T0237154

Compact Textbooks in Mathematics

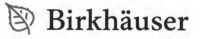

Compact Textbooks in Mathematics

This textbook series presents concise introductions to current topics in mathematics and mainly addresses advanced undergraduates and master students. The concept is to offer small books covering subject matter equivalent to 2- or 3-hour lectures or seminars which are also suitable for self-study. The books provide students and teachers with new perspectives and novel approaches. They may feature examples and exercises to illustrate key concepts and applications of the theoretical contents. The series also includes textbooks specifically speaking to the needs of students from other disciplines such as physics, computer science, engineering, life sciences, finance.

- **compact:** small books presenting the relevant knowledge
- **learning made easy:** examples and exercises illustrate the application of the contents
- **useful for lecturers:** each title can serve as basis and guideline for a semester course/lecture/seminar of 2–3 hours per week.

More information about this series at http://www.springer.com/series/11225

Piotr T. Chruściel

Elements of General Relativity

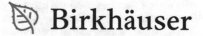 Birkhäuser

Piotr T. Chruściel
Faculty of Physics
University of Vienna
Vienna, Austria

ISSN 2296-4568   ISSN 2296-455X   (electronic)
Compact Textbooks in Mathematics
ISBN 978-3-030-28415-2   ISBN 978-3-030-28416-9   (eBook)
https://doi.org/10.1007/978-3-030-28416-9

Mathematics Subject Classification (2010): 83-01, 53Zxx

© Springer Nature Switzerland AG 2019
This work is subject to copyright. All rights are reserved by the Publisher, whether the whole
or part of the material is concerned, specifically the rights of translation, reprinting, reuse
of illustrations, recitation, broadcasting, reproduction on microfilms or in any other physical
way, and transmission or information storage and retrieval, electronic adaptation, computer
software, or by similar or dissimilar methodology now known or hereafter developed.
The use of general descriptive names, registered names, trademarks, service marks, etc. in this
publication does not imply, even in the absence of a specific statement, that such names are
exempt from the relevant protective laws and regulations and therefore free for general use.
The publisher, the authors, and the editors are safe to assume that the advice and information
in this book are believed to be true and accurate at the date of publication. Neither the
publisher nor the authors or the editors give a warranty, express or implied, with respect
to the material contained herein or for any errors or omissions that may have been made.
The publisher remains neutral with regard to jurisdictional claims in published maps and
institutional affiliations.

This book is published under the imprint Birkhäuser, www.birkhauser-science.com, by the
registered company Springer Nature Switzerland AG.
The registered company address is: Gewerbestrasse 11, 6330 Cham, Switzerland

Preface

This book arose from lecture notes for a one-semester undergraduate course in general relativity. It provides an introduction to the mathematics and physics of general relativity, its basic physical concepts, its observational implications, and the resulting insights into the nature of spacetime and the structure of the universe. It provides some of the most striking aspects of Einstein's theory of gravitation: black holes, gravitational waves, stellar models, and cosmology. It contains a self-contained introduction to tensor calculus and Riemannian geometry, which uses in parallel the language of modern differential geometry and the coordinate notation, more familiar to physicists. The author has strived to achieve mathematical rigour, with all notions given careful mathematical meaning, while trying to maintain the formalism to a minimum fit for purpose. Familiarity with special relativity is assumed.

The overall aim is to convey some of the main physical and geometrical properties of Einstein's theory of gravitation, providing a good entry point to further studies of the mathematics and physics of Einstein equations.

The material covered can be used for a fast-paced one-semester first course in general relativity, with 4 h of lectures per week. The usual heuristic motivations and introductory experimental justifications for general relativity have not been included, as in Vienna this is done in a different course. ▶ Chapter 1 can be dropped if the students are already familiar with elementary Riemannian geometry. An easy-going pace can be achieved by dropping one of the last three chapters, which are self-sufficient except for the contents of the introductory ▶ Chaps. 1 and 2.

The book should be useful to graduate and advanced undergraduate students in physics and mathematics wishing to obtain a solid background in mathematical general relativity.

Incidentally Indented text, typeset in smaller font, can be ignored by students in a first reading. Similarly for those proofs and Definitions, Remarks, etc. which are indented and typeset in a smaller font.

Acknowledgements I am grateful to Jacob Biamonte, David Fajman, Lukas Ifsits, and Maciek Maliborski for pointing out several mistakes in a previous version of these notes and Tim Paetz for making many corrections throughout.

Vienna, Austria   Piotr T. Chruściel

Contents

Introduction to Tensor Calculus and Riemannian Geometry

Piotr T. Chruściel

© Springer Nature Switzerland AG 2019
P. T. Chruściel, *Elements of General Relativity*, Compact Textbooks in Mathematics,
https://doi.org/10.1007/978-3-030-28416-9_1

In this chapter we introduce the reader to tensor calculus and Riemannian geometry, which are the main mathematical tools needed to study general relativity.

In Appendix A the reader will find some introductory material which makes contact with most introductory treatments of special relativity. There one can see at least two objects with rather different behavior under changes of special relativistic inertial frames—the frames themselves and the Maxwell tensor F. Tensor calculus, which we are about to present, provides a tool for a systematic treatment of such objects, and provides a mathematical background for general relativity.

We note that neither knowledge of, nor familiarity with, the topics covered in Appendix A is needed for what follows.

1.1 Introduction to Tensor Calculus

The right arena for tensor calculus is that of manifolds, so we start by shortly introducing the notion.

1.1.1 Manifolds

Manifolds can be thought of as well-behaved subsets of $\mathbb{R}^N$, for some (possibly very large) N, like curves, spheres, hyperboloids, and such. We will not delve into the nature and properties of manifolds in any length. However, it is appropriate to provide the definition, making clear what "well behaved" means:

Definition 1.1.1

A smooth n-dimensional manifold is a set M equipped with the following:

1. A topology: A "connected metrizable topological space."
2. Local charts: A collection of pairs $(\mathcal{U}, \phi)$, called coordinate patches, covering M, where $\mathcal{U}$ is an open subset of M, and $\phi : \mathcal{U} \to \mathbb{R}$ is continuous. One further requires that all the maps

$$M \supset \mathcal{U} \ni p \mapsto \phi(p) \equiv (x^1(p), \dots, x^n(p)) \in \mathcal{V} \subset \mathbb{R}^n$$

   are homeomorphisms.
3. Compatibility: Given two overlapping coordinate patches, $(\mathcal{U}, x^i)$ and $(\widetilde{\mathcal{U}}, \tilde{x}^i)$, with corresponding sets $\mathcal{V}, \widetilde{\mathcal{V}} \subset \mathbb{R}^n$, the maps $\tilde{x}^j \mapsto x^i(\tilde{x}^j)$ are smooth diffeomorphisms wherever defined. This means that they are bijections differentiable as many times as one wishes, and with

$$\det \left[\frac{\partial x^i}{\partial \tilde{x}^j} \right] \text{ nowhere vanishing} .$$

Remark 1.1.2 One can likewise talk about manifolds of finite differentiability, e.g., C^k, then the coordinate-change functions between coordinate patches, called transition functions, are required to be of C^k differentiability class. For example, a topological manifold will be a manifold where the transition functions are only required to be continuous homeomorphisms.

The collection of all pairs $(\mathcal{U}, \phi)$ as above is called an *atlas*. The maps ϕ provide local coordinates on M.

The main point of the definition of a manifold is that it allows one to talk about differentiable functions and maps, and therefore do analysis:

Definition 1.1.3

A function on M is smooth if it is smooth when expressed in terms of local coordinates. Similarly for tensor fields and, more generally, for maps between manifolds.

Remark 1.1.4 A "Hausdorff paracompact topological space" is the same as "Hausdorff sigma-compact topological space" or "metrizable topological space." Some references use these equivalent conditions in the definition.

The manifold structure allows us to define vectors and covectors, and to develop tensor calculus; we will return to these notions shortly.

Examples
1. $\mathbb{R}^n$ with the usual topology, and with a single global coordinate patch.
2. A sphere: One can use stereographic projection to obtain two overlapping coordinate systems (or use spherical coordinates, but then one must avoid borderline angles; one set of spherical coordinates does not provide well-behaved coordinates for the whole sphere!).
3. We will use several coordinate patches to describe the Schwarzschild black hole.
4. Let $f : \mathbb{R}^n \to \mathbb{R}$, and define $N := f^{-1}(0)$. If ∇f has no zeros on N, then every connected component of N is a smooth $(n-1)$-dimensional manifold. This construction leads to a plethora of examples. For example, if $f = \sqrt{(x^1)^2 + \ldots + (x^n)^2} - R$, with $R > 0$, then N is a sphere of radius R. In this context an interesting example is provided by the function $f = t^2 - x^2$ on $\mathbb{R}^2$ with coordinates (t, x): its zero-level-set is the light-cone $t = \pm x$, which is a manifold except at the origin. Note that $\nabla f = 0$ there, which shows that the requirement of a nonvanishing gradient is necessary in general.

1.1.2 Scalar Functions

The geometric definition of a scalar f on a manifold M is that f is a real-valued function on M. Equivalently, f is a map from M to $\mathbb{R}$.

One can likewise talk about complex-valued scalar functions.

Unless explicitly indicated otherwise, scalars are assumed to be as differentiable as the differentiable structure of the manifold allows.

Now, in physics one often defines scalars by their behavior under changes of coordinates: Under such a change, say $x \equiv (x^i) \to (y^j(x^i)) \equiv y(x)$, a scalar function f simply changes by composition of coordinates; so to a function $f(x)$ we associate a new function

$$\bar{f}(y) = f(x(y)).$$

In general relativity it is a common abuse of notation to write the same symbol f for what we denoted by $\bar{f}$, when we think that this is the same function but expressed in a different coordinate system. In this section, to make things clearer, we will make this notational distinction, but this will almost never be done in the remainder of the lectures. For example we will most of the time use the same symbol $g_{\mu\nu}$ for the metric components, whatever the coordinate system used.

1.1.3 Vector Fields

The notion of a vector field finds its roots in the notion of *tangents to a curve*, say $\mathbb{R} \supset I \ni s \to \gamma(s) \in M$.

If we use local coordinates x^i to write $\gamma(s)$ as $(\gamma^1(s), \gamma^2(s), \ldots, \gamma^n(s))$, the tangent to that curve at the point $\gamma(s)$ is defined as the set of numbers

$$(\dot\gamma^1(s), \dot\gamma^2(s), \ldots, \dot\gamma^n(s)),$$

where a dot denotes a derivative with respect to s.

Consider a curve $\gamma(s)$ given in a coordinate system x^i and let us perform a change of coordinates $x^i \to y^j(x^i)$. In the new coordinates y^j the curve γ is represented by the functions $y^j(\gamma^i(s))$, with new tangent

$$\frac{dy^j}{ds} = \frac{\partial y^j}{\partial x^i}\dot\gamma^i. \tag{1.1.1}$$

Here, and everywhere that follows, we are using the *summation convention*. This means that, unless explicitly stated otherwise, the summation symbol is omitted whenever two identical indices occur, one up and one down. So, (1.1.1) is the same as

$$\frac{dy^j}{ds} = \sum_i \frac{\partial y^j}{\partial x^i}\dot\gamma^i.$$

Equation (1.1.1) defines what is called *the transformation law of vectors*: given a point $x = (x^i)$ and a set of numbers $X = (X^i)$, the set (X^i) is called a *vector at x* if, under a change of coordinates $x^i \to y^j(x^i)$ the set (X^i) transforms as

$$X^i(x) \to \bar X^j(y(x)) = \frac{\partial y^j}{\partial x^i}(x)X^i(x); \tag{1.1.2}$$

equivalently

$$\bar X^j(y) = \frac{\partial y^j}{\partial x^i}(x(y))X^i(x(y)). \tag{1.1.3}$$

A very convenient way of representing vectors is using *first order homogeneous differential operators*. So, consider a vector field represented in a coordinate system $\{x^i\}$ by a set of functions (X^i), which transform using the transformation rule (1.1.2) under coordinate changes. Define the following differential operator

$$X := X^1\frac{\partial}{\partial x^1} + \ldots + X^n\frac{\partial}{\partial x^n}.$$

The point is that *the transformation rule* (1.1.2) *becomes implicit in the notation*. Indeed, consider a function f, so that the differential operator X acts on f by differentiation:

$$X(f)(x) := X^i\frac{\partial f(x)}{\partial x^i}. \tag{1.1.4}$$

If we make a coordinate change so that

$$x^j = x^j(y^k) \iff y^k = y^k(x^j),$$

keeping in mind that

$$\bar{f}(y) = f(x(y)) \quad \Longleftrightarrow \quad f(x) = \bar{f}(y(x)),$$

then

$$X(f)(x) := X^i(x)\frac{\partial f(x)}{\partial x^i}$$

$$= X^i(x)\frac{\partial \bar{f}(y(x))}{\partial x^i}$$

$$= X^i(x)\frac{\partial \bar{f}(y(x))}{\partial y^k}\frac{\partial y^k}{\partial x^i}(x)$$

$$= \bar{X}^k(y(x))\frac{\partial \bar{f}(y(x))}{\partial y^k}$$

$$= \left(\bar{X}^k\frac{\partial \bar{f}}{\partial y^k}\right)(y(x)), \tag{1.1.5}$$

with $\bar{X}^k$ given by (1.1.2). So, writing "iff" for "if and only if",

$X(f)$ is a scalar iff the coefficients X^i satisfy the transformation law of a vector.

Exercise 1.1.5
Check that this is a necessary and sufficient condition. ∎

We will often use the middle formula in (1.1.5) in the form

$$\frac{\partial}{\partial x^i} = \frac{\partial y^k}{\partial x^i}\frac{\partial}{\partial y^k}. \tag{1.1.6}$$

Note that the tangent to the curve $s \to (s, x^2, x^3, \ldots x^n)$, where $(x^2, x^3, \ldots x^n)$ are constants, is identified with the differential operator

$$\partial_1 \equiv \frac{\partial}{\partial x^1}.$$

Similarly the tangent to the curve $s \to (x^1, s, x^3, \ldots x^n)$, where $(x^1, x^3, \ldots x^n)$ are constants, is identified with

$$\partial_2 \equiv \frac{\partial}{\partial x^2}$$

and then $\dot{\gamma}$ is identified with

$$\dot{\gamma}(s) = \dot{\gamma}^i(s)\partial_i.$$

We note the following:

Proposition 1.1.6

Let $X : C^\infty(M) \to C^\infty(M)$ satisfy, for all smooth functions f and g,
1. *$X(f + g) = X(f) + X(g)$,*
2. *$X(cf) = cX(f)$ for all real numbers c, and*
3. *$X(fg) = X(f)g + X(g)f$.*

Then X is a vector field, in the sense that in local coordinates it takes the form $X = X^i \partial_i$ with smooth coefficients X^i.

The proof proceeds by using Taylor expansions and is left as an exercise.

Note that $X^i = X(x^i)$, where the right-hand side of the equation is meant as the action of a vector field X on the coordinate function x^i. Thus, X acting on x^i returns the ith component of x in the basis $\partial/\partial x^i$.

The *Lie bracket* of two vector fields is defined by its action on functions as

$$[X, Y](f) = X(Y(f)) - Y(X(f)) . \tag{1.1.7}$$

We leave it as another exercise to the reader to check that $[X, Y]$ is a vector field. A possible proof consists of checking the conditions of Proposition 1.1.6.

It is useful to introduce some notation. Let M be a manifold (see ▶ Sect. 1.1.1, p. 1). At any given point p of M the set of vectors forms a vector space, denoted by $T_p M$. See ◻ Fig. 1.1 for an example. The collection of all tangent spaces is called the *tangent bundle*, denoted by TM.

◻ **Fig. 1.1** The space of vectors tangent to a point on the sphere

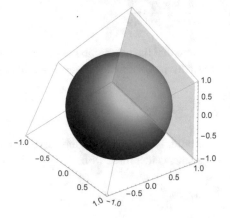

1.1.4 Covectors

Covector fields are defined as *maps from the space of vector fields to the space of functions which are linear under addition and under multiplication by functions.*

At a point $p \in M$, covectors at p are *maps from the space of vectors at p to $\mathbb{R}$ which are linear under addition and multiplication by real numbers.*

At any given point $p \in M$ the set of covectors forms a vector space, denoted by $T_p^* M$. The collection of all the tangent spaces is called the cotangent bundle to M, denoted by $T^* M$.

In this context the basic object is the *coordinate differential dx^i*, defined by its action on vectors as follows:

$$dx^i(X^j \partial_j) := X^i . \tag{1.1.8}$$

Equivalently,

$$dx^i(\partial_j) := \delta_j^i := \begin{cases} 1, \, i = j; \\ 0, \, \text{otherwise.} \end{cases}$$

The dx^i's form a basis for the space of covectors: indeed, let φ be a linear map on the space of vectors, then

$$\varphi(\underbrace{X}_{X^i \partial_i}) = \varphi(X^i \partial_i) \underset{\text{linearity}}{=} X^i \underbrace{\varphi(\partial_i)}_{\text{call this } \varphi_i} = \varphi_i dx^i(X) \underset{\text{def. sum of functions}}{=} (\varphi_i dx^i)(X) ,$$

hence

$$\varphi = \varphi_i dx^i ,$$

and every φ can indeed be written as a linear combination of the dx^i's.

Note that $\varphi(X)$ equals, by definition, the action of the linear map φ on the vector X, and therefore does not depend upon the choice of coordinates used to obtain an explicit formula for $\varphi(X)$. So, if we write φ as $\varphi_i dx^i$ and X as $X^i \frac{\partial}{\partial x^i}$ in the coordinates $\{x^i\}$, and $\bar{\varphi}_i dy^i$ and $\bar{X}^i \frac{\partial}{\partial y^i}$ in the coordinates $\{y^i\}$, then $\bar{\varphi}_i \bar{X}^i$ should be equal to $\varphi_i X^i$. Now, the formula for the transformation of components of a vector under changes of coordinates gives

$$\bar{\varphi}_i \bar{X}^i = \bar{\varphi}_i \frac{\partial y^i}{\partial x^k} X^k .$$

This will be equal to $\varphi_k X^k$ for all vector fields X if and only if the components φ_i of φ transform as

$$\varphi_k = \bar{\varphi}_i \frac{\partial y^i}{\partial x^k} . \tag{1.1.9}$$

Given a scalar f, we define its *differential df* as

$$df = \frac{\partial f}{\partial x^1} dx^1 + \ldots + \frac{\partial f}{\partial x^n} dx^n.$$

With this definition, dx^i *is the differential of the coordinate function x^i.*

As presented above, the differential of a function is a covector by definition. As an exercise, the reader can check directly that the collection of functions $\varphi_i := \partial_i f$ satisfies the transformation rule (1.1.9).

We have a formula which is often used in calculations

$$\boxed{dy^j = \frac{\partial y^j}{\partial x^k} dx^k,}$$

and which follows immediately from the fact that dy^j is the differential of the coordinate function y^j.

Example 1.1.7

Let (ρ, φ) be polar coordinates on $\mathbb{R}^2$, thus $x = \rho \cos \varphi$, $y = \rho \sin \varphi$, and then

$$dx = d(\rho \cos \varphi) = \cos \varphi d\rho - \rho \sin \varphi d\varphi,$$

$$dy = d(\rho \sin \varphi) = \sin \varphi d\rho + \rho \cos \varphi d\varphi. \qquad \blacksquare$$

Incidentally An elegant approach to the definition of differentials proceeds as follows: Given any function f, we define:

$$df(X) := X(f). \tag{1.1.10}$$

(Recall that here we are viewing a vector field X as a differential operator on functions, defined by (1.1.4).) The map $X \mapsto df(X)$ is linear under addition of vectors, and multiplication of vectors by numbers: if λ, μ are real numbers, and X and Y are vector fields, then

$$df(\lambda X + \mu Y) \underset{\text{by definition (1.1.10)}}{=} (\lambda X + \mu Y)(f)$$

$$\underset{\text{by definition (1.1.4)}}{=} \lambda X^i \partial_i f + \mu Y^i \partial_i f$$

$$\underset{\text{by definition (1.1.10)}}{=} \lambda df(X) + \mu df(Y).$$

Applying (1.1.10) to the function $f = x^i$ we obtain

$$dx^i(\partial_j) = \frac{\partial x^i}{\partial x^j} = \delta^i_j,$$

recovering (1.1.8).

1.1.5 Bilinear Maps, Two-Covariant Tensors

A map is said to be *multi-linear* if it is linear in every entry; e.g., g is *bilinear* if

$$g(aX + bY, Z) = ag(X, Z) + bg(Y, Z),$$

and

$$g(X, aZ + bW) = ag(X, Z) + bg(X, W).$$

A map g which is bilinear on the space of vectors can be represented by a "matrix with two indices down":

$$g(X, Y) = g(X^i \partial_i, Y^j \partial_j) = X^i Y^j \underbrace{g(\partial_i, \partial_j)}_{=:g_{ij}} = g_{ij} dx^i(X) dx^j(Y).$$

We say that g is a *covariant tensor of valence two*.

We say that g is *symmetric* if $g(X, Y) = g(Y, X)$ for all X, Y; equivalently, $g_{ij} = g_{ji}$.

A symmetric bilinear tensor field is said to be *nondegenerate* if $\det(g_{ij})$ has no zeros. Equivalently:

$$\forall Y \quad g(X, Y) = 0 \quad \Longrightarrow \quad X = 0. \tag{1.1.11}$$

By Sylvester's inertia theorem, there exists a basis $\{\theta^i\}$ of the space of covectors so that a symmetric bilinear map g can be written as

$$g(X, Y) = -\theta^1(X)\theta^1(Y) - \ldots - \theta^s(X)\theta^s(Y)$$
$$+\theta^{s+1}(X)\theta^{s+1}(Y) + \ldots + \theta^{s+r}(X)\theta^{s+r}(Y).$$

The pair (s, r) is called the *signature of* g; in geometry, unless explicitly indicated otherwise, one always assumes that the signature does not change from point to point.

If $r = N$, in dimension N, then g is said to be a *Riemannian metric tensor*. Thus, a Riemannian metric on a manifold M is a field of symmetric nondegenerate bilinear forms with signature $(0, \dim M)$.

A canonical example is provided by the *Euclidean* metric on $\mathbb{R}^n$,

$$g = (dx^1)^2 + \ldots + (dx^n)^2.$$

Thus, by definition, a *Riemannian metric* is a field of symmetric two-covariant tensors with signature $(+, \ldots, +)$.

Note that the last condition implies that $\det g_{ij}$ has no zeros anywhere, hence the matrix g_{ij} admits a matrix inverse everywhere.

Incidentally A Riemannian metric on M can be used to define the length of curves: if $\gamma : [a, b] \ni s \to \gamma(s) \in M$, then the length $\ell_g(\gamma)$ is defined as

$$\ell_g(\gamma) := \int_a^b \sqrt{g(\dot{\gamma}, \dot{\gamma})} ds \, .$$

One can then define the distance between points by minimizing the length of the curves connecting them. Compare Remark 6.4.2, p. 221.

If $s = 1$ and $r = N - 1$, in dimension N, then g is said to be a *Lorentzian metric tensor*.

For example, the *Minkowski metric* on $\mathbb{R}^{1+n}$ is

$$\eta = -(dx^0)^2 + (dx^1)^2 + \ldots + (dx^n)^2 \, .$$

1.1.6 Tensor Products

If φ and θ are covectors, we can define a bilinear map using the formula

$$(\varphi \otimes \theta)(X, Y) = \varphi(X)\theta(Y) \, . \tag{1.1.12}$$

For example

$$(dx^1 \otimes dx^2)(X, Y) = X^1 Y^2 \, .$$

Using this notation, if g is a bilinear map on vectors, then we have

$$g(X, Y) = g(X^i \partial_i, Y^j \partial_j) = \underbrace{g(\partial_i, \partial_j)}_{=:g_{ij}} \underbrace{X^i}_{dx^i(X)} \underbrace{Y^j}_{dx^j(Y)} = (g_{ij} dx^i \otimes dx^j)(X, Y) \, .$$
$$\underbrace{}_{(dx^i \otimes dx^j)(X,Y)}$$

$$\tag{1.1.13}$$

We will write $dx^i dx^j$ for the symmetric product,

$$dx^i dx^j := \frac{1}{2}(dx^i \otimes dx^j + dx^j \otimes dx^i) \, ,$$

so that (1.1.13) is most of the time written as

$$g(X, Y) = (g_{ij} dx^i dx^j)(X, Y) \quad \Longleftrightarrow \quad g = g_{ij} dx^i dx^j \, .$$

This formula allows one to read-off, without even having to think, the transformation law of a metric tensor under coordinate changes:

$$
g_{ij}(x) \underbrace{dx^i}_{\frac{\partial x^i}{\partial y^k} dy^k} \underbrace{dx^j}_{\frac{\partial x^j}{\partial y^\ell} dy^\ell} = g_{ij}(x) \frac{\partial x^i}{\partial y^k} \frac{\partial x^j}{\partial y^\ell} dy^k dy^\ell.
$$

So, if we write $\bar{g}_{k\ell}(y) dy^k dy^\ell$ for the metric in the coordinates y, we have

$$
\bar{g}_{kl}(y) = g_{ij}(x(y)) \frac{\partial x^i}{\partial y^k}(y) \frac{\partial x^j}{\partial y^\ell}(y).
$$

Example 1.1.8

Let (ρ, φ) be polar coordinates on $\mathbb{R}^2$:

$$
x = \rho \cos \varphi, \quad y = \rho \sin \varphi.
$$

We then have

$$
dx = d(\rho \cos \varphi) = \cos \varphi \, d\rho - \rho \sin \varphi \, d\varphi, \quad dy = d(\rho \sin \varphi) = \sin \varphi \, d\rho + \rho \cos \varphi \, d\varphi.
$$

From this, we find that the Euclidean metric $g = dx^2 + dy^2$ on the plane can also be written as

$$
\begin{aligned}
g &= dx^2 + dy^2 = (\cos \varphi \, d\rho - \rho \sin \varphi \, d\varphi)^2 + (\sin \varphi \, d\rho + \rho \cos \varphi \, d\varphi)^2 \\
&= d\rho^2 + \rho^2 \sin^2 \varphi \, d\varphi^2.
\end{aligned}
$$

∎

Incidentally Recall that, by definition, a Lorentzian metric is a symmetric two-covariant tensor field with signature $(-, +, \ldots, +)$. This implies that the determinant $\det g_{\mu\nu}$ of the matrix $g_{\mu\nu}$ representing g in some coordinate system is strictly negative; given a symmetric bilinear tensor, calculating this determinant provides a good hint whether or not this might be a Lorentzian metric.

One writes $dx^i \wedge dx^j$ for the antisymmetric tensor product:

$$
dx^i \wedge dx^j := dx^i \otimes dx^j - dx^j \otimes dx^i.
$$

It should be clear how this generalizes: the tensors $dx^i \otimes dx^j \otimes dx^k$, defined as

$$
(dx^i \otimes dx^j \otimes dx^k)(X, Y, Z) = X^i Y^j Z^k,
$$

form a basis for three-linear maps on the space of vectors. In other words, in local coordinates, every three-linear map X can be written in the form

$$
X = X_{ijk} dx^i \otimes dx^j \otimes dx^k.
$$

Here X is a tensor of valence $(0, 3)$. Under changes of coordinates, each index leads to a transformation factor as in a covector:

$$X = X_{ijk}dx^i \otimes dx^j \otimes dx^k = X_{ijk}\frac{\partial x^i}{\partial y^m}\frac{\partial x^j}{\partial y^\ell}\frac{\partial x^k}{\partial y^n}dy^m \otimes dy^\ell \otimes dy^n \,.$$

It is sometimes useful to think of vectors as linear maps on covectors, using a formula which looks funny when first met: if θ is a covector, and X is a vector, then

$$X(\theta) := \theta(X)$$

(the right-hand side is the value of θ on X, so this formula defines the left-hand side). So if $\theta = \theta_i dx^i$ and $X = X^i \partial_i$, then

$$\theta(X) = \theta_i X^i = X^i \theta_i = X(\theta)\,.$$

It then makes sense to define, e.g., $\partial_i \otimes \partial_j$ as a bilinear map on covectors: if θ and ψ are covectors, then

$$(\partial_i \otimes \partial_j)(\theta, \psi) := \partial_i(\theta)\,\partial_j(\psi) \equiv \theta_i \psi_j \,. \tag{1.1.14}$$

Here, as usual $\theta_i \equiv \theta(\partial_i)$, $\psi_j \equiv \psi(\partial_j)$, and we emphasize that in (1.1.14) $\partial_i(\theta)$ does *not* mean differentiation of θ with respect to x^i but the duality action of the vector ∂_i on the covector θ.

Next, one can define a map $\partial_i \otimes dx^j$ which is linear on forms in the first slot, and linear in vectors in the second slot as

$$(\partial_i \otimes dx^j)(\theta, X) := \partial_i(\theta)dx^j(X) = \theta_i X^j \,. \tag{1.1.15}$$

The $\partial_i \otimes dx^j$'s form the basis for the *space of tensors of rank* $(1, 1)$, or *valence* $(1, 1)$; if T is such a tensor, then in local coordinates it can be written as

$$T = T^i{}_j \partial_i \otimes dx^j \,.$$

Such a tensor transforms under coordinate changes in the obvious way:

$$T^i{}_j \underbrace{\partial_{x^i}}_{\frac{\partial y^j}{\partial x^i}\partial_{y^j}} \otimes \underbrace{dx^j}_{\frac{\partial x^j}{\partial y^k}dy^k} = T^i{}_j \frac{\partial y^j}{\partial x^i}\frac{\partial x^j}{\partial y^k}\partial_{y^j} \otimes dy^k \,.$$

Quite generally, a tensor of valence, or rank, (r, s) can be defined as an object which has r vector indices and s covector indices, so that it transforms as

$$S^{i_1 \ldots i_r}{}_{j_1 \ldots j_s} \to S^{m_1 \ldots m_r}{}_{\ell_1 \ldots \ell_s}\frac{\partial y^{i_1}}{\partial x^{m_1}} \cdots \frac{\partial y^{i_r}}{\partial x^{m_r}}\frac{\partial x^{\ell_1}}{\partial y^{j_1}} \cdots \frac{\partial x^{\ell_s}}{\partial y^{j_s}} \,.$$

For example, if X^i and Y^j are vectors, then $X^i Y^j$ forms a contravariant tensor of valence two.

Tensors of the same valence can be added in the obvious way: *e.g.*

$$(A + B)(X, Y) := A(X, Y) + B(X, Y) \quad \Longleftrightarrow \quad (A + B)_{ij} = A_{ij} + B_{ij}.$$

Tensors can be multiplied by scalars: *e.g.*, if f is a function,

$$(fA)(X, Y, Z) := fA(X, Y, Z) \quad \Longleftrightarrow \quad f \cdot (A_{ijk}) := (f A_{ijk}).$$

Finally, we have seen in (1.1.12) how to take tensor products for one-forms, and in (1.1.15) how to take a tensor product of a vector and a one-form, but this can also be done for higher order tensors: e.g., if S is of valence (a, b) and T is a multi-linear map of valence (c, d) (not to be confused with the signature!), then $S \otimes T$ is a multi-linear map of valence $(a + c, b + d)$, defined as

$$(S \otimes T)(\underbrace{\theta, \dots \qquad , \psi, \dots}_{a \text{ covectors and } b \text{ vectors} \quad c \text{ covectors and } d \text{ vectors}}) := S(\theta, \dots)T(\psi, \dots).$$

Incidentally A rather shorter way of saying what we said so far about tensor products, which includes all the cases we treated separately, is: suppose that V and W are vector spaces, and let α be a p-linear map on V, and β be a q-linear map on W, then, for $X_1, \dots, X_p \in V$ and $Y_1, \dots, Y_q \in W$, one defines

$$(\alpha \otimes \beta)(X_1, \dots, X_p, Y_1, \dots, Y_q) := \alpha(X_1, \dots, X_p)\beta(Y_1, \dots, Y_q).$$

1.1.6.1 Contractions

The simplest example of the *contraction* applies to tensor fields $S^i{}_j$ with one index down and one index up. One can then perform the sum

$$S^i{}_i.$$

This turns out to be a scalar; indeed, under changes of coordinates,

$$S^i{}_j \rightarrow \bar{S}^\ell{}_k = S^i{}_j \frac{\partial x^j}{\partial y^k} \frac{\partial y^\ell}{\partial x^i},$$

which leads to

$$\bar{S}^\ell{}_\ell = S^i{}_j \underbrace{\frac{\partial x^j}{\partial y^\ell} \frac{\partial y^\ell}{\partial x^i}}_{\delta^j_i} = S^i{}_i.$$

One can similarly do contractions on higher valence tensors, e.g.

$$S^{i_1 i_2 \ldots i_r i_{r+1}}{}_{j_1 j_2 j_3 \ldots j_s j_{s+1}} \;\to\; S^{i_1 i_2 \ldots i_r \ell}{}_{j_1 \ell j_3 \ldots j_s j_{s+1}} \,.$$

After contraction, a tensor of rank $(r+1, s+1)$ becomes a tensor of rank (r, s).

1.1.7 Raising and Lowering of Indices

Let g be a symmetric two-covariant tensor field on M, by definition such an object is the assignment to each point $p \in M$ of a bilinear map $g(p)$ from $T_p M \times T_p M$ to $\mathbb{R}$, with the additional property

$$g(X, Y) = g(Y, X)\,.$$

In this book the symbol g will be reserved to *nondegenerate symmetric two-covariant tensor fields*. It is usual to simply write g for $g(p)$, the point p being implicitly understood. We will sometimes write g_p for $g(p)$ when referencing p.

Sylvester's inertia theorem tells us that at each p the map g will have a well-defined signature; clearly this signature will be point-independent on a connected manifold when g is nondegenerate. A pair (M, g) is said to be a *Riemannian manifold* when the signature of g is $(0, \dim M)$; equivalently, when g is a positive definite bilinear form on every product $T_p M \times T_p M$. A pair (M, g) is said to be a *Lorentzian manifold* when the signature of g is $(1, \dim M - 1)$. One talks about *pseudo-Riemannian* metrics and manifolds whatever the signature of g, as long as g is nondegenerate, but we will only encounter Riemannian and Lorentzian metrics in these lectures.

Any pseudo-Riemannian metric g induces an isomorphism

$$\flat : T_p M \to T_p^* M$$

by the formula $X_\flat = g(X, \cdot)$, i.e.

$$\boxed{X_\flat(Y) := g(X, Y)\,,}$$

for $X, Y \in T_p M$. In local coordinates this gives

$$X_\flat = g_{ij} X^i dx^j =: X_j dx^j \,. \tag{1.1.16}$$

This last equality defines X_j—"the vector X^j with the index j lowered":

$$\boxed{X_j := g_{ij} X^i \,.} \tag{1.1.17}$$

The operation (1.1.17) is called the *lowering of indices* in the physics literature and, again in the physics literature, one does not make a distinction between the one-form $X_\flat$ and the vector X.

The inverse map is often denoted by $\sharp$ and is called the *raising of indices*; from (1.1.16) we obviously have

$$\alpha^\sharp = g^{ij}\alpha_i\partial_j =: \alpha^i\partial_i \quad \Longleftrightarrow \quad dx^i(\alpha^\sharp) = \boxed{\alpha^i = g^{ij}\alpha_j}, \tag{1.1.18}$$

where (g^{ij}) is the matrix inverse to (g_{ij}). For example,

$$(dx^i)^\sharp = g^{ik}\partial_k .$$

Clearly (g^{ij}), understood as the matrix of a bilinear map on T_p^*M, has the same signature as g, and can be used to define a scalar product $g^\sharp$ on $T_p^*(M)$:

$$g^\sharp(\alpha, \beta) := g(\alpha^\sharp, \beta^\sharp) \quad \Longleftrightarrow \quad g^\sharp(dx^i, dx^j) = g^{ij} .$$

This last equality is justified as follows:

$$g^\sharp(dx^i, dx^j) = g((dx^i)^\sharp, (dx^j)^\sharp) = g(g^{ik}\partial_k, g^{j\ell}\partial_\ell) = \underbrace{g^{ik}g_{k\ell}}_{=\delta_\ell^i}g^{j\ell} = g^{ji} = g^{ij} .$$

It is convenient to use the same letter g for $g^\sharp$—physicists do it all the time—or for scalar products induced by g on all the remaining tensor bundles, and we will often do so.

Incidentally One might wish to check by direct calculations that $g_{\mu\nu}X^\nu$ transforms as a one-form if X^μ transforms as a vector. The simplest way is to notice that $g_{\mu\nu}X^\nu$ is a contraction, over the last two indices, of the three-index tensor $g_{\mu\nu}X^\alpha$. Hence it is a one-form by the analysis at the end of the previous section. Alternatively, if we write $\bar{g}_{\mu\nu}$ for the transformed $g_{\mu\nu}$'s, and $\bar{X}^\alpha$ for the transformed X^α's, then

$$\underbrace{\bar{g}_{\alpha\beta}}_{g_{\mu\nu}\frac{\partial x^\mu}{\partial y^\alpha}\frac{\partial x^\nu}{\partial y^\beta}}\bar{X}^\beta = g_{\mu\nu}\frac{\partial x^\mu}{\partial y^\alpha}\frac{\partial x^\nu}{\partial y^\beta}\bar{X}^\beta = g_{\mu\nu}\underbrace{X^\nu}_{}\frac{\partial x^\mu}{\partial y^\alpha} ,$$

which is indeed the transformation law of a covector.

The *gradient* ∇f of a function f is a vector field obtained by raising the indices on the differential df:

$$g(\nabla f, Y) := df(Y) \quad \Longleftrightarrow \quad \nabla f := g^{ij}\partial_i f\partial_j . \tag{1.1.19}$$

1.2   Covariant Derivatives

When dealing with $\mathbb{R}^n$, or subsets thereof, there exists an obvious prescription how to differentiate tensor fields: In this case we have at our disposal the canonical "trivialization $\{\partial_i\}_{i=1,\dots,n}$ of $T\mathbb{R}^n$" (this means: a globally defined set of vectors which, at every point, form a basis of the tangent space), together with its dual trivialization $\{dx^j\}_{j=1,\dots,n}$ of $T^*\mathbb{R}^n$. We can expand a tensor field T of valence (k,ℓ) in terms of those bases,

$$
T = T^{i_1\dots i_k}{}_{j_1\dots j_\ell} \partial_{i_1} \otimes \dots \otimes \partial_{i_k} \otimes dx^{j_1} \otimes \dots \otimes dx^{j_\ell}
$$
$$
\iff \quad T^{i_1\dots i_k}{}_{j_1\dots j_\ell} = T(dx^{i_1},\dots,dx^{i_k},\partial_{j_1},\dots,\partial_{j_\ell}) , \tag{1.2.1}
$$

and differentiate each component $T^{i_1\dots i_k}{}_{j_1\dots j_\ell}$ of T separately:

$$
X(T)_{\text{in the coordinate system } x^i} :=
$$
$$
X^i \partial_{x^i}(T^{i_1\dots i_k}{}_{j_1\dots j_\ell})\partial_{x^{i_1}} \otimes \dots \otimes \partial_{x^{i_k}} \otimes dx^{j_1} \otimes \dots \otimes dx^{j_\ell} . \tag{1.2.2}
$$

The resulting object does, however, *not* behave as a tensor under coordinate transformations, in the sense that the above form of the right-hand side will not be preserved under coordinate transformations: As an example, consider the one-form $T := dx$ on $\mathbb{R}^n$, which has vanishing derivative as defined by (1.2.2). When expressed in spherical coordinates we have

$$
T = d(\rho\cos\varphi) = -\rho\sin\varphi d\varphi + \cos\varphi d\rho ,
$$

the partial derivatives of which are nonzero (both with respect to the original Cartesian coordinates (x,y) and to the new spherical ones (ρ,φ)). The notion of a *covariant derivative*, sometimes also referred to as a *connection*, is introduced precisely to obtain a notion of derivative which has tensorial properties. By definition, a covariant derivative is a map which to a vector field X and a tensor field T assigns a tensor field of the same type as T, denoted by $\nabla_X T$, with the following properties:

1. $\nabla_X T$ is linear with respect to addition both with respect to X and T:

$$
\nabla_{X+Y} T = \nabla_X T + \nabla_Y T , \qquad \nabla_X(T+Y) = \nabla_X T + \nabla_X Y ; \tag{1.2.3}
$$

2. $\nabla_X T$ is linear with respect to multiplication of X by functions f,

$$
\nabla_{fX} T = f\nabla_X T ; \tag{1.2.4}
$$

3. and, finally, $\nabla_X T$ satisfies the *Leibniz rule* under multiplication of T by a differentiable function f:

$$
\nabla_X(fT) = f\nabla_X T + X(f)T . \tag{1.2.5}
$$

It is natural to ask whether covariant derivatives do exist at all in general and, if so, how many of them can there be. First, it immediately follows from the axioms above that if D and ∇ are two covariant derivatives, then the map C defined as

$$C(X, T) := D_X T - \nabla_X T$$

is multi-linear both with respect to addition and multiplication by functions—the non-homogeneous terms $X(f)T$ in (1.2.5) cancel—and is thus a tensor field. Reciprocally, if D is a covariant derivative and $C(X, T)$ is bilinear with respect to addition and multiplication by functions, then

$$\nabla_X T := D_X T + C(X, T) \tag{1.2.6}$$

is a new covariant derivative. So, at least locally, on tensors of valence (r, s) there are as many covariant derivatives as tensors of valence $(r + s, r + s + 1)$.

1.2.1 Functions

The *canonical covariant derivative on functions* is defined as

$$\nabla_X(f) = X(f),$$

and we will always use the above. This has all the right properties, so obviously covariant derivatives of functions exist. From what has been said, any covariant derivative on functions is of the form

$$\nabla_X f = X(f) + \alpha(X) f, \tag{1.2.7}$$

where α is a one-form. Conversely, given any one-form α, (1.2.7) defines a covariant derivative on functions. The addition of the lower-order term $\alpha(X) f$ in (1.2.7) does not appear to be very useful for functions, but it plays a role in a geometric formulation of electrodynamics, or in geometric quantization. In any case such lower-order terms play an essential role when defining covariant derivatives for tensor fields.

1.2.2 Vectors

The simplest next case is that of a covariant derivative of vector fields. Let us not worry about existence at this stage, but assume that a covariant derivative exists, and work from there. (Anticipating, we will show shortly that any metric defines a covariant derivative, called the *Levi-Civita* covariant derivative, which is the unique covariant derivative operator satisfying a natural set of conditions.)

We will first assume that we are working on a set $\Omega \subset M$ over which we have a *trivialization* of the tangent bundle TM; as already mentioned, this means that there exist vector fields e_a, $a = 1, \ldots, \dim M$, such that at every point $p \in \Omega$ the fields $e_a|_p \in T_p M$ form a basis of $T_p M$.

A comment about existence of trivializations is in order. Suppose that Ω is a coordinate patch with coordinates (x^i), then the collection of vector fields $\{e_a\}_{a=1}^{\dim M}$ can be chosen to be equal to $\{\partial_i\}_{i=1}^{\dim M}$. Thus, trivializations always exist when Ω is part of the domain of definition of a system of coordinates. But trivializations do not exist globally in general, an example is provided by the two-dimensional sphere. We say that a manifold is *parallelizable* if a basis of TM can be chosen globally over M—in such a case Ω can be taken equal to M. We emphasize that we are *not* assuming that M is parallelizable, so that equations such as (1.2.8) below have only a local character in general.

Let θ^a denote the dual trivialization of T^*M—by definition the θ^a's satisfy

$$\boxed{\theta^a(e_b) = \delta_b^a}\ .$$

Given a covariant derivative ∇ on vector fields we set

$$\Gamma^a{}_b(X) := \theta^a(\nabla_X e_b) \iff \nabla_X e_b = \Gamma^a{}_b(X)e_a \,, \tag{1.2.8a}$$

$$\boxed{\Gamma^a{}_{bc} := \Gamma^a{}_b(e_c) = \theta^a(\nabla_{e_c} e_b)} \iff \nabla_X e_b = \Gamma^a{}_{bc} X^c e_a \,. \tag{1.2.8b}$$

The (locally defined) functions $\Gamma^a{}_{bc}$ are called *connection coefficients*. If $\{e_a\}$ is the coordinate basis $\{\partial_\mu\}$ we shall write

$$\Gamma^\mu{}_{\alpha\beta} := dx^\mu(\nabla_{\partial_\beta} \partial_\alpha) \quad \left(\iff \quad \nabla_{\partial_\mu} \partial_\nu = \Gamma^\sigma{}_{\nu\mu} \partial_\sigma \right), \tag{1.2.9}$$

etc. In this particular case the connection coefficients are usually called *Christoffel symbols*. We will sometimes write $\Gamma^\sigma{}_{\nu\mu}$ instead of $\Gamma^\sigma{}_{\nu\mu}$; note that the former convention is more common. Given vector fields X and Y, using the Leibniz rule (1.2.5) we find

$$\begin{aligned}
\nabla_X Y &= \nabla_X(Y^a e_a) \\
&= X(Y^a)e_a + Y^a \nabla_X e_a \\
&= X(Y^a)e_a + Y^a \Gamma^b{}_a(X)e_b \\
&= (X(Y^a) + \Gamma^a{}_b(X)Y^b)e_a \\
&= (X(Y^a) + \Gamma^a{}_{bc}Y^b X^c)e_a \,,
\end{aligned} \tag{1.2.10}$$

which gives various equivalent ways of writing $\nabla_X Y$. The (perhaps only locally defined) $\Gamma^a{}_b$'s are linear in X, and the collection $(\Gamma^a{}_b)_{a,b=1,\dots,\dim M}$ is sometimes referred to as the *connection one-form*. The one-covariant, one-contravariant tensor field ∇Y is defined as

$$\nabla Y := \nabla_a Y^b \theta^a \otimes e_b \iff \nabla_a Y^b := \theta^b(\nabla_{e_a} Y) \iff \boxed{\nabla_a Y^b = e_a(Y^b) + \Gamma^b{}_{ca}Y^c}\ . \tag{1.2.11}$$

We will often write ∇_a for ∇_{e_a}. Further, $\nabla_a Y^b$ will sometimes be written as $Y^b{}_{;a}$.

1.2.3 Transformation Law

Consider a coordinate basis ∂_{x^i}, it is natural to enquire about the transformation law of the connection coefficients $\Gamma^i{}_{jk}$ under a change of coordinates $x^i \to y^k(x^i)$. To make things clear, let us write $\Gamma^i{}_{jk}$ for the connection coefficients in the x-coordinates, and $\hat{\Gamma}^i{}_{jk}$ for the ones in the y-coordinates. We calculate:

$$
\Gamma^i{}_{jk} := dx^i\left(\nabla_{\frac{\partial}{\partial x^k}}\frac{\partial}{\partial x^j}\right)
$$

$$
= dx^i\left(\nabla_{\frac{\partial}{\partial x^k}}\frac{\partial y^\ell}{\partial x^j}\frac{\partial}{\partial y^\ell}\right)
$$

$$
= dx^i\left(\frac{\partial^2 y^\ell}{\partial x^k \partial x^j}\frac{\partial}{\partial y^\ell} + \frac{\partial y^\ell}{\partial x^j}\nabla_{\frac{\partial}{\partial x^k}}\frac{\partial}{\partial y^\ell}\right)
$$

$$
= \frac{\partial x^i}{\partial y^s}dy^s\left(\frac{\partial^2 y^\ell}{\partial x^k \partial x^j}\frac{\partial}{\partial y^\ell} + \frac{\partial y^\ell}{\partial x^j}\nabla_{\frac{\partial}{\partial x^k}\frac{\partial}{\partial y^r}}\frac{\partial}{\partial y^\ell}\right)
$$

$$
= \frac{\partial x^i}{\partial y^s}dy^s\left(\frac{\partial^2 y^\ell}{\partial x^k \partial x^j}\frac{\partial}{\partial y^\ell} + \frac{\partial y^\ell}{\partial x^j}\frac{\partial y^r}{\partial x^k}\nabla_{\frac{\partial}{\partial y^r}}\frac{\partial}{\partial y^\ell}\right)
$$

$$
= \frac{\partial x^i}{\partial y^s}\frac{\partial^2 y^s}{\partial x^k \partial x^j} + \frac{\partial x^i}{\partial y^s}\frac{\partial y^\ell}{\partial x^j}\frac{\partial y^r}{\partial x^k}\hat{\Gamma}^s{}_{\ell r}. \tag{1.2.12}
$$

Summarizing,

$$
\boxed{\Gamma^i{}_{jk} = \hat{\Gamma}^s{}_{\ell r}\frac{\partial x^i}{\partial y^s}\frac{\partial y^\ell}{\partial x^j}\frac{\partial y^r}{\partial x^k} + \frac{\partial x^i}{\partial y^s}\frac{\partial^2 y^s}{\partial x^k \partial x^j}.} \tag{1.2.13}
$$

Thus, the $\Gamma^i{}_{jk}$'s do *not* form a tensor; instead they transform as a tensor *plus* a nonhomogeneous term containing second derivatives.

1.2.4 Torsion

Because the inhomogeneous term in (1.2.13) is symmetric under the interchange of k and j, it follows from (1.2.13) that

$$
T^i{}_{jk} := \Gamma^i{}_{kj} - \Gamma^i{}_{jk}
$$

does transform as a tensor, called *the torsion tensor* of ∇.

> **Incidentally** An index-free definition of torsion proceeds as follows: Let ∇ be a covariant derivative defined for vector fields, the *torsion tensor* T is defined by the formula
>
> $$
> \boxed{T(X, Y) := \nabla_X Y - \nabla_Y X - [X, Y]}, \tag{1.2.14}
> $$

where $[X, Y]$ is the Lie bracket defined in (1.1.7). We obviously have

$$T(X, Y) = -T(Y, X).\tag{1.2.15}$$

Let us check that (1.2.14) defines a tensor field: multi-linearity with respect to addition is obvious. To check what happens under multiplication by functions, in view of (1.2.15) it is sufficient to do the calculation for the first slot of T. We then have

$$T(fX, Y) = \nabla_{fX}Y - \nabla_Y(fX) - [fX, Y]$$

$$= f\left(\nabla_X Y - \nabla_Y X\right) - Y(f)X - [fX, Y].\tag{1.2.16}$$

To work out the last commutator term we compute, for any function φ,

$$[fX, Y](\varphi) = fX(Y(\varphi)) - \underbrace{Y(fX(\varphi))}_{=Y(f)X(\varphi)+fY(X(\varphi))} = f[X, Y](\varphi) - Y(f)X(\varphi),$$

hence

$$[fX, Y] = f[X, Y] - Y(f)X,\tag{1.2.17}$$

and the last term here cancels the undesirable second-to-last term in (1.2.16), as required.

In a coordinate basis ∂_μ we have $[\partial_\mu, \partial_\nu] = 0$ and one finds from (1.2.9)

$$\boxed{T(\partial_\mu, \partial_\nu) = (\Gamma^\sigma{}_{\nu\mu} - \Gamma^\sigma{}_{\mu\nu})\partial_\sigma}\,,\tag{1.2.18}$$

which shows that—in coordinate frames—T is determined by "twice the antisymmetrization of the $\Gamma^\sigma{}_{\mu\nu}$'s over the lower indices." In particular, as already noticed by inspection of the transformation law for the Christoffel symbols, that last antisymmetrization produces a tensor field.

1.2.5 Covectors

Suppose that we are given a covariant derivative on vector fields, there is a natural way of inducing a covariant derivative on one-forms by imposing the condition that *the duality operation be compatible with the Leibniz rule*: given two vector fields X and Y together with a field of one-forms α, one sets

$$\boxed{(\nabla_X\alpha)(Y) := X(\alpha(Y)) - \alpha(\nabla_X Y)}\,.\tag{1.2.19}$$

Let us, first, check that (1.2.19) indeed defines a field of one-forms. The linearity, in the Y variable, with respect to addition is obvious. Next, for any function f we have

$$(\nabla_X\alpha)(fY) = X(\alpha(fY)) - \alpha(\nabla_X(fY))$$

$$= X(f)\alpha(Y) + fX(\alpha(Y)) - \alpha(X(f)Y + f\nabla_X Y)$$

$$= f(\nabla_X\alpha)(Y)\,,$$

as should be the case for one-forms. Next, we need to check that ∇ defined by (1.2.19) does satisfy the remaining axioms imposed on covariant derivatives. Again multi-linearity with respect to addition is obvious, as well as linearity with respect to multiplication of X by a function. Finally,

$$\nabla_X(f\alpha)(Y) = X(f\alpha(Y)) - f\alpha(\nabla_X Y)$$
$$= X(f)\alpha(Y) + f(\nabla_X\alpha)(Y),$$

as desired.

The duality pairing

$$T_p^*M \times T_pM \ni (\alpha, X) \to \alpha(X) \in \mathbb{R}$$

is a special case of the *contraction* operation. As already pointed out, the operation ∇ on one-forms has been defined in (1.2.19) so as to satisfy the *Leibniz rule under duality pairing*:

$$X(\alpha(Y)) = (\nabla_X\alpha)(Y) + \alpha(\nabla_X Y); \tag{1.2.20}$$

this follows directly from (1.2.19). This should not be confused with the Leibniz rule under multiplication by functions, which is part of the definition of a covariant derivative, and therefore always holds. It should be kept in mind that (1.2.20) does not necessarily hold for all covariant derivatives: if $^v\nabla$ is a random covariant derivative on vectors, and $^f\nabla$ is some covariant derivative on one-forms, typically one will have

$$X(\alpha(Y)) \neq (^f\nabla_X\alpha)(Y) + \alpha(^v\nabla_X Y).$$

Using the basis-expression (1.2.10) of $\nabla_X Y$ and the definition (1.2.19) we have

$$\nabla_X\alpha = X^a\nabla_a\alpha_b\,\theta^b,$$

with

$$\boxed{\nabla_a\alpha_b} := (\nabla_{e_a}\alpha)(e_b)$$
$$= e_a(\alpha(e_b)) - \alpha(\nabla_{e_a}e_b)$$
$$= \boxed{e_a(\alpha_b) - \Gamma^c{}_{ba}\alpha_c}.$$

1.2.6 Higher Order Tensors

It should now be clear how to extend ∇ to tensors of arbitrary valence: if T is r covariant and s contravariant one sets

$$(\nabla_X T)(X_1, \ldots, X_r, \alpha_1, \ldots, \alpha_s) := X\Big(T(X_1, \ldots, X_r, \alpha_1, \ldots, \alpha_s)\Big)$$

$$- T(\nabla_X X_1, \ldots, X_r, \alpha_1, \ldots, \alpha_s) - \ldots - T(X_1, \ldots, \nabla_X X_r, \alpha_1, \ldots, \alpha_s)$$

$$- T(X_1, \ldots, X_r, \nabla_X \alpha_1, \ldots, \alpha_s) - \ldots - T(X_1, \ldots, X_r, \alpha_1, \ldots, \nabla_X \alpha_s).$$

$$(1.2.21)$$

The verification that this defines a covariant derivative proceeds in a way identical to that for one-forms.

In a basis we write

$$\nabla_X T = X^a \nabla_a T_{a_1 \ldots a_r}{}^{b_1 \ldots b_s} \theta^{a_1} \otimes \ldots \otimes \theta^{a_r} \otimes e_{b_1} \otimes \ldots \otimes e_{b_s},$$

and (1.2.21) gives

$$\nabla_a T_{a_1 \ldots a_r}{}^{b_1 \ldots b_s} := (\nabla_{e_a} T)(e_{a_1}, \ldots, e_{a_r}, \theta^{b_1}, \ldots, \theta^{b_s})$$

$$= e_a(T_{a_1 \ldots a_r}{}^{b_1 \ldots b_s}) - \Gamma^c{}_{a_1 a} T_{c \ldots a_r}{}^{b_1 \ldots b_s} - \ldots - \Gamma^c{}_{a_r a} T_{a_1 \ldots c}{}^{b_1 \ldots b_s}$$

$$+ \Gamma^{b_1}{}_{ca} T_{a_1 \ldots a_r}{}^{c \ldots b_s} + \ldots + \Gamma^{b_s}{}_{ca} T_{a_1 \ldots a_r}{}^{b_1 \ldots c}.$$

$$(1.2.22)$$

Carrying over the last two lines of (1.2.21) to the left-hand side of that equation one obtains the Leibniz rule for ∇ under pairings of tensors with vectors or forms. It should be clear from (1.2.21) that ∇ defined by that equation is the *only covariant derivative which agrees with the original one on vectors, and which satisfies the Leibniz rule under tensor product and duality pairing*. We will only consider such covariant derivatives in this work.

1.2.7 The Levi-Civita Connection

One of the fundamental results in pseudo-Riemannian geometry is the existence of a torsion-free connection which preserves the metric:

Theorem 1.2.1

Let g be a two-covariant symmetric nondegenerate tensor field on a manifold M. Then there exists a unique connection ∇ such that

1. *$\nabla g = 0$,*
2. *the torsion tensor T of ∇ vanishes.*

Proof

Let us, first, suppose that we have a metric-compatible torsion-free connection ∇. Using the definition of $\nabla_i g_{jk}$ we have

$$0 = \nabla_i g_{jk} \equiv \partial_i g_{jk} - \Gamma^\ell_{ji} g_{\ell k} - \Gamma^\ell_{ki} g_{\ell j} \, ; \tag{1.2.23}$$

here we have written Γ^i_{jk} instead of $\Gamma^i{}_{jk}$, as is standard in the literature. We rewrite this equation making cyclic permutations of indices, and changing the overall sign:

$$0 = -\nabla_j g_{ki} \equiv -\partial_j g_{ki} + \Gamma^\ell_{kj} g_{\ell i} + \Gamma^\ell_{ij} g_{\ell k} \, ,$$

$$0 = -\nabla_k g_{ij} \equiv -\partial_k g_{ij} + \Gamma^\ell_{ik} g_{\ell j} + \Gamma^\ell_{jk} g_{\ell i} \, .$$

Adding the three equations and using symmetry of Γ^k_{ji} in ij one obtains

$$0 = \partial_i g_{jk} - \partial_j g_{ki} - \partial_k g_{ij} + 2\Gamma^\ell_{jk} g_{\ell i} \, .$$

Multiplying by g^{im} we obtain

$$\Gamma^m_{jk} = g^{mi} \Gamma^\ell_{jk} g_{\ell i} = \frac{1}{2} g^{mi} (\partial_j g_{ki} + \partial_k g_{ij} - \partial_i g_{jk}) \, . \tag{1.2.24}$$

This proves uniqueness.

To prove that (1.2.24) defines a connection, one can check that the formula (1.2.24) has the right transformation properties under coordinate transformations; we leave this as a (heavy) exercise to the reader.

The vanishing of the torsion is clear from (1.2.24).

Finally, it remains to check that the insertion of Γ^m_{jk}, as given by (1.2.24), into the right-hand side of (1.2.23), indeed gives zero, finishing the proof of existence. □

Incidentally Let us give a coordinate-free version of the above, which turns out to be somewhat messier in its initial part. However, it leads to *Koszul's formula* (1.2.27) below, which implies existence of the Levi-Civita connection without the even messier calculations mentioned, and skipped, in the paragraph following (1.2.24).

Suppose, first, that a connection satisfying 1. and 2. above is given. By the Leibniz rule we then have for all vector fields X, Y, and Z,

$$0 = (\nabla_X g)(Y, Z) = X(g(Y, Z)) - g(\nabla_X Y, Z) - g(Y, \nabla_X Z) \, . \tag{1.2.25}$$

One then rewrites the same equation applying cyclic permutations to X, Y and Z, with a minus sign for the last equation:

$$g(\nabla_X Y, Z) + g(Y, \nabla_X Z) = X(g(Y, Z)) \, ,$$

$$g(\nabla_Y Z, X) + g(Z, \nabla_Y X) = Y(g(Z, X)) \, ,$$

$$-g(\nabla_Z X, Y) - g(X, \nabla_Z Y) = -Z(g(X, Y)) \, . \tag{1.2.26}$$

As the torsion tensor vanishes, the sum of the left-hand sides of these equations can be manipulated as follows:

$$g(\nabla_X Y, Z) + g(Y, \nabla_X Z) + g(\nabla_Y Z, X) + g(Z, \nabla_Y X) - g(\nabla_Z X, Y) - g(X, \nabla_Z Y)$$

$$= g(\nabla_X Y + \nabla_Y X, Z) + g(Y, \nabla_X Z - \nabla_Z X) + g(X, \nabla_Y Z - \nabla_Z Y)$$

$$= g(2\nabla_X Y - [X, Y], Z) + g(Y, [X, Z]) + g(X, [Y, Z])$$

$$= 2g(\nabla_X Y, Z) - g([X, Y], Z) + g(Y, [X, Z]) + g(X, [Y, Z]).$$

This shows that the sum of the three equations (1.2.26) can be rewritten as

$$2g(\nabla_X Y, Z) = g([X, Y], Z) - g(Y, [X, Z]) - g(X, [Y, Z])$$

$$+ X(g(Y, Z)) + Y(g(Z, X)) - Z(g(X, Y)). \tag{1.2.27}$$

(As already mentioned, this equation is known as *Koszul's formula*.) Since Z is arbitrary and g is nondegenerate, the left-hand side of this equation determines the vector field $\nabla_X Y$ uniquely. Uniqueness of ∇ follows.

(Note that the calculation simplifies considerably when all the vector fields commute. This can be achieved by choosing $X = \partial_i$, $Y = \partial_j$, etc. Such a choice is often convenient when proving equality of tensor fields.)

To prove existence, let $S(X, Y)(Z)$ be defined as one half of the right-hand side of (1.2.27),

$$S(X, Y)(Z) = \frac{1}{2}\Big(X(g(Y, Z)) + Y(g(Z, X)) - Z(g(X, Y))$$

$$+ g(Z, [X, Y]) - g(Y, [X, Z]) - g(X, [Y, Z])\Big). \tag{1.2.28}$$

Clearly S is linear with respect to addition in all fields involved.
Let us check that it is linear with respect to multiplication of Z by a function:

$$S(X, Y)(fZ) = \frac{f}{2}\Big(X(g(Y, Z)) + Y(g(Z, X)) - Z(g(X, Y))$$

$$+ g(Z, [X, Y]) - g(Y, [X, Z]) - g(X, [Y, Z])\Big)$$

$$+ \frac{1}{2}\Big(X(f)g(Y, Z) + Y(f)g(Z, X) - g(Y, X(f)Z) - g(X, Y(f)Z)\Big)$$

$$= f S(X, Y)(Z). \tag{1.2.29}$$

Since g is nondegenerate, we conclude that there exists a unique vector field $W(X, Y)$ such that

$$S(X, Y)(Z) = g(W(X, Y), Z).$$

Calculations similar to those in (1.2.29) show that the assignment

$$(X, Y) \rightarrow \nabla_X Y := W(X, Y)$$

satisfies all the requirements imposed on a covariant derivative.

It is immediate from (1.2.27) that the connection ∇_X so defined is torsion-free: Indeed, the sum of all-but-first terms at the right-hand side of (1.2.27) is symmetric in (X, Y), while the first term is precisely what is needed to produce the torsion tensor when removing from (1.2.27) its counterpart with X and Y interchanged.

Finally, inserting $\nabla_X Y$ and $\nabla_X Z$, as defined by (1.2.27), into (1.2.25) one finds that ∇ is metric-compatible, which concludes the proof.

Let us check that (1.2.27) reproduces (1.2.24): Consider (1.2.27) with $X = \partial_\gamma$, $Y = \partial_\beta$, and $Z = \partial_\sigma$,

$$
\begin{aligned}
2g(\nabla_\gamma \partial_\beta, \partial_\sigma) &= 2g(\Gamma^\rho{}_{\beta\gamma} \partial_\rho, \partial_\sigma) \\
&= 2g_{\rho\sigma} \Gamma^\rho{}_{\beta\gamma} \\
&= \partial_\gamma g_{\beta\sigma} + \partial_\beta g_{\gamma\sigma} - \partial_\sigma g_{\beta\gamma} .
\end{aligned}
\tag{1.2.30}
$$

Multiplying this equation by $g^{\alpha\sigma}/2$ we then obtain the expected formula:

$$
\boxed{\Gamma^\alpha{}_{\beta\gamma} = \tfrac{1}{2} g^{\alpha\sigma} \{ \partial_\beta g_{\sigma\gamma} + \partial_\gamma g_{\sigma\beta} - \partial_\sigma g_{\beta\gamma} \}} .
\tag{1.2.31}
$$

1.2.8 Geodesics and Christoffel Symbols

A geodesic can be defined as the stationary point of the action

$$
I(\gamma) = \frac{1}{2} \int_a^b \underbrace{g(\dot\gamma, \dot\gamma)(s)}_{=:\mathscr{L}(\gamma, \dot\gamma)} ds ,
\tag{1.2.32}
$$

where $[a, b] \ni s \to \gamma(s) \in M$ is a differentiable curve, with $\dot\gamma := d\gamma/ds$. Thus,

$$
\mathscr{L}(x^\mu, \dot{x}^\nu) = \frac{1}{2} g_{\alpha\beta}(x^\mu) \dot{x}^\alpha \dot{x}^\beta .
$$

One readily finds the Euler-Lagrange equations for this Lagrange function:

$$
\frac{d}{ds}\left(\frac{\partial \mathscr{L}}{\partial \dot{x}^\mu}\right) = \frac{\partial \mathscr{L}}{\partial x^\mu} \quad \Longleftrightarrow \quad \frac{d^2 x^\mu}{ds^2} + \Gamma^\mu{}_{\alpha\beta} \frac{dx^\alpha}{ds} \frac{dx^\beta}{ds} = 0 .
\tag{1.2.33}
$$

This gives a very convenient way of calculating the Christoffel symbols: given a metric g, write down $\mathscr{L}$, work out the Euler-Lagrange equations, and identify the Christoffels as the coefficients of the first derivative terms in those equations.

(The Euler-Lagrange equations for (1.2.32) are identical with those of

$$
\tilde{I}(\gamma) = \int_a^b \sqrt{|g(\dot\gamma, \dot\gamma)(s)|} ds ,
\tag{1.2.34}
$$

but (1.2.32) is more convenient to work with. For example, $\mathscr{L}$ is differentiable at points where $\dot{\gamma}$ vanishes, while $\sqrt{|g(\dot{\gamma}, \dot{\gamma})(s)|}$ is not. The aesthetic advantage of (1.2.34), of being reparameterization-invariant, is more than compensated by the calculational convenience of $\mathscr{L}$.)

Example 1.2.2 As an example, consider a metric of the form

$$g = dr^2 + f(r)d\varphi^2 .$$

Special cases of this metric include the Euclidean metric on $\mathbb{R}^2$ (then $f(r) = r^2$), and the canonical metric on a sphere (then $f(r) = \sin^2 r$, with r actually being the polar angle θ). The Lagrangian (1.2.34) is thus

$$L = \frac{1}{2} \left(\dot{r}^2 + f(r)\dot{\varphi}^2 \right) .$$

The Euler-Lagrange equations read

$$\frac{d}{ds} \left(\frac{\partial L}{\partial \dot{\varphi}} \right) = \frac{d}{ds}(f(r)\dot{\varphi}) = \underbrace{\frac{\partial L}{\partial \varphi}}_{0} ,$$

so that

$$0 = f\ddot{\varphi} + f'\dot{r}\dot{\varphi} = f \left(\ddot{\varphi} + \Gamma^{\varphi}_{\varphi\varphi}\dot{\varphi}^2 + 2\Gamma^{\varphi}_{r\varphi}\dot{r}\dot{\varphi} + \Gamma^{\varphi}_{rr}\dot{r}^2 \right) ,$$

which implies

$$\Gamma^{\varphi}_{\varphi\varphi} = \Gamma^{\varphi}_{rr} = 0 , \quad \Gamma^{\varphi}_{r\varphi} = \frac{f'}{2f} .$$

Similarly

$$\frac{d}{ds} \left(\frac{\partial L}{\partial \dot{r}} \right) = \ddot{r} = \underbrace{\frac{\partial L}{\partial r}}_{f'\dot{\varphi}^2/2} ,$$

so that

$$\Gamma^{r}_{r\varphi} = \Gamma^{r}_{rr} = 0 , \quad \Gamma^{r}_{\varphi\varphi} = -\frac{f'}{2} .$$

This also shows that the curves of constant φ are geodesics of g.

1.3 Local Inertial Coordinates

A useful notion from the point of view of physics, but also as a simplifying tool for some calculations, is that of *local inertial coordinates*. These are the coordinates whose existence is asserted in the next result:

Proposition 1.3.1

1. Let *g* be a Lorentzian metric on *M*, for every *p* ∈ *M* there exists a neighborhood thereof with a coordinate system such that $g_{\mu\nu} = \eta_{\mu\nu} = \mathrm{diag}(-1, 1, \cdots, 1)$ at *p*.

2. If *g* is differentiable, then the coordinates can be further chosen so that

$$\partial_\sigma g_{\alpha\beta} = 0 \iff \Gamma^\alpha{}_{\beta\gamma} = 0 \qquad (1.3.1)$$

at *p*.

Proof

1. Let y^μ be any coordinate system around *p*, shifting by a constant vector we can assume that *p* corresponds to $y^\mu = 0$. Let $e_a = e_a{}^\mu \partial/\partial y^\mu$ be any frame at *p* such that $g(e_a, e_b) = \eta_{ab}$—such frames can be found by, *e.g.*, a Gram-Schmidt orthogonalization. Calculating the determinant of both sides of the equation

$$g_{\mu\nu} e_a{}^\mu e_b{}^\nu = \eta_{ab}$$

we obtain, at *p*,

$$\det(g_{\mu\nu}) \det(e_a{}^\mu)^2 = -1,$$

which shows that $\det(e_a{}^\mu)$ is nonvanishing. It follows that the map $y^\mu \mapsto x^a$ defined implicitly by the formula

$$y^\mu = e^\mu{}_a x^a$$

defines a (linear) diffeomorphism. Using the x^a as the new coordinates we have, again at *p*,

$$g\left(\frac{\partial}{\partial x^a}, \frac{\partial}{\partial x^b}\right) = e^\mu{}_a e^\nu{}_b \, g\left(\frac{\partial}{\partial y^\mu}, \frac{\partial}{\partial y^\nu}\right) = \eta_{ab}. \qquad (1.3.2)$$

2. We will use (1.2.13), which uses Latin indices, so let us switch to that notation. Let x^i be the coordinates described in point 1., recall that p lies at the origin of those coordinates. The new coordinates $\hat{x}^j$ will be implicitly defined by the equations

$$x^i = \hat{x}^i + \frac{1}{2}A^i{}_{jk}\hat{x}^j\hat{x}^k \,,$$

where $A^i{}_{jk}$ is a set of constants, symmetric with respect to the interchange of j and k. Recall (1.2.13),

$$\hat{\Gamma}^i{}_{jk} = \Gamma^s{}_{\ell r}\frac{\partial\hat{x}^i}{\partial x^s}\frac{\partial x^\ell}{\partial\hat{x}^j}\frac{\partial x^r}{\partial\hat{x}^k} + \frac{\partial\hat{x}^i}{\partial x^s}\frac{\partial^2 x^s}{\partial\hat{x}^k\partial\hat{x}^j}\,; \tag{1.3.3}$$

here we use the symbol $\hat{\Gamma}^s{}_{\ell r}$ to denote the Christoffel symbols of the metric in the hatted coordinates. At $x^i = 0$ this equation reads

$$\hat{\Gamma}^i{}_{jk} = \Gamma^s{}_{\ell r}\underbrace{\frac{\partial\hat{x}^i}{\partial x^s}}_{\delta^i_s}\underbrace{\frac{\partial x^\ell}{\partial\hat{x}^j}}_{\delta^\ell_j}\underbrace{\frac{\partial x^r}{\partial\hat{x}^k}}_{\delta^r_k} + \underbrace{\frac{\partial\hat{x}^i}{\partial x^s}}_{\delta^i_s}\underbrace{\frac{\partial^2 x^s}{\partial\hat{x}^k\partial\hat{x}^j}}_{A^s_{kj}}$$

$$= \Gamma^i{}_{jk} + A^i{}_{jk}\,.$$

Choosing A^i_{jk} as $-\Gamma^i{}_{jk}(0)$, the result follows. □

Incidentally If you do not like to remember formulae such as (1.2.13), proceed as follows: Let x^μ be the coordinates described in point 1. The new coordinates $\hat{x}^\alpha$ will be implicitly defined by the equations

$$x^\mu = \hat{x}^\mu + \frac{1}{2}A^\mu{}_{\alpha\beta}\hat{x}^\alpha\hat{x}^\beta\,,$$

where $A^\mu{}_{\alpha\beta}$ is a set of constants, symmetric with respect to the interchange of α and β. Set

$$\hat{g}_{\alpha\beta} := g\left(\frac{\partial}{\partial\hat{x}^\alpha}, \frac{\partial}{\partial\hat{x}^\beta}\right), \qquad g_{\alpha\beta} := g\left(\frac{\partial}{\partial x^\alpha}, \frac{\partial}{\partial x^\beta}\right).$$

Recall the transformation law

$$\hat{g}_{\mu\nu}(\hat{x}^\sigma) = g_{\alpha\beta}(x^\rho(\hat{x}^\sigma))\frac{\partial x^\alpha}{\partial\hat{x}^\mu}\frac{\partial x^\beta}{\partial\hat{x}^\nu}\,.$$

By differentiation one obtains at $x^\mu = \hat{x}^\mu = 0$,

$$\frac{\partial\hat{g}_{\mu\nu}}{\partial\hat{x}^\rho}(0) = \frac{\partial g_{\mu\nu}}{\partial x^\rho}(0) + g_{\alpha\beta}(0)\left(A^\alpha{}_{\mu\rho}\delta^\beta_\nu + \delta^\alpha_\mu A^\beta{}_{\nu\rho}\right)$$

$$= \frac{\partial g_{\mu\nu}}{\partial x^\rho}(0) + A_{\nu\mu\rho} + A_{\mu\nu\rho}\,, \tag{1.3.4}$$

where

$$A_{\alpha\beta\gamma} := g_{\alpha\sigma}(0) A^{\sigma}{}_{\beta\gamma} .$$

It remains to show that we can choose $A^{\sigma}{}_{\beta\gamma}$ so that the left-hand side can be made to vanish at p. An explicit formula for the $A_{\sigma\beta\gamma}$'s can be obtained from (1.3.4) by a cyclic permutation calculation similar to that in (1.2.26). After raising the first index, the final result is

$$A^{\alpha}{}_{\beta\gamma} = \frac{1}{2} g^{\alpha\rho} \left\{ \frac{\partial g_{\beta\gamma}}{\partial x^{\rho}} - \frac{\partial g_{\beta\rho}}{\partial x^{\gamma}} - \frac{\partial g_{\rho\gamma}}{\partial x^{\beta}} \right\} (0) .$$

The reader may wish to check directly that this does indeed lead to a vanishing right-hand side of (1.3.4).

1.4 Curvature

We have seen that we can get rid of first derivatives of the metric at any point by making a coordinate transformation. It turns out that second derivatives cannot be made to vanish in this way. This will follow from the study of a new object, called the curvature tensor.

Proposition 1.4.1

1. Let ∇ be torsion-free. There exists a tensor field $R^{d}{}_{abc}$ of type $(1, 3)$ such that

$$\nabla_a \nabla_b X^d - \nabla_b \nabla_a X^d = R^{d}{}_{cab} X^c . \tag{1.4.1}$$

2. It holds that

$$R^{d}{}_{cab} = \partial_a \Gamma^{d}_{bc} - \partial_b \Gamma^{d}_{ac} + \Gamma^{d}_{ae} \Gamma^{e}_{bc} - \Gamma^{d}_{be} \Gamma^{e}_{ac} . \tag{1.4.2}$$

Proof

We need to check that the derivatives of X cancel. Now,

$$\nabla_a \nabla_b X^d - \partial_a (\underbrace{\nabla_b X^d}_{\partial_b X^d + \Gamma^d_{be} X^e}) + \Gamma^d_{ac} \underbrace{\nabla_b X^c}_{\partial_b X^c + \Gamma^c_{be} X^e} - \Gamma^e_{ab} \nabla_e X^d$$

$$= \partial_a \partial_b X^d \underbrace{}_{=:1_{ab}} + \partial_a \Gamma^d_{be} X^e + \underbrace{\Gamma^d_{be} \partial_a X^e}_{=:2_{ab}} + \underbrace{\Gamma^d_{ac} \partial_b X^c}_{=:3_{ab}} + \Gamma^d_{ac} \Gamma^c_{be} X^e - \underbrace{\Gamma^e_{ab} \nabla_e X^d}_{=:4_{ab}} .$$

If we subtract $\nabla_b \nabla_a X^d$, then

1. 1_{ab} is symmetric in a and b, so will cancel out; similarly for 4_{ab} because ∇ has been assumed to have no torsion;

2. 2_{ab} will cancel out with 3_{ba}; similarly 3_{ab} will cancel out with 2_{ba}.

So the left-hand side of (1.4.1) is indeed linear in X^e. Since it is a tensor, the right-hand side also is. Since X^e is arbitrary, we conclude that $R^d{}_{cab}$ is a tensor of the desired type. This proves point 1.

To prove 2., from what has been said we have

$$\nabla_a \nabla_b X^d - \nabla_b \nabla_a X^d = \partial_a \Gamma^d_{be} X^e + \Gamma^d_{ac} \Gamma^c_{be} X^e - (a \leftrightarrow b)$$

$$= (\partial_a \Gamma^d_{bc} - \partial_b \Gamma^d{}_{ac} + \Gamma^d_{ae} \Gamma^e_{bc} - \Gamma^d_{be} \Gamma^e_{ac}) X^c ,$$

as desired. Here "$(a \leftrightarrow b)$" means "the same expression with a interchanged with b".   □

The calculation of the curvature tensor may be a very traumatic experience. There is one obvious case where things are painless, when all $g_{\mu\nu}$'s are constants: in this case the Christoffels vanish, and so does the curvature tensor. Metrics with the last property are called *flat*.

For more general metrics, one way out is to use symbolic computer algebra. MATHEMATICA packages to do this can be found at URLs http://www.math.washington.edu/~lee/Ricci, or http://grtensor.phy.queensu.ca/NewDemo, or http://luth.obspm.fr/~luthier/Martin-Garcia/xAct. This last package is least user-friendly as of today, but is the most flexible, especially for more involved computations.

Example 1.4.2
As an example less trivial than a metric with constant coefficients, consider the round two sphere, which we write in the form

$$g = d\theta^2 + e^{2f} d\varphi^2 , \qquad e^{2f} = \sin^2 \theta .$$

The Christoffel symbols are easily found from the Lagrangian for geodesics (see example 1.2.2):

$$\mathscr{L} = \frac{1}{2}(\dot{\theta}^2 + e^{2f} \dot{\varphi}^2) .$$

The Euler-Lagrange equations give

$$\Gamma^\theta{}_{\varphi\varphi} = -f' e^{2f} , \quad \Gamma^\varphi{}_{\theta\varphi} = \Gamma^\varphi{}_{\varphi\theta} = f' ,$$

with the remaining Christoffel symbols vanishing. Using the definition of the Riemann tensor we then immediately find that

$$R_{\theta\varphi\theta}{}^\varphi = -f'' - (f')^2 = 1 .$$

All remaining components of the Riemann tensor can be obtained from this one by raising and lowering of indices, together with symmetry operations.

Taking into account the symmetries of the Riemann tensor, the above calculation can be summarized neatly as

$$R_{abcd} = g_{ac}g_{bd} - g_{ad}g_{bc} \,. \tag{1.4.3}$$

Hence

$$R_{ab} := R^c{}_{acb} = g_{ab} \,, \qquad R := g^{ab} R_{ab} = 2 \,. \qquad \blacksquare$$

Incidentally A manifold (M, g) is called a *space-form* if there exists a constant κ such that the curvature tensor takes the *maximally symmetric* form, namely (compare(1.4.3))

$$R_{abcd} = \kappa(g_{ac}g_{bd} - g_{ad}g_{bc}) \,. \tag{1.4.4}$$

(Allowing κ to be a function would not lead to more general metrics in dimensions higher than two as the second Bianchi identity implies that κ must be constant anyway.) In dimension d this leads to

$$R_{ac} = \kappa(d-1)g_{ac} \,, \qquad R = d(d-1) \,. \tag{1.4.5}$$

Flat metrics obviously satisfy (1.4.4) with $\kappa = 0$. Example 1.4.2 shows that the two-dimensional sphere with the standard unit metric is a space-form, with $\kappa = 1$. Another two-dimension example is provided by the two-dimensional *projective space*, which is obtained from S^2 by identifying the antipodal points.

All higher-dimensional spheres with their canonical unit round metrics are space-forms, with $\kappa = 1$; the same is true of their quotients.

A negatively curved example, with $\kappa = -1$, is provided by the *hyperbolic space*. In the *half-space model* the manifold is $(0, \infty) \times \mathbb{R}^{n-1}$, with the metric taking the form

$$g = \frac{dz^2 + (dx^1)^2 + \ldots + (dx^{n-1})^2}{z^2} \,. \tag{1.4.6}$$

Exercise 1.4.3 As an exercise, we check that an equivalent, index-free definition of the Riemann tensor can be given as follows: Let ∇ be a *torsionless* covariant derivative defined for vector fields. Set

$$\boxed{R(X, Y)Z := \nabla_X \nabla_Y Z - \nabla_Y \nabla_X Z - \nabla_{[X,Y]}Z} \,, \tag{1.4.7}$$

where (as elsewhere) $[X, Y]$ is the Lie bracket defined in (1.1.7). Then the coordinate components

$$R^d{}_{cab} = dx^d(R(\partial_a, \partial_b)\partial_c) \quad \Longleftrightarrow \quad R(\partial_a, \partial_b)\partial_c = R^d{}_{cab}\partial_d$$

coincide with those defined by (1.4.1). To prove this, given a vector field Z, consider the tensor field S defined as

$$Y \longrightarrow S(Y) := \nabla_Y Z.$$

In local coordinates, S takes the form

$$S = \nabla_\mu Z^\nu \, dx^\mu \otimes \partial_\nu.$$

It follows from the Leibniz rule—or, equivalently, from the definitions in ▶ Sect. 1.2—that we have

$$(\nabla_X S)(Y) = \nabla_X(S(Y)) - S(\nabla_X Y)$$
$$= \nabla_X \nabla_Y Z - \nabla_{\nabla_X Y} Z.$$

The commutator of the derivatives can then be calculated as

$$(\nabla_X S)(Y) - (\nabla_Y S)(X) = \nabla_X \nabla_Y Z - \nabla_Y \nabla_X Z - \nabla_{\nabla_X Y} Z + \nabla_{\nabla_Y X} Z$$
$$= \nabla_X \nabla_Y Z - \nabla_Y \nabla_X Z - \nabla_{[X,Y]} Z$$
$$+ \nabla_{[X,Y]} Z - \nabla_{\nabla_X Y} Z + \nabla_{\nabla_Y X} Z$$
$$= R(X,Y)Z - \nabla_{T(X,Y)} Z. \tag{1.4.8}$$

Writing ∇S in the usual form

$$\nabla S = \nabla_\sigma S_\mu{}^\nu \, dx^\sigma \otimes dx^\mu \otimes \partial_\nu = \nabla_\sigma \nabla_\mu Z^\nu \, dx^\sigma \otimes dx^\mu \otimes \partial_\nu,$$

we are thus led to

$$\nabla_\mu \nabla_\nu Z^\alpha - \nabla_\nu \nabla_\mu Z^\alpha = R^\alpha{}_{\sigma\mu\nu} Z^\sigma - T^\sigma{}_{\mu\nu} \nabla_\sigma Z^\alpha. \tag{1.4.9}$$

In the important case of vanishing torsion, the coordinate-component equivalent of (1.4.7) is thus

$$\boxed{\nabla_\mu \nabla_\nu Z^\alpha - \nabla_\nu \nabla_\mu Z^\alpha = R^\alpha{}_{\sigma\mu\nu} Z^\sigma}. \tag{1.4.10}$$

A calculation identical to that in the proof of Proposition 1.4.1 gives, again for torsionless connections,

$$\nabla_\mu \nabla_\nu a_\alpha - \nabla_\nu \nabla_\mu a_\alpha = -R^\sigma{}_{\alpha\mu\nu} a_\sigma. \tag{1.4.11}$$

For a general tensor t and torsion-free connection, each tensor index comes with a corresponding Riemann tensor term:

$$\nabla_\mu \nabla_\nu t_{\alpha_1 \ldots \alpha_r}{}^{\beta_1 \ldots \beta_s} - \nabla_\nu \nabla_\mu t_{\alpha_1 \ldots \alpha_r}{}^{\beta_1 \ldots \beta_s} =$$
$$- R^\sigma{}_{\alpha_1 \mu\nu} t_{\sigma \ldots \alpha_r}{}^{\beta_1 \ldots \beta_s} - \ldots - R^\sigma{}_{\alpha_r \mu\nu} t_{\alpha_1 \ldots \sigma}{}^{\beta_1 \ldots \beta_s}$$
$$+ R^{\beta_1}{}_{\sigma\mu\nu} t_{\alpha_1 \ldots \alpha_r}{}^{\sigma \ldots \beta_s} + \ldots + R^{\beta_s}{}_{\sigma\mu\nu} t_{\alpha_1 \ldots \alpha_r}{}^{\beta_1 \ldots \sigma}. \tag{1.4.12}$$

From now on we assume a Levi-Civita connection.

One of the fundamental results in Riemannian geometry is:

Theorem 1.4.4
There exists a coordinate system in which the metric tensor field has vanishing second derivatives at p if and only if its Riemann tensor vanishes at p. Furthermore, there exists a coordinate system in which the metric tensor field has constant entries near p if and only if the Riemann tensor vanishes near p.

Proof
The condition is necessary, since $R^\delta{}_{\gamma\alpha\beta}$ is a tensor. The sufficiency will be admitted.   □

1.4.1 Symmetries

Here is a full list of algebraic symmetries of the curvature tensor of the Levi-Civita connection:

1. Directly from the definition, we obtain

$$\boxed{R^\delta{}_{\gamma\alpha\beta} = -R^\delta{}_{\gamma\beta\alpha}}\,. \tag{1.4.13}$$

2. The next symmetry, known as the *first Bianchi identity*, is less obvious:

$$\boxed{R^\delta{}_{\gamma\alpha\beta} + R^\delta{}_{\alpha\beta\gamma} + R^\delta{}_{\beta\gamma\alpha} = 0}\,. \tag{1.4.14}$$

An equivalent form is

$$R^\delta{}_{[\gamma\alpha\beta]} = 0\,, \tag{1.4.15}$$

where brackets over indices denote complete antisymmetrization, e.g.

$$A_{[\alpha\beta]} = \tfrac{1}{2}(A_{\alpha\beta} - A_{\beta\alpha})\,,$$
$$A_{[\alpha\beta\gamma]} = \tfrac{1}{6}(A_{\alpha\beta\gamma} - A_{\beta\alpha\gamma} + A_{\gamma\alpha\beta} - A_{\gamma\beta\alpha} + A_{\alpha\gamma\beta} - A_{\beta\gamma\alpha})\,,$$

with a similar definition for any number of indices. Equation (1.4.15) is justified using (1.4.13), which implies the straightforward identity

$$6R^\delta{}_{[\gamma\alpha\beta]} = 2(R^\delta{}_{\gamma\alpha\beta} + R^\delta{}_{\alpha\beta\gamma} + R^\delta{}_{\beta\gamma\alpha})\,.$$

3. Finally, we have the pair-interchange symmetry:

$$\boxed{R_{\alpha\beta\gamma\delta} = R_{\gamma\delta\alpha\beta}}\,. \tag{1.4.16}$$

Here, of course, $R_{\gamma\delta\alpha\beta} = g_{\gamma\sigma} R^{\sigma}{}_{\delta\alpha\beta}$.

It is not obvious, but true, that this list exhausts all independent algebraic identities satisfied by $R_{\alpha\beta\gamma\delta}$.

As a consequence of (1.4.13) and (1.4.16) we find

$$R_{\alpha\beta\delta\gamma} = R_{\delta\gamma\alpha\beta} = -R_{\delta\gamma\beta\alpha} = -R_{\beta\alpha\gamma\delta} \,,$$

and so the Riemann tensor is also antisymmetric in its first two indices:

$$R_{\alpha\beta\gamma\delta} = -R_{\beta\alpha\gamma\delta} \,. \tag{1.4.17}$$

The Ricci tensor is defined as

$$R_{\alpha\beta} := R^{\sigma}{}_{\alpha\sigma\beta} \,.$$

The pair-interchange symmetry implies that the Ricci tensor is symmetric:

$$R_{\alpha\beta} = g^{\sigma\rho} R_{\sigma\alpha\rho\beta} = g^{\sigma\rho} R_{\rho\beta\sigma\alpha} = R_{\beta\alpha} \,.$$

To prove

$$R_{\alpha\beta\gamma\delta} = R_{\gamma\delta\alpha\beta} \tag{1.4.18}$$

we suppose that the metric is twice-differentiable. By point 2. of Proposition 1.3.1, in a neighborhood of any point $p \in M$ there exists a coordinate system in which the connection coefficients $\Gamma^{\alpha}{}_{\beta\gamma}$ vanish at p. Equation (1.4.2) evaluated at p therefore reads

$$\begin{aligned}
R^{\alpha}{}_{\beta\gamma\delta} &= \partial_{\gamma} \Gamma^{\alpha}{}_{\beta\delta} - \partial_{\delta} \Gamma^{\alpha}{}_{\beta\gamma} \\
&= \frac{1}{2} \left\{ g^{\alpha\sigma} \partial_{\gamma} (\partial_{\delta} g_{\sigma\beta} + \partial_{\beta} g_{\sigma\delta} - \partial_{\sigma} g_{\beta\delta}) \right. \\
&\quad \left. - g^{\alpha\sigma} \partial_{\delta} (\partial_{\gamma} g_{\sigma\beta} + \partial_{\beta} g_{\sigma\gamma} - \partial_{\sigma} g_{\beta\gamma}) \right\} \\
&= \frac{1}{2} g^{\alpha\sigma} \left\{ \partial_{\gamma} \partial_{\beta} g_{\sigma\delta} - \partial_{\gamma} \partial_{\sigma} g_{\beta\delta} - \partial_{\delta} \partial_{\beta} g_{\sigma\gamma} + \partial_{\delta} \partial_{\sigma} g_{\beta\gamma} \right\} \,.
\end{aligned}$$

Equivalently,

$$R_{\sigma\beta\gamma\delta}(0) = \frac{1}{2} \left\{ \partial_{\gamma} \partial_{\beta} g_{\sigma\delta} - \partial_{\gamma} \partial_{\sigma} g_{\beta\delta} - \partial_{\delta} \partial_{\beta} g_{\sigma\gamma} + \partial_{\delta} \partial_{\sigma} g_{\beta\gamma} \right\}(0) \,. \tag{1.4.19}$$

The first term goes to the last one under interchange of $\sigma\beta$ with $\gamma\delta$; similarly for the second and the third, and (1.4.18) follows.

Further, the indices $\sigma\beta\gamma$ on the first term in (1.4.19) form a cyclic permutation of those in the second; similarly for the third and the fourth one. This proves the first Bianchi identity (1.4.14) in its equivalent form $R_{[\sigma\beta\gamma]\delta} = 0$.

Remark 1.4.5 In dimension two it holds that

$$R_{\alpha\beta\gamma\delta} = \frac{R}{2}(g_{\alpha\gamma}g_{\beta\delta} - g_{\alpha\delta}g_{\beta\gamma}), \quad R_{\alpha\beta} = \frac{R}{2}g_{\alpha\beta}, \tag{1.4.20}$$

which can be seen as follows: The symmetries of the Riemann tensor imply that the only possibly nonzero components thereof are

$$R_{1212} = -R_{2112} = R_{2121} = -R_{1221}.$$

Now, it is straightforward to check that the tensor $g_{\alpha\gamma}g_{\beta\delta} - g_{\alpha\delta}g_{\beta\gamma}$ enjoys an identical property. Hence there exists a function f such that

$$R_{\alpha\beta\gamma\delta} = f(g_{\alpha\gamma}g_{\beta\delta} - g_{\alpha\delta}g_{\beta\gamma}). \tag{1.4.21}$$

Taking traces one obtains (1.4.20).

In dimension three, a similar argument gives

$$R_{\alpha\beta\gamma\delta} = (P_{\alpha\gamma}g_{\beta\delta} - P_{\alpha\delta}g_{\beta\gamma} + g_{\alpha\gamma}P_{\beta\delta} - g_{\alpha\delta}P_{\beta\gamma}), \tag{1.4.22}$$

where

$$P_{\alpha\beta} := R_{\alpha\beta} - \frac{R}{4}g_{\alpha\beta}. \qquad \square$$

We close this section with a differential identity satisfied by the Riemann tensor, known as the *second Bianchi identity*:

$$\nabla_\alpha R_{\sigma\beta\gamma\delta} + \nabla_\sigma R_{\beta\alpha\gamma\delta} + \nabla_\beta R_{\alpha\sigma\gamma\delta} = 0. \tag{1.4.23}$$

The proof is again simplest in coordinates in which the derivatives of the metric vanish at p: Indeed, a calculation very similar to the one leading to (1.4.19) gives

$$\nabla_\alpha R_{\sigma\beta\gamma\delta}(0) = \partial_\alpha R_{\upsilon\beta\gamma\delta}(0) -$$

$$\frac{1}{2}\left\{\partial_\alpha\partial_\gamma\partial_\beta g_{\sigma\delta} - \partial_\alpha\partial_\gamma\partial_\sigma g_{\beta\delta} - \partial_\alpha\partial_\delta\partial_\beta g_{\sigma\gamma} + \partial_\alpha\partial_\delta\partial_\sigma g_{\beta\gamma}\right\}(0), \tag{1.4.24}$$

and the result follows by inspection.

Curved Spacetime

Piotr T. Chruściel

© Springer Nature Switzerland AG 2019
P. T. Chruściel, *Elements of General Relativity*, Compact Textbooks in Mathematics,
https://doi.org/10.1007/978-3-030-28416-9_2

In this chapter we introduce the main ideas of general relativity.

2.1 The Heuristics

Having realized that special relativity provides a better description of our world than Newtonian mechanics, it is necessary to find a theory of gravitation which will be compatible with the basic facts of special-relativity. Recall that Newtonian gravity is a theory where gravitational interaction proceeds via the gravitational potential ϕ, as obtained by solving the Laplace equation,

$$\Delta_\delta \phi = -4\pi G\mu \,, \tag{2.1.1}$$

where G is Newton's constant, μ is the density of matter fields, and Δ_δ is the Laplace operator associated with the flat Euclidean metric. This equation is certainly not compatible with special relativity, as it requires preferred constant-time-slices on which ϕ is determined.[1] In particular it is certainly not covariant under Lorentz transformations.

The question thus arises, how to formulate a theory which would include (2.1.1) in the limit of small velocities and weak fields, and which would be compatible with special relativity. A hint how such a theory should look like is given by the observed equality between the inertial mass and the passive gravitational mass: Indeed, let us denote by m_i the parameter which arises in the Newtonian equation relating the force and the acceleration,

$$\vec{F} = m_i \vec{a} \,, \tag{2.1.2}$$

---

[1]It is rather ironic that general relativistic cosmology actually predicts existence of a preferred cosmic time. This does, of course, not contradict the basic principle that the theory should be formulated so that no preferred coordinates exist.

and let us call this parameter the *inertial mass*. Here the index i on m stands for "inertial". Let us denote by m_g the parameter which appears in the Newtonian equation asserting that the gravitational force is proportional to the gradient of the potential ϕ of (2.1.1)

$$\vec{F} = -m_g \nabla \phi .$$
(2.1.3)

It has been measured to high precision that

$$m_i = m_g \quad \Longrightarrow \quad \vec{a} = -\nabla \phi .$$
(2.1.4)

Incidentally The history of modern measurements constraining the ratio $(m_i - m_g)/(m_i + m_g)$ is usually thought to start with the torsion-balance experiments of Eötvös, carried out in the period 1895–1906 and leading to an upper bound for the ratio of the order of 10^{-9}. The most recent, at the time of writing of these notes, pre-official-release upper bound of 10^{-14} has been obtained from the satellite experiment MICROSCOPE. The satellite mission ended in October 2018, and the final results are expected to be published around the end of 2019.

Equation (2.1.3) has the consequence that in Newton's theory, and in the absence of other forces, *acceleration cannot be distinguished from a gravitational field*. Equivalently, local effects of gravity can be made to disappear by introducing a new coordinate system which is accelerated with respect to the original one. Yet another way of saying this is, that at every point of spacetime there exists a class of preferred coordinate systems in which no gravitational forces are experienced locally.

The above observation, arising within the framework of Newtonian gravity, has been elevated to a universal principle, independent of the model, the *equivalence principle*:

> no local experiment can distinguish between acceleration and gravity.   (2.1.5)

Einstein's key observation is that something similar happens when considering local inertial frames associated with a general metric with Lorentzian signature. Indeed, at every point p of every Lorentzian manifold, the metric g defines a preferred class of coordinate systems, the *inertial frames at* p, in which $g_{\mu\nu}$ takes the Minkowskian form $\eta_{\mu\nu}$ and where the partial derivatives of the metric tensor vanish. His working proposal, how to do physics on Lorentzian manifolds $(\mathcal{M}, g)$, is to carry over the equations of special-relativistic physics to $(\mathcal{M}, g)$ by requiring that these equations take their usual form at the center of local inertial coordinates. Letting p be a point in $\mathcal{M}$ with associated inertial coordinates x^μ, we thus have:

1. In Minkowski space-time a vector X is called

$$\begin{cases} \text{timelike} \\ \text{null} \\ \text{causal} \\ \text{spacelike} \end{cases} \text{if} \quad \begin{cases} \eta(X, X) < 0 , \\ \eta(X, X) = 0 , \; X \neq 0 , \\ \eta(X, X) \leq 0 , \; X \neq 0 , \\ \text{otherwise.} \end{cases}$$

At p and in local inertial coordinates this is the same as:

$$
\begin{cases}
\text{timelike} \\
\text{null} \\
\text{causal} \\
\text{spacelike}
\end{cases}
\text{if}
\begin{cases}
g(X, X) < 0, \\
g(X, X) = 0, \ X \neq 0, \\
g(X, X) \leq 0, \ X \neq 0, \\
\text{otherwise.}
\end{cases}
\tag{2.1.6}
$$

Since (2.1.6) is point- and coordinate-independent, one adopts it as the relevant definition in general relativity.

One also has the obvious generalization of this definition to curves: a curve $\mathbb{R} \supset I \ni \lambda \mapsto \gamma(\lambda) \in \mathcal{M}$, where I is an interval, with tangent $\dot{\gamma} := d\gamma/d\lambda$ is

$$
\begin{cases}
\text{timelike} \\
\text{null} \\
\text{causal} \\
\text{spacelike}
\end{cases}
\text{if}
\begin{cases}
g(\dot{\gamma}, \dot{\gamma}) < 0, \\
g(\dot{\gamma}, \dot{\gamma}) = 0, \ \dot{\gamma} \neq 0 \\
g(\dot{\gamma}, \dot{\gamma}) \leq 0, \ \dot{\gamma} \neq 0 \\
\text{otherwise.}
\end{cases}
\text{for all } \lambda \in I
\tag{2.1.7}
$$

The proper-time coordinate s along a timelike curve is obtained by an obvious similar generalization of the Minkowskian definition:

$$
s(\lambda_2) - s(\lambda_1) = \int_{s_1}^{s_2} \sqrt{\left| g\left(\frac{d\gamma}{d\lambda}, \frac{d\gamma}{d\lambda} \right) \right|} \, d\lambda \,.
\tag{2.1.8}
$$

2. The special-relativistic requirement that physical objects follow causal curves, $\eta(\dot{x}, \dot{x}) \leq 0$, reads at p

$$
\eta(\dot{x}, \dot{x})|_{x(s)=p} \leq 0 \,,
\tag{2.1.9}
$$

which, in local inertial coordinates, is the same as

$$
g(\dot{x}, \dot{x})|_{x(s)-p} \leq 0 \,.
\tag{2.1.10}
$$

Similarly the special relativistic requirement that massive physical objects move along worldlines, i.e., curves with timelike tangent $\dot{x}$, becomes at p

$$
g(\dot{x}, \dot{x})|_{x(s)=p} < 0 \,.
\tag{2.1.11}
$$

3. The special relativistic free-fall equation $d^2 x^\mu / ds^2 = 0$ becomes at p

$$
\left. \frac{d^2 x^\mu}{ds^2} \right|_{x(s)=p} = 0 \,,
\tag{2.1.12}
$$

which is the same as the geodesic equation in a coordinate system in which the Christoffel symbols vanish at p:

$$\frac{d^2 x^\mu}{ds^2} + \underbrace{\Gamma^\mu_{\alpha\beta}\Big|_{x(s)=p}}_{0} \frac{dx^\alpha}{ds} \frac{dx^\beta}{ds} = 0. \tag{2.1.13}$$

4. The special-relativistic divergence identity for an energy-momentum tensor, $\partial_\mu T^{\mu\nu} = 0$, becomes at p

$$\partial_\mu T^{\mu\nu}|_p = 0, \tag{2.1.14}$$

which, as before, is the same as the tensorial equation

$$\nabla_\mu T^{\mu\nu}|_p = 0. \tag{2.1.15}$$

5. The special-relativistic source-free Maxwell equations, when rewritten in inertial coordinates at p,

$$\partial_\mu F^{\mu\nu}|_p = 0, \quad \partial_{[\alpha} F_{\beta\gamma]}|_p = 0, \quad \text{where } F_{\beta\gamma}|_p = \eta_{\beta\mu}\eta_{\gamma\nu} F^{\mu\nu}|_p, \tag{2.1.16}$$

coincide, at p, with the tensorial equations

$$\nabla_\mu F^{\mu\nu}|_p = 0, \quad \nabla_{[\alpha} F_{\beta\gamma]}|_p = 0, \quad \text{where } F_{\beta\gamma}|_p = g_{\beta\mu} g_{\gamma\nu} F^{\mu\nu}|_p. \tag{2.1.17}$$

Having written the equations above, one realizes that (2.1.10), (2.1.11), (2.1.13), (2.1.15), and (2.1.17) make sense without invoking local inertial frames. This leads one the conclusion that it is convenient, whenever possible, to write all equations without reference to a specific coordinate system or frame.

2.2   Summary of Basic Ideas

1. "Special relativity holds over small distances and short times in local inertial frames." This is implemented by allowing the Minkowski metric

$$\eta = -dt^2 + dx^2 + dy^2 + dz^2$$

to be replaced by a "metric tensor" with Lorentzian signature

$$g = g_{\mu\nu} dx^\mu dx^\nu,$$

where the coefficients $g_{\mu\nu}$ are now allowed to be functions.

In particular it follows that physical observers move along curves with $g(\dot\gamma, \dot\gamma) < 0$—such curves are called *timelike*.

2. Gravity appears as the relative acceleration of nearby local inertial frames.
3. **Principle of general covariance:** the theory should be formulated in a way which does not give any special role to any particular coordinate system.
4. Vacuum Einstein equations: in vacuum, the Ricci tensor $R_{\mu\nu} := R_{\mu\alpha\nu}{}^{\alpha}$ vanishes.
5. Freely falling test particles move along timelike geodesics.[2]
6. Space-time is a manifold!

2.2.1 Geodesic Deviation (Jacobi Equation) and Tidal Forces

How are extended bodies, as opposed to point objects, affected by the gravitational field? To understand this, let us look at families of geodesics.

Suppose that we have a one parameter family of geodesics

$$\gamma(s, \lambda) \text{ (in local coordinates, } (\gamma^{\alpha}(s, \lambda))),$$

where s is the parameter along the geodesic, and λ is a parameter which distinguishes the geodesics. Set

$$Z(s, \lambda) := \frac{\partial \gamma(s, \lambda)}{\partial \lambda} \equiv \frac{\partial \gamma^{\alpha}(s, \lambda)}{\partial \lambda} \partial_{\alpha};$$

for each λ this defines a vector field Z along $\gamma(s, \lambda)$, which measures how nearby geodesics deviate from each other, since, to first order, using a Taylor expansion,

$$\gamma^{\alpha}(s, \lambda) = \gamma^{\alpha}(s, \lambda_0) + Z^{\alpha}(\lambda - \lambda_0) + O((\lambda - \lambda_0)^2).$$

To measure how a vector field W changes along $s \mapsto \gamma(s, \lambda)$, one introduces the differential operator D/ds, defined as

$$\frac{DW^{\mu}}{ds} := \frac{\partial(W^{\mu} \circ \gamma)}{\partial s} + \Gamma^{\mu}{}_{\alpha\beta} \dot{\gamma}^{\alpha} W^{\beta} \tag{2.2.1}$$

$$= \dot{\gamma}^{\beta} \frac{\partial W^{\mu}}{\partial x^{\beta}} + \Gamma^{\mu}{}_{\alpha\beta} \dot{\gamma}^{\alpha} W^{\beta} \tag{2.2.2}$$

$$= \dot{\gamma}^{\beta} \nabla_{\beta} W^{\mu}. \tag{2.2.3}$$

(It would perhaps be more logical to write $\frac{DW^{\mu}}{\partial s}$ in the current context, but people never do that.) The last two lines only make sense if W is defined in a whole neighborhood of γ, but for the first it suffices that $W(s)$ be defined along $s \mapsto \gamma(s, \lambda)$. (One possible way of making sense of the last two lines is to extend W^{μ} to any smooth vector field defined in a neighborhood of $\gamma^{\mu}(s, \lambda)$, and note that the result is independent of the particular choice of extension because the equation involves only derivatives tangential to $s \mapsto \gamma^{\mu}(s, \lambda)$.)

---

[2]There is actually a sense in which this follows from non-vacuum Einstein equations [31, 37, 90], compare ▶ Sect. 2.3.2 below, but this is usually admitted as an axiom.

Analogously one sets

$$\frac{DW^\mu}{d\lambda} := \frac{\partial(W^\mu \circ \gamma)}{\partial\lambda} + \Gamma^\mu{}_{\alpha\beta}\frac{\partial\gamma^\alpha}{\partial\lambda}W^\beta \qquad (2.2.4)$$

$$= \frac{\partial\gamma^\beta}{\partial\lambda}\frac{\partial W^\mu}{\partial x^\beta} + \Gamma^\mu{}_{\alpha\beta}\frac{\partial\gamma^\alpha}{\partial\lambda}W^\beta \qquad (2.2.5)$$

$$= Z^\beta \nabla_\beta W^\mu . \qquad (2.2.6)$$

Note that since $s \to \gamma(s,\lambda)$ is a geodesic we have from (2.2.1) and (2.2.3)

$$\frac{D^2\gamma^\mu}{ds^2} := \frac{D\dot\gamma^\mu}{ds} = \frac{\partial^2\gamma^\mu}{\partial s^2} + \Gamma^\mu{}_{\alpha\beta}\dot\gamma^\alpha\dot\gamma^\beta = 0 . \qquad (2.2.7)$$

Remark 2.2.1 Equation (2.2.7) is sometimes written as

$$\dot\gamma^\alpha \nabla_\alpha \dot\gamma^\mu = 0 , \qquad (2.2.8)$$

which is an abuse of notation since typically we will only know $\dot\gamma^\mu$ as a function of s, and so there is no such thing as $\nabla_\alpha \dot\gamma^\mu$. However, this equation makes sense in the following context: Suppose that we have a family of geodesics $s \mapsto \gamma(s)$, each of which is an *integral curve* of a vector field u. By definition, this means that

$$\frac{d\gamma^\mu}{ds}(s) = u^\mu(\gamma(s)) , \qquad (2.2.9)$$

which is often shortly written as

$$\dot\gamma = u . \qquad (2.2.10)$$

We then have

$$\frac{d^2\gamma^\mu}{ds^2}(s) = \frac{du^\mu(x(s))}{ds} = \frac{\partial u^\mu(x(s))}{\partial x^\alpha}\frac{d\gamma^\alpha(s)}{ds} ,$$

so that

$$\frac{D^2\gamma^\mu}{ds^2} \equiv \frac{d^2\gamma^\mu}{ds^2} + \Gamma^\mu{}_{\alpha\beta}\frac{d\gamma^\alpha}{ds}\frac{d\gamma^\beta}{ds} = \frac{\partial u^\mu}{\partial x^\alpha}u^\alpha + \Gamma^\mu{}_{\alpha\beta}u^\alpha u^\beta = u^\alpha \nabla_\alpha u^\mu .$$

We see that the fact that the integral curves of u are geodesics, $\frac{D^2\gamma^\mu}{ds^2} = 0$, is equivalent to

$$u^\alpha \nabla_\alpha u^\mu = 0 , \qquad (2.2.11)$$

which can be rewritten as (2.2.8) when using the notation (2.2.10).

Now,

$$\frac{DZ^\mu}{ds} \underset{(2.2.1)}{=} \frac{\partial^2 \gamma^\mu}{\partial s \partial \lambda} + \Gamma^\mu{}_{\alpha\beta} \dot\gamma^\alpha \frac{\partial \gamma^\beta}{\partial \lambda} \underset{(2.2.4)}{=} \frac{D\dot\gamma^\mu}{d\lambda} , \tag{2.2.12}$$

(The abuse-of-notation derivation of the same formula proceeds as:

$$\nabla_{\dot\gamma} Z^\mu = \dot\gamma^\nu \nabla_\nu Z^\mu = \dot\gamma^\nu \nabla_\nu \frac{\partial\gamma}{\partial\lambda}^\mu \underset{(2.2.3)}{=} \frac{\partial^2\gamma^\mu}{\partial s \partial\lambda} + \Gamma^\mu{}_{\alpha\beta}\dot\gamma^\alpha \frac{\partial\gamma^\beta}{\partial\lambda} \underset{(2.2.6)}{=} Z^\beta \nabla_\beta \dot\gamma^\mu = \nabla_Z \dot\gamma^\mu ,$$

$$\tag{2.2.13}$$

which can then be written as

$$\nabla_{\dot\gamma} Z = \nabla_Z \dot\gamma \; .) \tag{2.2.14}$$

One can now repeat the calculation leading to (1.4.1) to obtain, for any vector field W defined along $\gamma^\mu(s, \lambda)$,

$$\frac{D}{ds}\frac{D}{d\lambda} W^\mu - \frac{D}{d\lambda}\frac{D}{ds} W^\mu = R^\mu{}_{\delta\alpha\beta} \dot\gamma^\alpha Z^\beta W^\delta . \tag{2.2.15}$$

If $W^\mu = \dot\gamma^\mu$ the second term at the left-hand side is zero, and from $\frac{D}{d\lambda}\dot\gamma = \frac{D}{ds}Z$ we obtain

$$\frac{D^2 Z^\mu}{ds^2}(s) = R^\mu{}_{\delta\alpha\beta} \dot\gamma^\alpha Z^\beta \dot\gamma^\delta . \tag{2.2.16}$$

We have obtained an equation known as the *Jacobi equation*, or as the *geodesic deviation equation*.

In the "index-free notation" of (1.4.7), p. 31 this equation reads

$$\boxed{\frac{D^2 Z}{ds^2} = R(\dot\gamma, Z)\dot\gamma} . \tag{2.2.17}$$

Solutions of (2.2.17) are called *Jacobi fields* along γ.

Equation (2.2.17) shows that curvature causes relative acceleration between neighboring geodesics. Keeping in mind that gravitational force and acceleration are indistinguishable, we say that curvature produces a "gravitational tidal force" between freely falling nearby observers.

> **Incidentally** The advantage of the abuse-of-notation equations above, which can be justified by
> the extension-artifact already mentioned, is that one can invoke the result of Proposition 1.4.1,
> instead of repeating its calculations, to obtain (2.2.15):
>
> $$\frac{D^2 Z^\mu}{ds^2}(s) = \dot{\gamma}^\alpha \nabla_\alpha (\dot{\gamma}^\beta \nabla_\beta Z^\mu)$$
>
> $$= \dot{\gamma}^\alpha \nabla_\alpha (Z^\beta \nabla_\beta \dot{\gamma}^\mu)$$
>
> $$= (\dot{\gamma}^\alpha \nabla_\alpha Z^\beta)\nabla_\beta \dot{\gamma}^\mu + Z^\beta \dot{\gamma}^\alpha \nabla_\alpha \nabla_\beta \dot{\gamma}^\mu$$
>
> $$= (\dot{\gamma}^\alpha \nabla_\alpha Z^\beta)\nabla_\beta \dot{\gamma}^\mu + Z^\beta \dot{\gamma}^\alpha (\nabla_\alpha \nabla_\beta - \nabla_\beta \nabla_\alpha)\dot{\gamma}^\mu + Z^\beta \dot{\gamma}^\alpha \nabla_\beta \nabla_\alpha \dot{\gamma}^\mu$$
>
> $$= (\dot{\gamma}^\alpha \nabla_\alpha Z^\beta)\nabla_\beta \dot{\gamma}^\mu + Z^\beta \dot{\gamma}^\alpha R^\mu{}_{\sigma\alpha\beta}\dot{\gamma}^\sigma + Z^\beta \dot{\gamma}^\alpha \nabla_\beta \nabla_\alpha \dot{\gamma}^\mu$$
>
> $$= (\dot{\gamma}^\alpha \nabla_\alpha Z^\beta)\nabla_\beta \dot{\gamma}^\mu + Z^\beta \dot{\gamma}^\alpha R^\mu{}_{\sigma\alpha\beta}\dot{\gamma}^\sigma + Z^\beta \nabla_\beta \underbrace{(\dot{\gamma}^\alpha \nabla_\alpha \dot{\gamma}^\mu)}_{0}$$
>
> $$- (Z^\beta \nabla_\beta \dot{\gamma}^\alpha)\nabla_\alpha \dot{\gamma}^\mu. \tag{2.2.18}$$
>
> After renaming the indices on the first and the last term one finds
>
> $$(\dot{\gamma}^\alpha \nabla_\alpha Z^\beta)\nabla_\beta \dot{\gamma}^\mu - (Z^\beta \nabla_\beta \dot{\gamma}^\alpha)\nabla_\alpha \dot{\gamma}^\mu = (\dot{\gamma}^\alpha \nabla_\alpha Z^\beta - Z^\alpha \nabla_\alpha \dot{\gamma}^\beta)\nabla_\beta \dot{\gamma}^\mu,$$
>
> which is zero by (2.2.14). This leads again to (2.2.16).

2.3 Einstein Equations and Matter

We have already seen in the lectures that Einstein's equations in vacuum read

$$R_{\mu\nu} = 0,$$

where $R_{\mu\nu}$ is the Ricci tensor. Anticipating, in the presence of matter the right-hand side
will not be zero, but will involve an object describing the density of energy of matter
fields.

The idea is: energy produces curvature. So we need a tensor with two indices which
will describe the energy contents of the matter fields.

This requires examining the matter models. We start with the simplest one, that of
dust.

2.3.1 Dust in Special and General Relativity

By definition, *dust* is a cloud of noninteracting particles, whose velocities vary smoothly
from point to point in spacetime.

So at each point we have a scalar ρ which represents the density of the dust: this
is the mass per unit volume measured in a frame in which the particles are at rest. For
example, if there are n particles per unit volume and each has rest mass m, then $\rho = nm$.

Let us first assume that we are in special relativity. We wish to calculate the energy density of the dust in a general inertial frame.

By definition, a rest frame is a frame in which the particles do not move, so that their space velocity is zero, and therefore their velocity four-vector is

$$u = u^\mu \partial_\mu = \partial_t \quad \Longleftrightarrow \quad (u^\mu) = (1, \vec{0}) \,.$$

Let an observer move with space-velocity $\vec{v}$ with respect to the dust, so she has a four-velocity vector

$$(v^\mu) = (v^0 = \gamma := \frac{1}{\sqrt{1 - \vec{v}^2}}, \gamma\vec{v}) \,.$$

Choosing a coordinate system so that the velocity is aligned along the x axis and pointing in the positive direction, the observer has velocity v along the x-axis.

Let there be n particles of rest mass m in a box with sides dx, dy, and dz in the reference frame of the dust.

The observer sees n particles of rest mass m in a box with sides $\gamma^{-1}dx$ (Lorentz contraction factor!), dy and dz, with space velocity $-\vec{v}$, and therefore energy

$$mn\gamma$$

in a volume

$$\gamma^{-1}dx\,dy\,dz \,,$$

and hence a density

$$mn\gamma^2 = \rho(u_\mu v^\mu)^2 = \underbrace{\rho u_\mu u_\nu}_{=:T_{\mu\nu}} v^\mu v^\nu \,. \tag{2.3.1}$$

The tensor field

$$T_{\mu\nu} = \rho u_\mu u_\nu \tag{2.3.2}$$

is called the *energy-momentum tensor of dust with energy density ρ and four-velocity u^μ*, and is used to measure the energy density of dust in general frames, in a sense made clear by Eq. (2.3.1).

The above carries over immediately to general relativity, using the correspondence between physics in special relativity with that in local inertial frames. What needs to be done is thus to replace the special relativistic normalization $\eta_{\mu\nu}u^\mu u^\nu = -1$ of the four-velocity vector with $g_{\mu\nu}u^\mu u^\nu = -1$. (And, in any relevant equations, indices are raised and lowered with the metric g rather than with the Minkowski metric η, while partial derivatives are replaced with covariant ones.)

Energy-momentum tensors are the most basic objects in general relativity, as they provide the source-part of the Einstein equations.

Another example of energy-momentum tensor is given by the Maxwell energy-momentum tensor

$$T_{\mu\nu} = \frac{1}{4\pi} \left(F_{\alpha\mu} F^{\alpha}_{\ \nu} - \frac{1}{4} g_{\mu\nu} F_{\alpha\beta} F^{\alpha\beta} \right) , \tag{2.3.3}$$

see, e.g., Section 3.3 of [89].

2.3.2 The Continuity Equation

An important property of energy-momentum tensors *in special relativity* is that they satisfy a *conservation identity*:

$$\partial_{\nu} T^{\mu\nu} = 0 . \tag{2.3.4}$$

As an exercise, you can check that the divergence identity (2.3.4) for the Maxwell energy-momentum tensor (2.3.3) follows from the Maxwell equations.

In order to verify (2.3.4) for dust, we need first to know what the equations are. Since we assume that the particles are noninteracting, in special relativity each of them moves along a straight line. Now, straight lines are geodesics in Minkowski spacetime, so if u^{μ} is the field of vectors tangent to geodesics followed by the particles, normalized so that $g(u, u) = -1$, we have seen in Remark 2.2.1 that

$$u^{\mu} \nabla_{\mu} u^{\nu} = 0 . \tag{2.3.5}$$

Since the number of particles is conserved we also have the *conservation equation*

$$\nabla_{\mu}(\rho u^{\mu}) = 0 . \tag{2.3.6}$$

These are thus the equations for our model.

These are local equations, and one uses the correspondence principle to carry them over to general relativity, where the covariant derivative becomes the covariant derivative of a possibly curved metric.

Whether in curved spacetime or not, if we calculate the divergence of the energy-momentum tensor we obtain

$$\nabla_{\mu} T^{\mu\nu} \equiv \nabla_{\mu}(\rho u^{\mu} u^{\nu}) = \nabla_{\mu}(\rho u^{\mu}) u^{\nu} + \rho u^{\mu} \nabla_{\mu} u^{\nu} = 0 , \tag{2.3.7}$$

where the first term is zero by the continuity equation (2.3.6), and the second vanishes by the geodesic equation (2.3.5). We see that $\nabla_{\mu} T^{\mu\nu} = 0$, as required.

In fact, the vanishing of the divergence of $T^{\mu\nu}$ is equivalent to (2.3.5)–(2.3.6) in regions where ρ does not vanish if we remember the condition that

$$u_\nu u^\nu = -1 \,. \tag{2.3.8}$$

Indeed, by (2.3.8) we have

$$0 = \nabla_\mu(u_\nu u^\nu) = \nabla_\mu(g_{\alpha\nu}u^\alpha u^\nu) = 2g_{\alpha\nu}u^\alpha\nabla_\mu u^\nu = 2u_\nu\nabla_\mu u^\nu \,.$$

If we multiply (2.3.7) with u_ν and use the last equation we recover the conservation equation $\nabla_\mu(\rho u^\mu) = 0$, but then the geodesic character of u^μ follows.

2.3.3 Einstein Equations with Sources

The energy-momentum tensor $T_{\mu\nu}$ provides a good candidate for the source term in Einstein's theory of gravitation. Keeping in mind the special relativity correspondence principle, and our analysis of dust and of the Maxwell field, the energy-momentum of matter fields will thus be described by a symmetric tensor satisfying

$$\nabla_\mu T^{\mu\nu} = 0 \,, \tag{2.3.9}$$

or, equivalently,

$$\nabla^\mu T_{\mu\nu} = 0 \,. \tag{2.3.10}$$

We thus need to write an equation which is compatible with this restriction. For this, we note the important identity:

$$\nabla^\mu\left(R_{\mu\nu} - \frac{1}{2}Rg_{\mu\nu}\right) = 0 \,, \tag{2.3.11}$$

where

$$R := R^\alpha{}_\alpha = R^{\alpha\beta}{}_{\alpha\beta} \,.$$

To prove (2.3.11), recall the second Bianchi identity

$$\nabla_\mu R_{\nu\rho\alpha\beta} + \nabla_\nu R_{\rho\mu\alpha\beta} + \nabla_\rho R_{\mu\nu\alpha\beta} = 0 \,. \tag{2.3.12}$$

Multiplying by $g^{\mu\alpha}g^{\nu\beta}$ we obtain

$$\nabla^\alpha R^\beta{}_{\rho\alpha\beta} + \nabla^\beta R_\rho{}^\alpha{}_{\alpha\beta} + \nabla_\rho R^{\alpha\beta}{}_{\alpha\beta} = 0 \,, \tag{2.3.13}$$

which is another way of writing (2.3.11).

Recalling that for any constant Λ we have

$$\nabla^{\mu}(\Lambda g_{\mu\nu}) = 0 \,,$$

we are led to an equation which will be compatible with (2.3.10):

$$\boxed{R_{\mu\nu} - \tfrac{1}{2}Rg_{\mu\nu} + \Lambda g_{\mu\nu} = \kappa T_{\mu\nu} \,.} \tag{2.3.14}$$

We will see in ▶ Sect. 4.1.5, p. 135 that the constant κ can be determined by analyzing the Newtonian limit, where $g_{\mu\nu}$ is very close to the Minkowski metric, and all velocities are very small compared with the speed of light, leading to

$$\kappa = \frac{8\pi G}{c^4} \,.$$

The numerical values of G and c depend upon choices of units. It is convenient to use units where $G = 1 = c$, and we shall often do so.

The constant Λ is called *the cosmological constant*, and current state-of-the art observations [73, 88] indicate strongly that Λ is nonzero: Indeed, the current standard model of cosmology requires a cosmological constant which is of the order

$$\Lambda \approx 10^{-35}\mathrm{s}^{-2} \approx 10^{-47}\mathrm{Gev}^4 \approx 10^{-29}\mathrm{g\,cm}^{-3} \approx 10^{-121}\mathrm{Planck\ units} \,,$$

admittedly a rather small but nonzero number. This makes a difference for cosmology, but not for describing black holes on a galactical scale, or planets within the solar system. So from now on we will mostly assume that $\Lambda = 0$, and use units $G = c = 1$, so that (2.3.14) becomes

$$G_{\mu\nu} := R_{\mu\nu} - \frac{1}{2}Rg_{\mu\nu} = 8\pi T_{\mu\nu} \,. \tag{2.3.15}$$

When writing his Eq. (2.3.15), Einstein was not aware of the following result of Lovelock which, in view of (2.3.10), shows that (2.3.14) is the only reasonable tensor equation in which the energy-momentum tensor appears as a source:

Theorem 2.3.1 (Lovelock [59])

In four spacetime dimensions, all symmetric tensors $E_{\mu\nu}$ built-out of the metric, its first and its second derivatives, and satisfying identically $\nabla^{\mu}E_{\mu\nu} = 0$ are of the form

$$E_{\mu\nu} = \alpha G_{\mu\nu} + \beta g_{\mu\nu} \,, \tag{2.3.16}$$

where α and β are constants.

One might be interested in deriving the Einstein equations from a variational principle. In this context another enlightening result of Lovelock reads:

Theorem 2.3.2 (Lovelock [58])
In spacetime dimension four, all coordinate-invariant Lagrange functions $\mathscr{L}$ which lead to second-order field equations for a metric are of the form

$$\mathscr{L} = (\alpha R + \beta)\sqrt{|\det g_{\mu\nu}|}, \tag{2.3.17}$$

where α and β are constants.

As first observed by Hilbert [45], the Lagrangians (2.3.17) with $\alpha \neq 0$ lead to the vacuum Einstein equations with a cosmological constant.

Lovelock also provides a classification of all such tensors and Lagrangians in higher dimensions, see [59] and references therein.

3

The Schwarzschild Metric

Piotr T. Chruściel

© Springer Nature Switzerland AG 2019
P. T. Chruściel, *Elements of General Relativity*, Compact Textbooks in Mathematics,
https://doi.org/10.1007/978-3-030-28416-9_3

A basic black-hole solution of the Einstein equations is the Schwarzschild metric. The aim of this chapter is to introduce the reader to its geometry.

3.1 The Metric, Birkhoff's Theorem

The simplest stationary solutions describing compact isolated objects are the spherically symmetric ones. The flagship example is the *Schwarzschild metric*:

$$g = -(1-\frac{2m}{r})dt^2 + \frac{dr^2}{1 - \frac{2m}{r}} + r^2 d\Omega^2 \,, \tag{3.1.1}$$

$$t \in \mathbb{R} \,, \ r \neq 2m, 0 \,. \tag{3.1.2}$$

Here $d\Omega^2$ denotes the metric of the round unit 2-sphere,

$$d\Omega^2 = d\theta^2 + \sin^2 \theta d\varphi^2 \,.$$

(In SI units,

$$g = -(1 - \frac{2Gm}{c^2 r})c^2 dt^2 + \frac{dr^2}{1 - \frac{2Gm}{c^2 r}} + r^2 d\Omega^2 \,.) \tag{3.1.3}$$

A theorem due to Jebsen [49], but usually attributed to Birkhoff [20], shows that:

Theorem 3.1.1
In a vacuum region, away from the set $\{r = 2m\}$ any spherically symmetric metric can locally be written in the Schwarzschild form, for some mass parameter m.

See Woodhouse [89] or Besse [16] for two very different verifications that this metric satisfies the vacuum Einstein equations.

We conclude that the hypothesis of spherical symmetry implies in vacuum, at least locally, the existence of two further symmetries: translations in t and t-reflections $t \to -t$. More precisely, we obtain *time translations* and *time-reflections* in the region where $1 - 2m/r > 0$ (a metric with those two properties is called *static*). However, in the region where $r < 2m$ the notation "t" for the coordinate appearing in (3.1.1) is misleading, as t is then a space-coordinate, and r is a time one. So in this region t–translations are actually translations in space.

When $m = 0$ the metric reduces to the Minkowski metric in spherical coordinates,

$$g|_{m=0} = \eta := -dt^2 + dr^2 + r^2 d\Omega^2 \equiv -dt^2 + dx^2 + dy^2 + dz^2 .$$

We assume that the reader is familiar with this metric from a course in special relativity and therefore this case will not be discussed any further.

3.1.1 $r = 0$

The metric (3.1.1) has problems when the sets $\{r = 0\}$ and $\{r = 2m\}$ are approached. We start with an analysis of the former, which is called a "singularity." A calculation gives (see, *e.g.*, http://grtensor.phy.queensu.ca/NewDemo)

$$R_{\alpha\beta\gamma\delta} R^{\alpha\beta\gamma\delta} = \frac{48m^2}{r^6} , \tag{3.1.4}$$

which shows that the scalar $R_{\alpha\beta\gamma\delta} R^{\alpha\beta\gamma\delta}$, called *Kretschmann scalar*, tends to infinity as r tends to zero.

This is true regardless of the sign of $m \neq 0$, but the sign of m makes a difference: Namely, if $m > 0$, any continuous curve starting in the region $r < 2m$ has to cross $r = 2m$ before reaching "the exterior world," where r is allowed to grow without bound. But the value $r = 2m$ is not allowed at this stage of our analysis. (What happens at $r = 2m$ will be studied in detail shortly). But when $m < 0$ nothing prevents a continuous curve starting near $r = 0$ to reach any value of r.

Since $R_{\alpha\beta\gamma\delta} R^{\alpha\beta\gamma\delta}$ is a scalar, we see that the curvature of the metric grows without bounds when $\{r = 0\}$ is approached *independently of the coordinate system used*. In other words, there is no coordinate system in which the metric remains twice differentiable (as needed to define the curvature tensor) and in which all components of the Riemann tensor would remain bounded when approaching the set $\{r = 0\}$.

Incidentally   It turns out that *there exist no continuous Lorentzian metrics extending* the Schwarzschild metric across the set $\{r = 0\}$ [76], which is much more difficult to prove.

Throughout the remainder of this chapter, unless explicitly indicated otherwise we will assume

$$m > 0 \, .$$

The point of this assumption is the occurrence, as will be seen shortly, of an *event horizon* located at $\{r = 2m\}$: the singular set $\{r = 0\}$ is then "hidden" behind this event horizon. This is considered to be less unpleasant than the situation where $m < 0$, where no such horizons occur.

3.2 Stationary Observers and the Parameter *m*

An observer is called stationary if her coordinates (r, θ, φ) are fixed, so that she is described by a world line

$$t \mapsto \gamma(t) = (t, r, \theta, \varphi) \, . \tag{3.2.1}$$

The tangent is $\dot{\gamma} = \partial_t$. It is timelike in the region $r > 2m$, and the four velocity u there (normalized to unit length) is

$$u := \frac{1}{\sqrt{-g(\dot{\gamma}, \dot{\gamma})}} \dot{\gamma} = \frac{1}{\sqrt{1 - \frac{2m}{r}}} \partial_t \, . \tag{3.2.2}$$

Clearly there would have been a problem with such a definition in the region $r \le 2m$, since the square-root is purely imaginary or vanishes:

$$\boxed{\text{there are no stationary observers on or under the horizon } r = 2m.} \tag{3.2.3}$$

Equivalently, the paths defined by (3.2.1) are *not* timelike for $r \le 2m$.

Incidentally Strictly speaking, the last two sentences do not make sense on the horizon at this stage of the analysis, since $\{r = 2m\}$ is not part of our manifold yet. Now, the key element in the definition above is the vector field ∂_t, which equals ∂_v in the Eddington–Finkelstein coordinates to be introduced shortly. This vector field is well defined throughout the extended manifold, and null on the horizon, which gives a precise sense to (3.2.3)

Let s be a proper-time parameter along the world-lines of the stationary observers, so that (3.2.1) reads instead

$$s \mapsto \gamma(s) = (\frac{1}{\sqrt{1 - \frac{2m}{r}}} s, r, \theta, \varphi) \, . \tag{3.2.4}$$

The acceleration four-vector a^μ is defined as

$$a^\mu := \frac{Du^\mu}{ds} = u^\nu \nabla_\nu u^\mu = \frac{1}{\sqrt{1 - \frac{2m}{r}}} \nabla_0 u^\mu = \frac{1}{1 - \frac{2m}{r}} \Gamma^\mu{}_{00} \,.$$

We need to calculate the relevant Christoffel symbols

$$\Gamma^\mu{}_{00} = \frac{1}{2} g^{\mu\nu} (2\partial_0 g_{\nu 0} - \partial_\nu g_{00}) = -\frac{1}{2} g^{\mu r} \partial_r g_{00} = \delta^\mu_r \frac{1}{2} (1 - \frac{2m}{r}) \partial_r (1 - \frac{2m}{r})$$

$$= \delta^\mu_r (1 - \frac{2m}{r}) \frac{m}{r^2} \,.$$

Thus, the four-acceleration vector $a := a^\mu \partial_\mu$ takes the form

$$a = \frac{m}{r^2} \partial_r \,,$$

which looks identical as the force needed to counteract the Newtonian gravitational force, when the parameter m is identified with the mass of the object producing the gravitational field.

Some care has to be taken with the above: We have calculated a vector in some coordinate system, which a priori has no clear invariant meaning. One can, however, consider the length

$$\sqrt{|g(a,a)|} = \sqrt{g(a,a)} = \sqrt{g_{rr}} \frac{m}{r^2} = \frac{1}{\sqrt{1 - \frac{2m}{r}}} \frac{m}{r^2} \,,$$

which provides an invariant description of the strength of the gravitational acceleration. For very large r this approximates very well the Newtonian formula for the gravitational acceleration, which confirms the interpretation of *active gravitational mass* for the parameter m, as experienced by distant observers.

Strikingly, the invariant length $\sqrt{g(a,a)}$ diverges as the horizon $\{r = 2m\}$ is approached.

An alternative justification of the interpretation of the parameter m as the Newtonian mass will be done in ▶ Sect. 3.9.1; cf. (3.9.18), p. 79.

3.3 Time Functions and Causality

In order to understand what happens at $r = 2m$, some elementary notions of *Lorentzian causality theory* will be needed.

We start by defining *time orientation*: This is the decision about which timelike vectors are future-pointing, and which ones are past-pointing. In special relativity this is taken for granted: in coordinates where the Minkowski metric η takes the form

$$\eta = -dt^2 + dx^2 + dy^2 + dz^2 \,, \tag{3.3.1}$$

a timelike vector $X^\mu \partial_\mu$ is said to be past-pointing if $X^0 < 0$. But, it should be realized that this is a question of conventions: we could very well agree that past-pointing vectors are those with positive X^0. We will shortly meet a situation where such a decision will have to be made, namely when trying to distinguish black holes from white holes.

The above special-relativistic notion of time-orientation carries over to a general manifold as follows: at every point $p \in \mathcal{M}$ the set of timelike vectors, defined as

$$\{X \in T_p\mathcal{M} \mid g(X, X) < 0\},$$

splits into two components. A simple way of seeing this is to introduce local inertial coordinates near a given point p, then the metric at p coincides with the Minkowski one. In this coordinate system timelike vectors $X = (X^0, \vec{X})$ at p satisfy *either* $X^0 > |\vec{X}|$, *or* $X^0 < -|\vec{X}|$. The *time orientation* is defined as the choice, which timelike vectors will be called *future*, and which will be called *past*. If this can be done continuously throughout the manifold, we say that the manifold is *time orientable*. There are Lorentzian manifolds for which this cannot be done, but we will only consider time-orientable ones. Such Lorentzian manifolds are called *spacetimes*.

Recall that one defines *null* vectors as nonzero vectors X such that $g(X, X) = 0$. A vector is *causal* if it is timelike or null. One can likewise talk about past- or future-directed null or causal vectors. All remaining vectors are called *spacelike*.

A function f will be called a *time function* if ∇f is everywhere timelike past pointing. (The choice here of "past" rather than "future" is related to our signature $(-, +, \ldots, +)$. In the opposite signature ∇f would have been required to be future-pointing.) A coordinate, say x^0, will be said to be a *time coordinate* if x^0 is a time function.

So, for example, $f = t$ on Minkowski spacetime is a time function: indeed, in canonical coordinates as in (3.3.1)

$$\nabla t = \eta^{\mu\nu} \partial_\mu t \, \partial_\nu = \eta^{0\nu} \partial_\nu = -\partial_t \,,$$

and so

$$\eta(\nabla t, \nabla t) = \eta(-\partial_t, -\partial_t) = -1 \,.$$

On the other hand, consider $f = t$ in the Schwarzschild metric: the inverse metric now reads

$$g^{\mu\nu} \partial_\mu \partial_\nu = -\frac{1}{1 - \frac{2m}{r}} \partial_t^2 + \left(1 - \frac{2m}{r}\right)\partial_r^2 + r^{-2}\left(\partial_\theta^2 + \sin^{-2}\theta \, \partial_\varphi^2\right), \tag{3.3.2}$$

and so

$$\nabla t = g^{\mu\nu} \partial_\mu t \, \partial_\nu = g^{0\nu} \partial_\nu = -\frac{1}{1 - \frac{2m}{r}} \partial_t \,.$$

The length-squared of ∇t is thus

$$g(\nabla t, \nabla t) = \frac{g(\partial_t, \partial_t)}{\left(1 - \frac{2m}{r}\right)^2} = -\frac{1}{1 - \frac{2m}{r}} \begin{cases} < 0, \ r > 2m; \\ > 0, \ r < 2m. \end{cases}$$

We conclude that t *is a time function in the region* $\{r > 2m\}$, but is *not* on the manifold $\{r < 2m\}$.

A similar calculation gives $\nabla r = \left(1 - \frac{2m}{r}\right)\partial_r$ and

$$g(\nabla r, \nabla r) = \left(1 - \frac{2m}{r}\right)^2 g(\partial_r, \partial_r) = \left(1 - \frac{2m}{r}\right) \begin{cases} > 0, \ r > 2m; \\ < 0, \ r < 2m. \end{cases}$$

So r is a *time function* in the region $\{r < 2m\}$, *if* the time orientation is chosen so that r is increasing towards the future. On the other hand, the alternative choice of time-orientation implies that *the negative of* r is a time function in this region.

Recall, next, that a differentiable curve is called *timelike* if its tangent vector is timelike everywhere. There are obvious corresponding definitions of *null, causal* or *spacelike* curves. Causal curves can further be *future directed*, or *past directed*, according to the time orientation of their tangents.

It is a basic axiom of general relativity that massive physical objects move along *timelike future directed* curves.

The usefulness of time functions in our context stems from the following:

Proposition 3.3.1
Time functions are strictly increasing along future directed causal curves.

Before passing to the proof, we need a lemma:

Lemma 3.3.2
Let X be timelike and Y causal. Then $g(X, Y) < 0$ if both X and Y are consistently time-oriented, while $g(X, Y) > 0$ if they have opposite time orientations.

Proof
Given a point p in the spacetime we can find a coordinate system so that $g_{\mu\nu}$ is diagonal at p with entries $(-1, 1, \ldots, 1)$, and in which X is proportional to ∂_0, thus $X = X^0 \partial_0$. Then

$$g(X, Y) = -X^0 Y^0 .$$

Since Y is causal we have $Y^0 \neq 0$, with the same sign as X^0 if both X and Y are consistently time-oriented, which leads to $g(X, Y) < 0$. Opposite time-orientation of X and Y corresponds to opposite signs of X^0 and Y^0, and leads to a positive scalar product. $\square$

Proof of Proposition 3.3.1

Let $\gamma(s)$ be a future directed timelike curve and let f be a time function, then

$$\frac{d(f \circ \gamma)}{ds} = \dot{\gamma}^\mu \partial_\mu f = \dot{\gamma}^\mu g_{\mu\nu} g^{\sigma\nu} \partial_\sigma f = g_{\mu\nu} \dot{\gamma}^\mu \nabla^\nu f = g(\nabla f, \dot{\gamma}).$$

Since ∇f is timelike and $\dot{\gamma}$ is causal but oppositely time-directed, their scalar product is positive. Hence $d(f \circ \gamma)/ds$ is positive, and so f is strictly increasing along γ. $\qquad\square$

3.4 The Eddington–Finkelstein Extensions

The metric (3.1.1) is singular as $r = 2m$ is approached. It turns out that this singularity is related to a poor choice of coordinates (one talks about "a coordinate singularity"). Indeed, we can replace t by a new coordinate v, chosen to cancel out the singularity in g_{rr}: if we set

$$v = t + f(r),$$

we find $dv = dt + f'dr$, so that

$$\left(1 - \frac{2m}{r}\right)dt^2 = \left(1 - \frac{2m}{r}\right)(dv - f'dr)^2$$

$$= \left(1 - \frac{2m}{r}\right)(dv^2 - 2f'dv\,dr + (f')^2 dr^2).$$

The offending g_{rr} terms in (3.1.1) will go away if we choose f to satisfy

$$\left(1 - \frac{2m}{r}\right)(f')^2 = \frac{1}{1 - \frac{2m}{r}}.$$

There are two possibilities for the sign; we choose

$$f' = \frac{1}{1 - \frac{2m}{r}} = \frac{r}{r - 2m} = \frac{r - 2m + 2m}{r - 2m} = 1 + \frac{1}{\frac{r}{2m} - 1}, \qquad (3.4.1)$$

leading to

$$v = t + r + 2m \ln\left(\frac{r}{2m} - 1\right).$$

(The alternative choice amounts to introducing another coordinate

$$u = t - f(r), \qquad (3.4.2)$$

with f still as in (3.4.1), we will return to this possibility shortly.) This brings g to the form

$$g = -\left(1 - \frac{2m}{r}\right) dv^2 + 2dv\,dr + r^2 d\Omega^2 . \tag{3.4.3}$$

Now, all coefficients of g in the new coordinate system are smooth. Further,

$$\det g = -r^4 \sin^2 \theta ,$$

which is nonzero for $r > 0$ except at the north and south pole, where we have the usual spherical coordinates singularity. Since g has signature $(-, +, +, +)$ for $r > 2m$, the signature cannot change across $r = 2m$, as for this the determinant would have to vanish there. We conclude that g is a well-defined smooth Lorentzian metric on the set

$$\{v \in \mathbb{R}, r \in (0, \infty)\} \times S^2 . \tag{3.4.4}$$

More precisely, (3.4.3)–(3.4.4) define an analytic extension of the original spacetime (3.1.1).

The coordinates (v, r, θ, φ) are called *"retarded Eddington–Finkelstein coordinates"*.

For further reference, note that increasing t while keeping the remaining coordinates fixed corresponds to increasing v.

We claim the following:

Theorem 3.4.1
The region $\{r \leq 2m\}$ for the metric (3.4.3) is a black hole region, *in the sense that*

$$\text{observers, or signals, can enter this region, but can never leave it.} \tag{3.4.5}$$

Proof
We have already seen that either r or *minus r* is a time function on the region $\{r < 2m\}$. Now, as already mentioned, observers in general relativity always move on *future directed timelike curves*, that is, curves with timelike future directed tangent vector. But time functions are strictly monotonous along future directed causal curves in view of Proposition 3.3.1. It follows that, along a future directed causal curve, either r or $-r$ is strictly increasing in the region $\{r < 2m\}$.

Suppose that there exists at least one future directed causal curve γ_0 which enters from $r > 2m$ to $r < 2m$. Then $\dot{r}$ must have been negative somewhere along γ_0 in the region $\{r < 2m\}$. This implies that *the time orientation has to be chosen so that $-r$ is a time function*. But then r is decreasing along every future directed causal curve. So no such curve passing through $\{r < 2m\}$ can cross $\{r = 2m\}$ again, when followed to the future.

To finish the proof, it remains to exhibit one future directed causal γ_0 which enters $\{r < 2m\}$ from the region $\{r > 2m\}$. For this, consider the radial curve

$$(-\infty, 0) \ni s \mapsto \gamma_0(s) = \big(v(s), r(s), \theta(s), \varphi(s)\big) = \Big(0, -s, \frac{\pi}{2}, 0\Big).$$

Then $\dot{\gamma}_0 = -\partial_r$, hence

$$g(\dot{\gamma}_0, \dot{\gamma}_0) = g_{rr} = 0$$

in the (v, r, θ, φ) coordinates, see (3.4.3). We see that γ_0 lies in the region $\{r > 2m\}$ for $s < -2m$, is null (hence causal), and crosses $\{r = 2m\}$ at $s = -2m$. Next, we have

$$t(s) = v(s) - f(r(s)) = -f(r(s)),$$

hence

$$\frac{dt(s)}{ds} = -f'(r(s))\frac{dr}{ds} = f'(r(s)) > 0 \ \text{ for } r(s) > 2m,$$

which shows that t is increasing along γ_0 in the region $\{r > 2m\}$, hence γ_0 is future directed there, which concludes the proof. $\qquad\square$

The last theorem motivates the name *black hole event horizon* for the hypersurface $\{r = 2m, v \in \mathbb{R}\} \times S^2$.

Incidentally The analogous construction using the coordinate u instead of v leads to a *white hole* spacetime, with $\{r = 2m\}$ being a *white hole event horizon*. The latter can only be crossed by those future directed causal curves which originate in the region $\{r < 2m\}$; this can be seen by a rather straightforward adaptation of the arguments above. In either case, $\{r = 2m\}$ is a causal membrane which prevents future directed causal curves to go *back and forth*. The whole picture will become clearer in ▶ Sect. 3.5.

While this is irrelevant for the considerations so far, it should be pointed out that the Eddington-Finkelstein coordinates (v, r, θ, φ) are constructed using *ingoing radial geodesics*. To see this, we start by noting that the metric inverse to (3.4.3) reads

$$g^{\mu\nu}\partial_\mu\partial_\nu = 2\partial_v\partial_r + (1 - \frac{2m}{r})\partial_r^2 + r^{-2}\partial_\theta^2 + r^{-2}\sin^{-2}\theta\,\partial_\varphi^2 \ . \qquad (3.4.6)$$

In particular

$$0 = g^{vv} = g(\nabla v, \nabla v),$$

which implies that the integral curves of

$$\nabla v = \partial_r$$

are null, affinely parameterized geodesics: Indeed, let $X = \nabla v$, then

$$X^\alpha \nabla_\alpha X^\beta = \nabla^\alpha v \nabla_\alpha \nabla^\beta v = \nabla^\alpha v \nabla^\beta \nabla_\alpha v = \frac{1}{2} \nabla^\beta (\nabla^\alpha v \nabla_\alpha v) = 0. \tag{3.4.7}$$

So if γ is an integral curve of X (thus, by definition,

$$\dot{\gamma}^\mu = X^\mu), \tag{3.4.8}$$

we obtain the geodesic equation:

$$X^\alpha \nabla_\alpha X^\beta = \dot{\gamma}^\alpha \nabla_\alpha \dot{\gamma}^\beta = 0. \tag{3.4.9}$$

We also have

$$g(\nabla r, \nabla r) = g^{rr} = -\left(1 - \frac{2m}{r}\right), \tag{3.4.10}$$

and since this vanishes at $r = 2m$ we say that the surface $r = 2m$ is *null*. It is reached by all the radial null geodesics $v = $ const, $\theta = $ const$'$, $\varphi = $ const$''$, in finite affine time.

Incidentally The calculation leading to (3.4.9) generalizes to functions f such that ∇f satisfies an equation of the form

$$g(\nabla f, \nabla f) = \psi(f), \tag{3.4.11}$$

for some function ψ; note that $f = r$ satisfies this, in view of (3.4.10). Repeating the calculation done in (3.4.7) with $X = \nabla f$ we instead recover

$$X^\alpha \nabla_\alpha X^\beta = \frac{1}{2} \nabla^\beta (\nabla^\alpha f \nabla_\alpha f) = \frac{1}{2} \psi' \nabla^\beta f = \frac{1}{2} \psi' X^\beta. \tag{3.4.12}$$

So if γ satisfies (3.4.8) we obtain

$$\dot{\gamma}^\alpha \nabla_\alpha \dot{\gamma}^\beta = \frac{1}{2} \psi' \dot{\gamma}^\beta. \tag{3.4.13}$$

This is again a geodesic, except that now the parameter along γ that results from the defining equation (3.4.8) is not affine; however, a suitable reparameterization of γ will lead to the usual affinely parameterized form of the geodesic equation, as in the second equality of (3.4.9).

3.5 The Kruskal–Szekeres Extension

The transition from

$$g = -(1 - \frac{2m}{r})dt^2 + \frac{dr^2}{1 - \frac{2m}{r}} + r^2 d\Omega^2 \tag{3.5.1}$$

to

$$g = -(1 - \frac{2m}{r})dv^2 + 2dvdr + r^2d\Omega^2 \qquad (3.5.2)$$

using the coordinate

$$v = t + f(r), \quad f' = \frac{1}{1 - \frac{2m}{r}}, \qquad (3.5.3)$$

so that

$$v = t + r + 2m \ln\left(\frac{r}{2m} - 1\right),$$

is not the end of the story, as further extensions are possible, which will be clear from the calculations that we will do shortly. For the metric (3.5.1) a maximal analytic extension has been found independently by Kruskal [56], Szekeres [84], and Fronsdal [36]; for some obscure reason Fronsdal is almost never mentioned in this context. This extension is visualized[1] in ◼ Fig. 3.1. The region *I* there corresponds to $r > 2m$, the extension

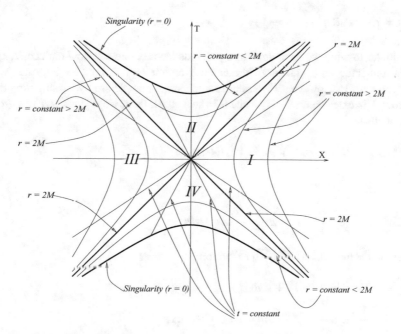

◼ **Fig. 3.1** The Kruskal–Szekeres extension of the Schwarzschild solution, reprinted with permission from [61]

---

[1]I am grateful to J.-P. Nicolas for allowing me to use his electronic figures [61].

constructed using the (u, r, θ, φ) coordinates corresponds to the regions I and IV, while the extension constructed using the (v, r, θ, φ) coordinates corresponds to the regions I and II.

The general construction of extensions as in ◘ Fig. 3.1 for static spherically symmetric metrics, which we write in the form

$$g = -V dt^2 + \frac{dr^2}{V} + r^2 d\Omega^2 \,, \tag{3.5.4}$$

proceeds as follows: We generalize (3.5.3) by writing

$$v = t + f(r) \,, \quad f' = \frac{1}{V} \,. \tag{3.5.5}$$

We introduce yet another coordinate u defined by changing a sign in the equation for v,

$$u = t - f(r) \,, \quad f' = \frac{1}{V} \,. \tag{3.5.6}$$

In the Schwarzschild case where $V = 1 - 2m/r$ this gives

$$u = t - r - 2m \ln \left(\frac{r}{2m} - 1 \right) .$$

Similar to what we have done in a previous lecture, one can now replace (t, r) by (u, r), obtaining an extension of the exterior region I of ◘ Fig. 3.1 into the "white hole" region IV. However, it is easier to understand this picture by passing directly to the complete extension, which proceeds in two steps. First, we replace (t, r) by (u, v). We note that

$$V \, du = V \, dt - dr \,, \quad V \, dv = V \, dt + dr \,,$$

which gives

$$V \, dt = \frac{V}{2}(du + dv) \,, \quad dr = \frac{V}{2}(dv - du) \,.$$

Inserting this into (3.5.4) brings g to the form

$$\begin{aligned} g &= -V dt^2 + V^{-1} dr^2 + r^2 d\Omega^2 \\ &= -\frac{V}{4} \Big((du + dv)^2 - (du - dv)^2 \Big) + r^2 d\Omega^2 \\ &= -V du \, dv + r^2 d\Omega^2 \,. \end{aligned} \tag{3.5.7}$$

The metric so obtained is still degenerate at $\{V = 0\}$. The desingularization is now obtained by setting

$$\hat{u} = -\exp(-cu), \quad \hat{v} = \exp(cv), \tag{3.5.8}$$

with an appropriately chosen c: since

$$d\hat{u} = c\exp(-cu)\,du, \quad d\hat{v} = c\exp(cv)dv,$$

we obtain

$$V\,du\,dv = \frac{V}{c^2}\exp(c(u-v))d\hat{u}\,d\hat{v}$$

$$= \frac{V}{c^2}\exp(-2cf(r))d\hat{u}\,d\hat{v}. \tag{3.5.9}$$

In the Schwarzschild case this reads

$$\frac{V}{c^2}\exp(-2cf(r)) = \frac{r-2m}{c^2r}\exp\left(-2c(r+2m\ln(\frac{r}{2m}-1))\right)$$

$$= \frac{\exp(-2cr)}{c^2r}(r-2m)\exp\left(-4mc\ln(\frac{r}{2m}-1)\right),$$

and with the choice

$$4mc = 1$$

the term $r - 2m$ cancels out, leading to a factor in front of $d\hat{u}\,d\hat{v}$ in (3.5.9) which has no zeros near $r = 2m$. Thus, the desired coordinate transformation is

$$\hat{u} = -\exp(-cu) = -\exp(\frac{r-t}{4m})\sqrt{\frac{r}{2m}-1}, \tag{3.5.10}$$

$$\hat{v} = \exp(cv) = \exp(\frac{r+t}{4m})\sqrt{\frac{r}{2m}-1}, \tag{3.5.11}$$

with g taking the form

$$g = -V\,du\,dv + r^2d\Omega^2$$

$$= -\frac{32m^3\exp(-\frac{r}{2m})}{r}d\hat{u}\,d\hat{v} + r^2d\Omega^2. \tag{3.5.12}$$

Here r should be viewed as a function of $\hat{u}$ and $\hat{v}$ defined implicitly by the equation

$$-\hat{u}\hat{v} = \underbrace{\exp(\frac{r}{2m})\left(\frac{r}{2m}-1\right)}_{=:G(r)}. \tag{3.5.13}$$

In order to check that this defines a smooth function $r = r(\hat{u}\hat{v})$, we note that

$$\left(\exp(\frac{r}{2m})(\frac{r}{2m} - 1)\right)' = \frac{r}{4m^2}\exp(\frac{r}{2m}) > 0.$$

This shows that the function $G(r)$ defined by the right-hand side of (3.5.13) is strictly increasing for $r > 0$. We have $G(0) = -1$, and G tends to infinity as r does, so G defines a bijection of $(0, \infty)$ with $(-1, \infty)$. Monotonicity guarantees existence of the inverse G^{-1}, while the implicit function theorem guarantees smoothness. We conclude that there exists indeed a smooth function

$$r = G^{-1}(-\hat{u}\hat{v})$$

solving (3.5.13) on the set $\hat{u}\hat{v} \in (-\infty, 1)$.

Note that so far we had $r > 2m$, but there are *a priori* no reasons for the function $r(u, v)$ defined above to satisfy this constraint. In fact, we already know from our experience with the (v, r, θ, φ) coordinate system that a restriction $r > 2m$ would lead to a spacetime with poor global properties.

We have

$$\det g = -16^2 m^6 \exp(-\frac{r}{m})r^2 \sin^2\theta\,,$$

with all coefficients of g smooth, which shows that (3.5.12) defines a smooth Lorentzian metric on the set

$$\{\hat{u},\ \hat{v} \in \mathbb{R},\ r > 0\} \times S^2\,. \tag{3.5.14}$$

It follows from (3.5.13) that this is the same as the set

$$\{\hat{u},\ \hat{v} \in \mathbb{R},\ \hat{u}\hat{v} < 1\} \times S^2\,. \tag{3.5.15}$$

The set (3.5.15) together with the metric (3.5.12) is, by definition, the *Kruszkal-Szekeres* extension of the original Schwarzschild spacetime.

It should be emphasized that (3.5.10)–(3.5.11) is only valid in the exterior world $\{r > 2m\}$, which should not bother us since we are constructing an *extension* of the exterior region. So while the initial metric (3.5.12) was only defined for $\hat{u} < 0$ and $\hat{v} > 0$, now we allow $\hat{u}$ and $\hat{v}$ to range over $\mathbb{R}$ except for the restriction $\hat{u}\hat{v} \in (-\infty, 1)$.

◼ Figure 3.1 gives a representation of the extended spacetime in coordinates

$$X := (\hat{v} - \hat{u})/2, \quad T := (\hat{v} + \hat{u})/2\,.$$

In these coordinates the metric takes the form

$$g = \frac{32m^3\exp(-\frac{r}{2m})}{r}(-dT^2 + dX^2) + r^2 d\Omega^2\,. \tag{3.5.16}$$

Here one should keep in mind that, as already mentioned, the metric cannot be extended across the set $r = 0$ in the class of continuous Lorentzian metrics.

Let us discuss some features of ◻ Fig. 3.1:

1. The singular set $r = 0$ corresponds to the spacelike hyperboloids

$$(X^2 - T^2)|_{r=0} = -\hat{u}\hat{v}|_{r=0} = -1 < 0.$$

2. More generally, the sets $r = $ const are hyperboloids $X^2 - T^2 = $ const', which are timelike in the regions I and III (since $X^2 - T^2 < 0$ there), and which are spacelike in the regions II and IV.

3. The vector field ∇T satisfies

$$g(\nabla T, \nabla T) = g^{\sharp}(dT, dT) = \frac{1}{4}g^{\sharp}(d\hat{u} + d\hat{v}, d\hat{u} + d\hat{v}) = \frac{1}{2}g^{\sharp}(d\hat{u}, d\hat{v}) < 0,$$

which shows that T is a time coordinate. Similarly X is a space-coordinate, so that ◻ Fig. 3.1 respects our implicit convention of representing time along the vertical axis and space along the horizontal one.

4. The map

$$(\hat{u}, \hat{v}) \to (-\hat{u}, -\hat{v}) \tag{3.5.17}$$

is clearly an isometry: Indeed, (3.5.17) leaves both $d\hat{u}\, d\hat{v}$ and $\hat{u}\hat{v}$ unchanged, and thus also r remains unchanged. It follows that the region I is isometric to region III, and region II is isometric to region IV. In particular the extended manifold has two asymptotically flat regions, the original region I, and region III which is an identical copy of I.

5. The hypersurface $t = 0$ from the region I corresponds to $\hat{u} = -\hat{v} < 0$, equivalently it is the subset $X > 0$ of the hypersurface $T = 0$. This can be smoothly continued to negative X, which corresponds to a second copy of this hypersurface. The resulting geometry is often referred to as the *Einstein–Rosen bridge*.

Before continuing, a definition is in order. Given a hypersurface $\mathscr{S} := \{t = f(\vec{x})\}$ in a spacetime $(\mathscr{M}, g)$, the *metric induced by g on* $\mathscr{S}$, say γ, is defined as the tensor obtained by replacing in g every occurrence of dt by $df \equiv \frac{\partial f}{\partial x^i}dx^i$, and substituting $f(\vec{x})$ for t in all metric functions. (See (4.3.13), p. 145 for a more geometric definition.)

Armed with this definition, we can carry out the construction of the Einstein–Rosen bridge directly from the Riemannian metric γ induced by g on $t = $ const:

$$\gamma = \frac{dr^2}{1 - \frac{2m}{r}} + r^2 d\Omega^2, \quad r > 2m. \tag{3.5.18}$$

A convenient coordinate ρ is given by

$$\rho = \sqrt{r^2 - 4m^2} \iff r = \sqrt{\rho^2 + 4m^2}.$$

We then have

$$r\,dr = \rho\,d\rho \quad\Longrightarrow$$

$$dr^2 = \frac{\rho^2}{r^2}d\rho^2 = \frac{(r-2m)(r+2m)}{r^2}d\rho^2 = \left(1-\frac{2m}{r}\right)\left(1+\frac{2m}{r}\right)d\rho^2\,.$$

Hence γ takes the form

$$\gamma = \left(1+\frac{2m}{\sqrt{\rho^2+4m^2}}\right)d\rho^2 + (\rho^2+4m^2)d\Omega^2\,, \tag{3.5.19}$$

which can be smoothly continued from the original range $\rho > 0$ to $\rho \in \mathbb{R}$. Equation (3.5.19) further explicitly exhibits asymptotic flatness of both asymptotic regions $\rho \to \infty$ and $\rho \to -\infty$. Indeed,

$$\gamma \sim d\rho^2 + \rho^2 d\Omega^2$$

to leading order, for large $|\rho|$, which is the flat metric in radial coordinates with radius $|\rho|$.

Remark 3.5.1 A geometrically natural procedure to regularize the metric (3.5.18) is to introduce a new coordinate x by the formula

$$dx = \frac{dr}{\sqrt{1-\frac{2m}{r}}}\,,$$

so that γ becomes

$$\gamma = dx^2 + r^2 d\Omega^2\,. \tag{3.5.20}$$

The resulting function x measures the geometric distance from the surface $\{r = 2m\}$. Integration gives

$$x = 2m\ln\left(\sqrt{\frac{r}{2m}-1}+\sqrt{\frac{r}{m}}\right) + \sqrt{r(r-2m)}\,. \tag{3.5.21}$$

The drawback is that an explicit formula for r as a function of x is not available.

---

6. In order to understand how the Eddington–Finkelstein extension using the v coordinate fits into ◼ Fig. 3.1, we need to express $\hat{u}$ and $\hat{v}$ in terms of v and r. For this we have

$$u = t - f(r) = v - 2f(r) = v - 2r - 4m\ln\left(\frac{r}{2m}-1\right),$$

hence

$$\hat{u} = -e^{-\frac{u}{4m}} = -e^{-\frac{v-2r}{4m}}\left(\frac{r}{2m} - 1\right), \quad \hat{v} = e^{\frac{v}{4m}}.$$

So $\hat{v}$ remains positive but $\hat{u}$ is allowed to become negative as r crosses $r = 2m$ from above. This corresponds to the region above the diagonal $T = -X$ in the coordinates (X, T) of ☐ Fig. 3.1.

A similar calculation shows that the Eddington–Finkelstein extension using the coordinate u corresponds to the region $\hat{u} < 0$ within the Kruskal–Szekeres extension, which is the region below the diagonal $T = X$ in the coordinates of ☐ Fig. 3.1.

7. Vector fields generating isometries are called *Killing vector fields*. Since time-translations are isometries in our case, the vector field $K = \partial_t$ is a Killing vector field. In the Kruskal–Szekeres coordinate system K takes the form

$$K = \partial_t = \frac{\partial \hat{u}}{\partial t}\partial_{\hat{u}} + \frac{\partial \hat{v}}{\partial t}\partial_{\hat{v}}$$

$$= \frac{1}{4m}\left(\hat{v}\partial_{\hat{v}} - \hat{u}\partial_{\hat{u}}\right). \tag{3.5.22}$$

More precisely, the Killing vector field ∂_t defined on the original Schwarzschild region extends to a Killing vector field K defined throughout the Kruskal–Szekeres manifold by the right-hand side of (3.5.22).

Exercise 3.5.2 A geodesic is said to be *complete* if it is defined for all values of an affine parameter. Show that the geodesics in the Kruskal–Szekeres manifold are either complete or asymptote to the singularity $\{r = 0\}$ in finite affine time.

Incidentally The Kruskal–Szekeres extension is *inextendible* in the class of C^2 metrics, which can be proved as follows: first, (3.1.4) shows that the *Kretschmann scalar* $R_{\alpha\beta\gamma\delta}R^{\alpha\beta\gamma\delta}$ diverges as r approaches zero. As already pointed out, this implies that no C^2 extension of the metric is possible across the set $\{r = 0\}$. The result of Exercise 3.5.2 implies then inextendibility.

It can be shown that the Kruskal–Szekeres extension is singled out by being maximal in the vacuum, analytic, simply connected class, with all maximally extended geodesics γ either complete, or with the curvature scalar $R_{\alpha\beta\gamma\delta}R^{\alpha\beta\gamma\delta}$ diverging along γ in finite affine time.

3.6 The Flamm Paraboloid

In this section we will try to understand the geometry of the hypersurfaces $\{t = \text{const}\}$ in the Schwarzschild spacetime. One way of doing this is to try to embed these hypersurfaces in four-dimensional Euclidean space. It is not clear that this can be done for general metrics (and, in fact, there are no necessary and sufficient conditions known for this in general); here it works because of spherical symmetry.

To proceed, recall that the metric, say γ, induced by the Schwarzschild metric g on the hypersurfaces of constant t is obtained by discarding the dt^2 term in g:

$$\gamma = \frac{dr^2}{1 - \frac{2m}{r}} + r^2 d\Omega^2. \tag{3.6.1}$$

Let us write

$$\overset{\circ}{g} = dz^2 + (dx^1)^2 + (dx^2)^2 + (dx^3)^2 = dz^2 + dr^2 + r^2 d\Omega^2,$$

to denote the Euclidean metric on $\mathbb{R}^4$. Now, the metric induced by $\overset{\circ}{g}$ on the surface $z = z(r)$, and which we will denote by h, is obtained by replacing in $\overset{\circ}{g}$ the differential dz by $\frac{dz}{dr} dr$:

$$h = \left(\left(\frac{dz}{dr} \right)^2 + 1 \right) dr^2 + r^2 d\Omega^2.$$

This will coincide with (3.6.1) if we require that

$$\frac{dz}{dr} = \pm \sqrt{\frac{2m}{r - 2m}}.$$

Integration gives

$$z = z_0 \pm 4m \sqrt{\frac{r}{2m} - 1}, \quad \text{where } r > 2m. \tag{3.6.2}$$

Choosing $z_0 = 0$ and solving for $r = r(z)$ lead to

$$r = 2m + \frac{z^2}{8m}.$$

This is a paraboloid, as first noted by Flamm. The embedding is visualized in ◻ Fig. 3.2.

The above gives alternative insight to the already-made observation, that the manifolds $t = \text{const}, r > m$, can be "doubled" by attaching another copy of the manifold to itself across the boundary $\{r = 2m\}$. We recover the already-mentioned *Einstein–Rosen bridge*.

The positive sign in (3.6.2) corresponds to our usual black hole exterior, while the negative sign corresponds to a second asymptotically flat region, on the "other side" of the Einstein–Rosen bridge.

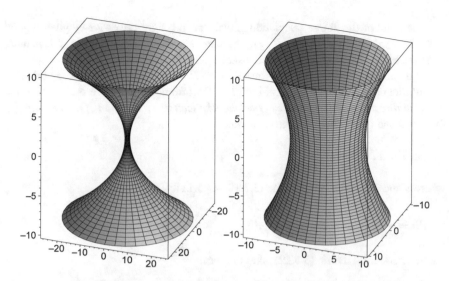

◻ **Fig. 3.2** Isometric embedding of the space-geometry of a Schwarzschild black hole into four-dimensional Euclidean space, near the throat of the Einstein–Rosen bridge $r = 2m$, with $2m = 1$ (left) and $2m = 6$ (right)

3.7   Conformal Carter-Penrose Diagrams

We return to the form (3.5.16) of the metric, namely

$$g = \underbrace{\frac{32m^3 \exp(-\frac{r}{2m})}{r} (-dT^2 + dX^2)}_{=:^2g} + r^2 d\Omega^2 \,. \tag{3.7.1}$$

Because the angular part of g is positive-definite, the (T, X) part of any vector $V = V^\mu \partial_\mu$ which is causal for g is a causal vector for the two-dimensional Minkowski metric ^2g. Indeed we have

$$g(V, V) = \frac{32m^3 \exp(-\frac{r}{2m})}{r} (-(V^T)^2 + (V^X)^2) + r^2((V^\theta)^2 + \sin^2(\theta)(V^\varphi)^2)$$

$$\geq \frac{32m^3 \exp(-\frac{r}{2m})}{r} (-(V^T)^2 + (V^X)^2) \,. \tag{3.7.2}$$

This shows that

$$g(V, V) \leq 0 \quad \Longrightarrow \quad -(V^T)^2 + (V^X)^2 \leq 0 \,,$$

as claimed. Moreover, this implication is an equivalence for vectors which have no angular components.

The point of the above calculation is that drawing light-cones for ^2g gives a good idea of the causal structure of $(\mathscr{M}, g)$. This can likewise be used in ▣ Fig. 3.1, making clear the black hole character of the Kruskal–Szekeres spacetime.

The *Penrose diagram* for the Kruskal–Szekeres spacetime is obtained by bringing the infinite domain of definition of the (T, X)-coordinates to a finite one, so that *one can visualize the global structure of the whole spacetime on a region of final extent.* For this, let $\bar{u}$ and $\bar{v}$ be defined by the equations

$$\tan \bar{u} = \hat{u}, \quad \tan \bar{v} = \hat{v},$$

where $\hat{v}$ and $\hat{u}$ have been defined in (3.5.10)–(3.5.11). Using

$$d\hat{u} = \frac{1}{\cos^2 \bar{u}} d\bar{u}, \quad d\hat{v} = \frac{1}{\cos^2 \bar{v}} d\bar{v},$$

the Schwarzschild metric (3.5.12) takes the form

$$g = -\frac{32m^3 \exp(-\frac{r}{2m})}{r} d\hat{u}\, d\hat{v} + r^2 d\Omega^2$$

$$= -\frac{32m^3 \exp(-\frac{r}{2m})}{r \cos^2 \bar{u} \cos^2 \bar{v}} d\bar{u}\, d\bar{v} + r^2 d\Omega^2. \tag{3.7.3}$$

Introducing new time- and space-coordinates $\bar{t} = (\bar{u} + \bar{v})/2$, $\bar{x} = (\bar{v} - \bar{u})/2$, so that

$$\bar{u} = \bar{t} - \bar{x}, \quad \bar{v} = \bar{t} + \bar{x},$$

one obtains a more familiar-looking form

$$g = \frac{32m^3 \exp(-\frac{r}{2m})}{r \cos^2 \bar{u} \cos^2 \bar{v}} (-d\bar{t}^2 + d\bar{x}^2) + r^2 d\Omega^2.$$

This is regular except at $\cos \bar{u} = 0$, and $\cos \bar{v} = 0$, and $r = 0$. The first set corresponds to the straight lines $\bar{u} = \bar{t} - \bar{x} \in \{\pm\pi/2\}$, while the second is the union of the lines $\bar{v} = \bar{t} + \bar{x} \in \{\pm\pi/2\}$.

The analysis of $\{r = 0\}$ requires some work: recall that $r = 0$ corresponds to $\hat{u}\hat{v} = 1$, which is equivalent to

$$\tan(\bar{u}) \tan(\bar{v}) = 1.$$

Using the identity

$$\tan(\bar{u}) \tan(\bar{v}) = \frac{\cos(\bar{u} - \bar{v}) - \cos(\bar{u} + \bar{v})}{\cos(\bar{u} - \bar{v}) + \cos(\bar{u} + \bar{v})}$$

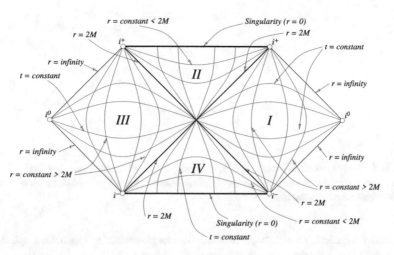

◻ Fig. 3.3 The Carter-Penrose diagram for the Kruskal–Szekeres spacetime with mass M. There are actually two asymptotically flat regions, with corresponding event horizons defined with respect to the second region. Each point in this diagram represents a two-dimensional sphere, and coordinates are chosen so that intersections of the light-cones with the plane of the figure have slopes plus minus one. Regions are numbered as in ◻ Fig. 3.1. I am grateful to J.P. Nicolas and M. Maliborski for providing the figure

we find that

$$\tan(\bar{u}) \tan(\bar{v}) = 1 \quad \Longleftrightarrow \quad \cos(\bar{u} + \bar{v}) = 0,$$

which is further equivalent to

$$\underbrace{\bar{u} + \bar{v}}_{2\bar{t}} = \pm\pi/2.$$

So the $(\hat{u}, \hat{v})$-part of the Kruskal–Szekeres metric is proportional to the two-dimensional Minkowski metric restricted to the set displayed in ◻ Fig. 3.3.

Exercise 3.7.1
Consider the four-dimensional Minkowski metric η with the usual spherical coordinates (t, r, θ, φ). Introduce new coordinates $\hat{u} = t - r$ and $\hat{v} = t + r$. Set

$$\hat{u} = \tan\bar{u}, \quad \hat{v} = \tan\bar{v}, \quad \bar{t} = \frac{\bar{v} + \bar{u}}{2}, \quad \bar{r} = \frac{\bar{v} - \bar{u}}{2}. \tag{3.7.4}$$

Rewrite η in the coordinates $(\bar{t}, \bar{r}, \theta, \varphi)$. Find the set on which the map

$$(t, r) \mapsto (\bar{t}, \bar{r})$$

◻ Fig. 3.4 The Penrose diagram of $(1+n)$-dimensional Minkowski spacetime, $n \geq 2$. In the left figure every point, except the dotted line, is an $(n-1)$-dimensional sphere. The right picture is obtained by rotating the left one around the dotted line

is a diffeomorphism. One thus obtains the *Penrose diagram* for Minkowski space-time, shown in ◻ Fig. 3.4. [*Hint: Do not forget that $r > 0$.*]   ∎

3.8   Keplerian Orbits

We will shortly start the analysis of geodesics in Schwarzschild, concentrating mainly on timelike geodesics, which describe the motion of test bodies in the Schwarzschild gravitational field. It will be instructive to compare the results to their Newtonian counterparts, i.e., the Keplerian orbits. Having this in mind, in this section we review the Newtonian situation.

Consider thus a body of mass m_0 moving in the Newtonian gravitational field created by a spherically symmetric star of mass m. The corresponding Newtonian potential is $-\frac{Gm}{r}$, leading to the equation of motion

$$m_0 \ddot{\vec{x}} = \nabla\left(\frac{Gm_0 m}{r}\right).$$ (3.8.1)

Let $\vec{J}_N$ denote the angular-momentum of the moving body,

$$\vec{J}_N = m_0 \vec{x} \times \dot{\vec{x}}.$$ (3.8.2)

The vector $\vec{J}_N$ is conserved:

$$\frac{d\vec{J}_N}{dt} = m_0\big(\vec{x} \times \underbrace{\ddot{\vec{x}}}_{\sim\vec{x}} + \dot{\vec{x}} \times \dot{\vec{x}}\big) = 0.$$

This shows that the position vector $\vec{x}(t)$ is orthogonal to the *time-independent* vector $\vec{J}_N$. If the latter is nonzero, we deduce that the motion takes place in the plane orthogonal to $\vec{J}_N$, hence is planar. If $\vec{J}_N$ vanishes, the velocity is proportional to the position vector, which implies that the motion takes places along a straight line through the origin. We conclude that the motion is planar for all $\vec{J}_N \in \mathbb{R}^3$.

If $\vec{J}_N \neq 0$ we choose the coordinate system so that $\vec{J}_N = (0, 0, J_N)$, then the motion takes place in the equatorial plane $\theta = \pi/2$. Similarly, for orbits with $\vec{J}_N = 0$, we can also rotate the coordinate system so that the straight line describing the motion lies in the equatorial place. Using polar coordinates

$$(x, y) = (r \cos \varphi, r \sin \varphi)$$

on the plane $z = 0$, a standard calculation gives

$$J_N = m_0 r^2 \dot{\varphi}, \tag{3.8.3}$$

where a dot denotes a derivative with respect to t.

Conservation of energy, say E_N, and of J_N gives the following expression for the Newtonian energy:

$$
\begin{aligned}
E_N &= \frac{m_0 \dot{r}^2}{2} + \frac{m_0 r^2 \dot{\varphi}^2}{2} - \frac{Gmm_0}{r} \\
&= \frac{m_0 \dot{r}^2}{2} + \frac{J_N^2}{2m_0 r^2} - \frac{Gmm_0}{r}.
\end{aligned} \tag{3.8.4}
$$

Assuming $J_N \neq 0$, the trick for solving the equations of motion is to consider

$$u := \frac{m}{r}$$

as a function of φ. One obtains:

$$\frac{du}{d\varphi} = -\frac{m \frac{dr}{dt}}{r^2 \frac{d\varphi}{dt}} = -\frac{m_0 m}{J_N} \frac{dr}{dt}. \tag{3.8.5}$$

Inserting into (3.8.4) we are led to

$$
\begin{aligned}
\left(\frac{du}{d\varphi} \right)^2 &= 2\frac{m_0 m^2}{J_N^2} \left(E_N - \frac{J_N^2}{2m_0 r^2} + \frac{Gmm_0}{r} \right) \\
&= \frac{2m_0 m^2 E_N}{J_N^2} - u^2 + \frac{2Gm_0^2 m^2}{J_N^2} u.
\end{aligned} \tag{3.8.6}
$$

Differentiating with respect to φ and dividing by $2du/d\varphi$, one finds the desired second-order equation of motion for u,

$$\frac{d^2 u}{d\varphi^2} = -u + \mathring{u}, \quad \text{where} \quad \boxed{\mathring{u} := \frac{Gm_0^2 m^2}{J_N^2}}. \tag{3.8.7}$$

> **Incidentally** Anticipating, for a particle of unit mass $m_0 = 1$, and using units in which $G = 1$, this coincides with the Schwarzschildian equation (3.9.15), p. 78 below with $\lambda = 1$ *if* one identifies J_N with J there, and if one neglects the quadratic terms in (3.9.15).

Equation (3.8.7) is elementary to solve, the general solution can be written as

$$\frac{m}{r} \equiv u = \mathring{u}\left(1 + e\cos(\varphi - \varphi_0)\right), \tag{3.8.8}$$

for some constants $e > 0$ and φ_0.

Let d be the distance of closest approach, this occurs when the right-hand side of (3.8.8) is largest, hence at $\varphi = \varphi_0$, and then $r = d$. This determines e:

$$e = \frac{m}{\mathring{u}d} - 1. \tag{3.8.9}$$

Rotating the (x, y)-plane so that $\varphi_0 = 0$, and keeping in mind that $r\cos(\varphi) = x$, straightforward algebra leads to the following form of (3.8.8):

$$\frac{m}{\mathring{u}} - ex = r. \tag{3.8.10}$$

Squaring both sides of (3.8.10), we see that the orbit lies on a quadratic curve in the equatorial plane:

$$\left(1 - e^2\right)x^2 + 2\frac{me}{\mathring{u}}x + y^2 = \left(\frac{m}{\mathring{u}}\right)^2. \tag{3.8.11}$$

If e equals one we obtain a parabola, if e is smaller—an ellipse and, finally, a hyperbola whenever e is larger than one.

In what follows we will mainly be interested in elliptical orbits. Consider, then, an ellipse, the equation of which can always be brought to the form

$$\frac{(x - x_0)^2}{a^2} + \frac{(y - y_0)^2}{b^2} = 1. \tag{3.8.12}$$

Assuming $b^2 \leq a^2$, the *eccentricity* $\mathring{e}$ is defined by the equation

$$\mathring{e}^2 := 1 - \frac{b^2}{a^2}.$$

Noting that (3.8.12) can be rewritten as

$$\frac{b^2}{a^2}(x - x_0)^2 + (y - y_0)^2 = b^2, \tag{3.8.13}$$

we see that the ratio b^2/a^2 can be read from the coefficient in front of x^2 when the coefficient in front of y^2 equals one. Comparing with (3.8.11) we conclude that

$$e = \overset{\circ}{e} .$$

This provides the geometric interpretation of the constant e.

Summarizing, (3.8.8) can be rewritten as

$$\boxed{u = \overset{\circ}{u}(e\cos\varphi + 1) = \frac{Gm_0^2 m^2}{J_N^2}(e\cos\varphi + 1)} , \tag{3.8.14}$$

where e is the eccentricity of the orbit when $e < 1$.

For further reference we note that on a circular orbit (hence, $e = 0$) of radius R we have

$$u(t) = \frac{m}{R} = \frac{Gm_0^2 m^2}{J_N^2} \iff J_N = m_0\sqrt{GmR} , \tag{3.8.15}$$

which gives the following expression for the angular velocity Ω

$$\boxed{\Omega := \frac{d\varphi}{dt} = \frac{J_N}{m_0 R^2} = \sqrt{\frac{Gm}{R^3}}} . \tag{3.8.16}$$

3.9 Geodesics

The Lagrangian for geodesics of the metric (3.5.4), parameterized by a parameter s, is:

$$\mathcal{L} = \frac{1}{2}\left(-V\left(\frac{dt}{ds}\right)^2 + V^{-1}\left(\frac{dr}{ds}\right)^2 + r^2\left(\frac{d\theta}{ds}\right)^2 + r^2\sin^2\theta\left(\frac{d\varphi}{ds}\right)^2\right) .$$

The Euler–Lagrange equation for the coordinate function $s \mapsto t(s)$ reads

$$\frac{d}{ds}\left(V\frac{dt}{ds}\right) = 0 , \tag{3.9.1}$$

which gives a conservation law. Similarly

$$\frac{d}{ds}\left(r^2\sin^2\theta\frac{d\varphi}{ds}\right) = 0 . \tag{3.9.2}$$

Those Euler–Lagrange equations which are not already covered by the conservation laws read:

$$\frac{d}{ds}\left(V^{-1}\frac{dr}{ds}\right) = -\frac{1}{2}\partial_r V\left(\left(\frac{dt}{ds}\right)^2 + V^{-2}\left(\frac{dr}{ds}\right)^2\right)$$

$$+ r\left[\left(\frac{d\theta}{ds}\right)^2 + \sin^2\theta\left(\frac{d\varphi}{ds}\right)^2\right], \tag{3.9.3}$$

$$\frac{d}{ds}\left(r^2\frac{d\theta}{ds}\right) = r^2\sin\theta\cos\theta\left(\frac{d\varphi}{ds}\right)^2. \tag{3.9.4}$$

As an exercise, one can read the Christoffel symbols of g from the above: e.g., from (3.9.1),

$$\Gamma^t{}_{tr} = \frac{V'}{2V},$$

etc.

Since $\mathcal{L}$ is s-independent, the Hamiltonian H is conserved:

$$H := \underbrace{\frac{\partial\mathcal{L}}{\partial\dot{x}^\mu}\dot{x}^\mu}_{=2\mathcal{L}} - \mathcal{L} = \mathcal{L}.$$

This gives one more constant of motion

$$V\left(\frac{dt}{ds}\right)^2 - V^{-1}\left(\frac{dr}{ds}\right)^2 - r^2\left(\frac{d\theta}{ds}\right)^2 - r^2\sin^2\theta\left(\frac{d\varphi}{ds}\right)^2 = \lambda, \tag{3.9.5}$$

and without loss of generality, making an affine transformation of s, we can choose

$$\boxed{\lambda \in \{0, \pm 1\},}$$

where $\lambda = 0$ for null geodesics and $\lambda = 1$ for timelike ones.

To simplify things somewhat, let us start by showing that the motion is *planar*. Consider any geodesic, and think of the coordinates (r, θ, φ) as spherical coordinates on $\mathbb{R}^3$. Then the initial position vector (which is assumed *not* to be the origin, as the metric is singular at $r = 0$) and the initial velocity vector, which is assumed *not* to be radial (otherwise the geodesic will be radial, and the claim is immediately true) define a unique plane in $\mathbb{R}^3$. We can then choose the spherical coordinates so that this plane is the plane $\theta = \pi/2$. This leads to $\theta(0) = \pi/2$ and $\dot{\theta}(0) = 0$, and then the function $\theta(s) \equiv \pi/2$ is a solution of (3.9.4) satisfying the initial values. By uniqueness of solutions, this is *the* solution.

We have thus shown that

> without loss of generality we can assume $\theta = \pi/2$.

In this case the equations of motion are

$$\frac{d}{ds}\left(V\frac{dt}{ds}\right) = 0 \implies \frac{dt}{ds} = \frac{E}{1 - \frac{2m}{r}}, \tag{3.9.6}$$

$$\frac{d}{ds}\left(r^2\frac{d\varphi}{ds}\right) = 0 \implies \frac{d\varphi}{ds} = \frac{J}{r^2}, \tag{3.9.7}$$

$$\frac{d}{ds}\left(V^{-1}\frac{dr}{ds}\right) = -\partial_r V\left(\left(\frac{dt}{ds}\right)^2 + V^{-2}\left(\frac{dr}{ds}\right)^2\right) + r\left(\frac{d\varphi}{ds}\right)^2. \tag{3.9.8}$$

We will ignore this last equation, combining instead (3.9.6)–(3.9.7) with (3.9.5),

$$\underbrace{V\left(\frac{dt}{ds}\right)^2}_{E^2 V^{-1}} - V^{-1}\left(\frac{dr}{ds}\right)^2 - \underbrace{r^2\left(\frac{d\varphi}{ds}\right)^2}_{J^2/r^2} = \lambda \in \{0, \pm1\}, \tag{3.9.9}$$

to conclude that

$$\left(\frac{dr}{ds}\right)^2 = -\left(\lambda + \frac{J^2}{r^2}\right)(1 - \frac{2m}{r}) + E^2. \tag{3.9.10}$$

The second order equation of motion could be obtained from (3.9.8), but at this stage it is simpler to realize that differentiation of (3.9.10) with respect to s *must* produce the second order equation of motion, obtaining

$$2\frac{d^2r}{ds^2} = \frac{d}{dr}\left(E^2 - (1 - \frac{2m}{r})(\lambda + \frac{J^2}{r^2})\right). \tag{3.9.11}$$

For further reference we note the following:

First, we see that φ is constant along those geodesics on which $J = 0$. From what has been said we see that all trajectories with constant θ and φ are geodesics.

Next, for $J \neq 0$, we have $\dot{\varphi} \neq 0$, and so we can invert the function $s \mapsto \varphi(s)$ to obtain a function $\varphi \mapsto s(\varphi)$. Set

$$u(\varphi) = \frac{m}{r(s(\varphi))},$$

then

$$\frac{du}{d\varphi} = -\frac{m}{r^2}\frac{dr}{ds}\frac{ds}{d\varphi} = -\frac{m}{r^2}\frac{\frac{dr}{ds}}{\frac{d\varphi}{ds}} = -\frac{m}{J}\frac{dr}{ds}. \tag{3.9.12}$$

From (3.9.10) and (3.9.7) one then obtains, for $J \neq 0$,

$$\left(\frac{du}{d\varphi}\right)^2 = \frac{m^2 E^2}{J^2} - u^2(1 - 2u) - \frac{\lambda m^2(1 - 2u)}{J^2}.$$ (3.9.13)

Similarly to the derivation of (3.9.11), the second order equation of motion for u can be obtained by differentiating with respect to φ, leading to

$$2\frac{d^2u}{d\varphi^2}\frac{du}{d\varphi} = \frac{d}{du}\left(-u^2(1 - 2u) - \frac{\lambda m^2(1 - 2u)}{J^2}\right)\frac{du}{d\varphi},$$ (3.9.14)

so either r is constant or

$$\frac{d^2u}{d\varphi^2} = -u + 3u^2 + \frac{\lambda m^2}{J^2}.$$ (3.9.15)

3.9.1 The Interpretation of E

In this section we momentarily suspend the convention that $c = 1$, and use the coordinate

$$x^0 := ct.$$

Consider two observers in special relativity, with unit four-velocity vectors u^μ and v^μ, there exists a frame in which (u^μ) equals $(1, \vec{0}) = (1, 0, 0, 0)$, while $v^\mu \partial_\mu$ has the general form $(v^\mu) = (\gamma, \gamma\vec{v})$, so that the scalar product $u^\mu v_\mu$ equals

$$u^\mu v_\mu = -\gamma := -\frac{1}{\sqrt{1 - |\vec{v}|^2 c^{-2}}}.$$ (3.9.16)

In curved spacetime we can use local inertial frames of two observers passing through the same spacetime point to similarly obtain (3.9.16), where $\vec{v}$ is the space-velocity of the local inertial frames with respect to each other:

$$\frac{1}{\sqrt{1 - |\vec{v}|^2 c^{-2}}} \equiv \gamma = -g(u, v).$$ (3.9.17)

We turn our attention now to a test observer moving in a Schwarzschild gravitational field. We want to relate the parameter m appearing in the metric to Newtonian physics. To avoid any preconceived ideas, let us call this parameter α, writing thus the metric as

$$g = -\left(1 - \frac{2\alpha}{r}\right)\underbrace{c^2 dt^2}_{(dx^0)^2} + \frac{dr^2}{1 - \frac{2\alpha}{r}} + r^2 d\Omega^2.$$

Let u^μ be the four-velocity vector of the *stationary* observers of ▶ Sect. 3.2 (with m there replaced by α) hence, as seen in (3.2.2),

$$u^\mu \partial_\mu = \frac{1}{\sqrt{1 - \frac{2\alpha}{r}}} \partial_0 .$$

Taking for v the four-velocity of a freely falling observer, i.e., $v = v^\mu \partial_\mu = \dot{x}^\mu \partial_\mu$ where $\dot{x}$ is tangent to a future directed timelike geodesic, we find

$$-\frac{1}{\sqrt{1 - |\vec{v}|^2 c^{-2}}} = g_{\mu\nu} u^\mu v^\nu = g_{00} \underbrace{u^0}_{1/\sqrt{1-\frac{2\alpha}{r}}} v^0 = -\sqrt{1 - \frac{2\alpha}{r}} \underbrace{\dot{t}}_{\frac{E}{1-\frac{2\alpha}{r}}}$$

$$= -\frac{E}{\sqrt{1 - \frac{2\alpha}{r}}} .$$

Hence

$$E = \frac{\sqrt{1 - \frac{2\alpha}{r}}}{\sqrt{1 - |\vec{v}|^2 c^{-2}}} .$$

Taylor expanding, for large r and small v/c, and multiplying by the rest energy, say $m_0 c^2$, of the geodesic observer we obtain

$$E m_0 c^2 \approx m_0 c^2 + \frac{m_0}{2} |\vec{v}|^2 - \frac{m_0 \alpha c^2}{r} . \tag{3.9.18}$$

This leads to the obvious interpretation of each term, as rest energy, Newtonian kinetic energy, and Newtonian gravitational energy in a gravitational field generated by a spherically symmetric object of mass

$$m = \frac{\alpha c^2}{G} .$$

In other words, for m to be the Newtonian mass seen at large distances, the Schwarzschild metric should read

$$g = -\left(1 - \frac{2Gm}{c^2 r}\right) c^2 dt^2 + \frac{dr^2}{1 - \frac{2Gm}{c^2 r}} + r^2 d\Omega^2 . \tag{3.9.19}$$

These calculations further show that, for timelike geodesics, we can interpret $E c^2$ as the *general relativistic energy per unit mass of the geodesic test objects*. The interpretation is consistent with the Newtonian theory for small velocities at large distances.

The property that E is constant along timelike geodesics becomes thus the law of conservation of energy for freely falling observers in a Schwarzschild field.

So far we have considered timelike geodesics, $\lambda = 1$. The situation is different for null geodesics because there is no preferred affine parameter along them. Indeed, given an affine parameter s and a positive constant a, we can introduce a new affine parameter $\bar{s} = as$. Under such a rescaling of a null geodesic $s \mapsto x(s)$ the new constant of motion, say $\bar{E}$, equals

$$\bar{E} := -g\left(\partial_t, \frac{dx}{d\bar{s}}\right) = -\frac{1}{a}g\left(\partial_t, \frac{dx}{ds}\right) = \frac{E}{a}.$$

This shows that, for future directed null geodesics, E can be changed to any positive value by a rescaling of the affine parameter.

One can provide prescriptions for a unique choice in specific situations, but there does not seem to be a clearly preferred universal choice.

3.9.2 Gravitational Redshift

In special relativity photons move along straight lines with null tangent $\eta(\dot{\gamma}, \dot{\gamma}) = 0$; these are affinely parameterized geodesics of the Minkowski metric η. In view of the correspondence principle we require that

> test photons in general relativity move along null geodesics.

Here, a *test photon* is a photon, the gravitational field of which can be ignored at the scale at which experiments are carried out.

In this section we wish to derive the frequency shift along radial null geodesics. So let a wave of light with frequency ω_1 be emitted radially at r_1, and let

$$\Delta s_1 := 2\pi/\omega_1$$

be the proper time between two consecutive maxima of the wave. In the "geometric optics approximation," the wave then travels outwards on radial null geodesics.

On such geodesics we have $\lambda = 0$, while θ and φ are constant. Therefore

$$-\left(1 - \frac{2m}{r}\right)dt^2 + \frac{dr^2}{1 - 2m/r} = 0,$$

leading to

$$\frac{dt}{dr} = \frac{r}{r - 2m}. \tag{3.9.20}$$

By integrating (3.9.20) we find that the coordinate time t_1 at which the crest of the wave leaves r_1 is related to the coordinate time t_2 at which it arrives at r_2 by

$$t_2 - t_1 = \int_{r_1}^{r_2} \frac{r\,dr}{r - 2m}. \tag{3.9.21}$$

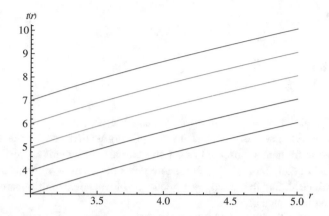

□ Fig. 3.5 Crests of a light-wave emitted radially at $r = 3$ propagate to $r = 5$ in a Schwarzschild metric of $m = 1$

Since the right-hand side of (3.9.21) is independent of t_1, we see that the coordinate time interval Δt_1 between the emission times of two successive crests at r_1 is the same as the coordinate time interval Δt_2 between their observations by a stationary observer at r_2; cf. □ Fig. 3.5.

Recall that the four-velocity vector

$$u \equiv u^\mu \partial_\mu := \frac{dx^\mu}{ds} \partial_\mu$$

of a stationary observer at coordinate radius r takes the form (3.2.2):

$$u = \frac{1}{\sqrt{1 - \frac{2m}{r}}} \partial_t = \frac{dt}{ds} \partial_t \quad \Longleftrightarrow \quad \frac{dt}{ds} = \frac{1}{\sqrt{1 - \frac{2m}{r}}}. \tag{3.9.22}$$

Since the emitter of light has been assumed to be stationary, (3.9.22) shows that the proper time interval Δs_1 is related to the coordinate time interval Δt_1 by

$$\Delta t_1 = \frac{dt}{ds} \Delta s_1 = \frac{1}{\sqrt{1 - \frac{2m}{r_1}}} \Delta s_1 \,,$$

with a similar formula relating Δt_2 with Δs_2.

As we have seen that $\Delta t_1 = \Delta t_2$, we obtain

$$\frac{\Delta s_1}{\Delta s_2} = \frac{\sqrt{1 - 2m/r_1}}{\sqrt{1 - 2m/r_2}}. \tag{3.9.23}$$

Subsequently,

$$\omega_2 = \frac{\sqrt{1 - \frac{2m}{r_1}}}{\sqrt{1 - \frac{2m}{r_2}}} \omega_1 \,. \tag{3.9.24}$$

This is the *gravitational redshift formula*, as observed, e.g., in the Pound-Rebka experiment in 1959, see, e.g., http://en.wikipedia.org/wiki/Pound-Rebka_experiment. We see that the frequency observed by an observer at infinity will be smaller than the energy emitted at any radius $r_1 > 2m$. More generally, if $r_2 > r_1$, then the observed spectrum will be shifted to the red, by a frequency-independent multiplicative factor, as compared to the emitted one.

As another application of (3.9.24), imagine that you send a beacon emitting at constant frequency towards a black-hole. You will see its frequency shifting away towards the red as the beacon approaches the event horizon $r = 2m$, with the frequency tending to zero as the event horizon is approached. As a result your eye, or your measuring apparatus, will stop seeing the beacon once the observation-threshold frequency of your instrument has been reached. This will happen even before the beacon has crossed the horizon.

Example 3.9.1 As yet another application, consider a wealthy owner of two watches, one worn on the wrist and one on the ankle. When m is small, the multiplicative factor at the right-hand side of (3.9.23) can be approximated as

$$1 - \frac{m}{r_1} + \frac{m}{r_2} \approx 1 - \frac{mh}{r_2^2} \,,$$

where $h = r_2 - r_1$, with second-order terms in m/r and h/r neglected. In SI units, this reads

$$\Delta s_1 = \Delta s_2 \left(1 - \frac{Gmh}{r_2^2 c^2}\right) = \Delta s_2 \left(1 - \frac{gh}{c^2}\right) \,,$$

where $g = Gm/r_2^2$ is the Newtonian gravitational acceleration at r_2. After 1 year, so that $\Delta s_1 \approx \Delta s_2 \approx 3 \times 10^7$ s, using the approximate values $g = 10 \, \text{m/s}^2$, $h = 1 \, \text{m}$, $c = 3 \times 10^8$ m/s, (3.9.23) reads

$$\frac{\Delta s_1 - \Delta s_2}{\Delta s_2} \approx -3 \times 10^{-9} \, \text{s} \,.$$

This shows that the watch on the ankle (at $r = r_1$) loses 3×10^{-9} s, each second, relative to the watch on the wrist (at $r = r_2 = r_1 + h$) every year. The watches must be pretty expensive, and cumbersome to wear, to notice the difference.

Note, however, that the effect becomes nontrivial at a height difference of 10 km. This has been measured by Hafele and Keating in 1972, who flew precise clocks in an aircraft around Earth. In this experiment one observes a superposition of the special relativistic time dilation, due to the motion of the aircraft, and a relativistic time correction. The results can be found in Table 3.1.

The effect becomes even more pronounced for GPS satellites. In fact, the most accurate measurement of the gravitational time shift so far arose from an accidental failure to put the satellites GSAT-0201 and GSAT-0202 of the European Global Navigational Satellite System Galileo (which orbit at about 20,000 km) into their originally planned circular orbits [27, 44]. The resulting ellipticity of the orbits allows to test the predictions of general relativity at the level of fractions of nanoseconds, see ◻ Fig. 3.6, with the fractional deviation of the gravitational redshift from the general relativistic prediction being smaller than 10^{-4} at 1σ.

Exercise 3.9.2 Explain the difference between the signs in the time-dilation balance of the westward and eastward measurements in Table 3.1.

Table 3.1 Time differences, in nanoseconds, measured in the Hafele–Keating experiment [39, 40]

| | Predicted | | | | |
	General rel.	Special rel.	Total	Measured	Difference
Eastward	$+144 \pm 14$	-184 ± 18	-40 ± 23	-59 ± 10	0.76σ
Westward	$+179 \pm 18$	$+96 \pm 10$	$+275 \pm 21$	$+273 \pm 7$	0.09σ

◻ **Fig. 3.6** Gravitational time shift for an eccentric Galileo satellite, reprinted with permission from [27]. The lower plot shows the residuals between the data and the general relativistic prediction. © 2019 by the American Physical Society

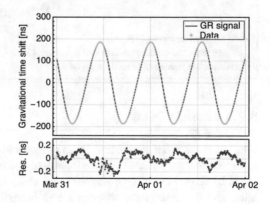

Incidentally An alternative derivation of the redshift effect uses the *frequency four vector* K of a photon, seen in the special relativity lectures: If a photon moves along a Minkowski spacetime null geodesic γ, then K is a constant multiple of $\dot{\gamma}$ such that an observer moving with four-velocity U observes a frequency ω given by

$$\omega = -U^\mu K_\mu.$$

In general relativity the frequency vector will thus again be a constant multiple of the tangent $\dot{\gamma}$ to a null geodesic.

So, consider a static observer O_1 at $r = r_1$, with velocity four-vector U_1, who sends a photon with frequency ω_1 to a static observer O_2 at $r = r_2$, with velocity four-vector U_2. We want to find the frequency ω_2 observed by O_2. If $K = \kappa\dot{\gamma}$, for some constant $\kappa > 0$, then the frequencies at $O_a, a = 1, 2$, equal

$$\omega_a = -g(U_a, K) = -g_{00}U_a^0 K^0 = -g_{00}U_a^0\kappa\dot{t} = \kappa\sqrt{1 - \frac{2m}{r_a}}\,\dot{t}.$$

But $(1 - 2m/r)\dot{t}$ is constant along the null geodesic followed by the photon. So

$$\omega_2 = \kappa\sqrt{1 - \frac{2m}{r_2}}\,\dot{t} = \kappa\frac{1}{\sqrt{1 - \frac{2m}{r_2}}}\underbrace{\left(1 - \frac{2m}{r_2}\right)\dot{t}}_{=\left(1 - \frac{2m}{r_1}\right)\dot{t}} = \frac{\sqrt{1 - \frac{2m}{r_1}}}{\sqrt{1 - \frac{2m}{r_2}}}\underbrace{\kappa\dot{t}\sqrt{1 - \frac{2m}{r_1}}}_{\omega_1},$$

and we have recovered (3.9.24).

As such, the calculation here is more general than the one which led to (3.9.24), as here any geodesic, not necessarily radial, is allowed.

3.9.3 Shapiro Delay

Suppose that a radar signal, sent by an observer at radius r_A, is reflected back at radius r_B to return to the observer. We are interested in the time needed for the signal to return.

In Minkowski spacetime photons move along null geodesics, which are null straight lines, and the time in the observer's own inertial frame will be twice the Euclidean distance between the observer and the reflecting object.

Consider the same problem in Schwarzschild spacetime. Clearly null geodesics in the Schwarzschild geometry will not be straight lines in general, thus the trajectories will differ from the Minkowskian ones. Similarly the time needed to hear the echo will be different. The difference between the Schwarzschildian time and the Minkowskian time is called *Shapiro delay*, for reasons that will become apparent shortly.

To illustrate the effect, we calculate the Shapiro delay for the simplest possible configuration, when the reflecting object at radius r_A has identical angular coordinates as the emitter located at r_B, with $r_B > r_A$. The signal moves thus first on an incoming, and then on an outgoing, null radial geodesic. Equation (3.9.21) applies and gives

$$t_B - t_A = \int_{r_A}^{r_B} \frac{r \, dr}{r - 2m} = r_B - r_A + 2m \ln\left(\frac{r_B - 2m}{r_A - 2m}\right). \tag{3.9.25}$$

It immediately follows from the first equality in (3.9.25) that both trips will take the same time, leading to a return coordinate-time

$$t_A + 2\left(r_B - r_A + 2m \ln\left(\frac{r_B - 2m}{r_A - 2m}\right)\right). \tag{3.9.26}$$

The associated proper-time at r_B equals

$$2\sqrt{1 - \frac{2m}{r_B}}\left(r_B - r_A + 2m \ln\left(\frac{r_B - 2m}{r_A - 2m}\right)\right). \tag{3.9.27}$$

The analogous expression in Minkowski spacetime is $2(r_B - r_A)$, leading to a difference, say δs, of proper times equal to

$$\delta s = 2\left(\sqrt{1 - \frac{2m}{r_B}} - 1\right)(r_B - r_A) + 4m \sqrt{1 - \frac{2m}{r_B}} \ln\left(\frac{r_B - 2m}{r_A - 2m}\right). \tag{3.9.28}$$

It turns out that δs is strictly positive: for positive m the arrival time is *delayed*, as compared to the Minkowskian case. To prove this, note that (3.9.27) equals

$$2\sqrt{1 - \frac{2m}{r_B}} \int_{r_A}^{r_B} \frac{dr}{1 - \frac{2m}{r}} = 2 \int_{r_A}^{r_B} \frac{\sqrt{1 - \frac{2m}{r_B}}}{1 - \frac{2m}{r}} dr$$

$$> 2 \int_{r_A}^{r_B} \sqrt{\frac{1 - \frac{2m}{r_B}}{1 - \frac{2m}{r}}} dr > 2 \int_{r_A}^{r_B} dr = 2(r_B - r_A). \tag{3.9.29}$$

In 1962 Irwin Shapiro suggested to measure this time difference for radio signals sent to Venus (◘ Fig. 3.7). When in 1964 the 120 foot Haystack antenna in Westford (see ◘ Fig. 3.8) was left by the military to MIT, Shapiro and his team began plans to carry out the experiment, with measurements taking place in 1966 and 1997. The resulting time-delay, as a function of time, is shown in ◘ Fig. 3.7. Shapiro improved the precision of his initial measurements to less than 1% in subsequent years. An interesting discussion of the experiment can be found on http://www.extinctionshift.com/SignificantFindings06B.htm.

The analysis of signals sent back and forth to the Mars Viking lander by Reasenberg et al. [71] was carried out in 1979, giving agreement with the general relativity prediction to about 0.2 %.

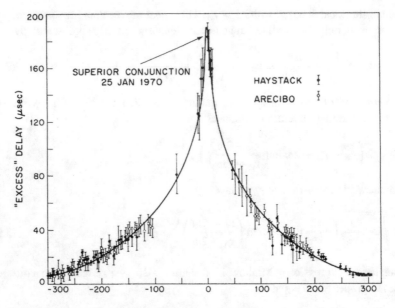

◘ Fig. 3.7 Time-delay for radar signals bouncing off Venus near the opposition, reprinted with permission from [78]. © 2019 by the American Physical Society

◘ Fig. 3.8 © MIT Haystack Observatory, reproduced with kind permission

More recently, the Shapiro effect has been used for accurate determination of masses in a binary millisecond radio pulsar [28].

An illustrative animation of the Shapiro effect can be viewed on https://en.wikipedia.org/wiki/File:Shapiro_delay.gif.

3.9.4 Circular Causal Geodesics

We consider the question of existence of *causal* geodesics (thus, $\lambda \in \{0, 1\}$) such that $\dot{r} \equiv 0, \theta \equiv \pi/2$. Clearly

$$J \neq 0$$

for such geodesics. Next, from (3.9.10) we must have

$$0 = \left(\frac{dr}{ds}\right)^2 = -\left(\lambda + \frac{J^2}{r^2}\right)\left(1 - \frac{2m}{r}\right) + E^2 \,, \tag{3.9.30}$$

which determines E:

$$E^2 = \left(\lambda + \frac{J^2}{r^2}\right)\left(1 - \frac{2m}{r}\right). \tag{3.9.31}$$

Now, (3.9.30) is only a *necessary condition* for a circular orbit; it would also hold, e.g., at a local extremum of r along any orbit. So, in addition, we need to make sure that the geodesic equation is satisfied with $r = $ const. In particular (3.9.11) must hold, leading to

$$0 = 2\frac{d^2r}{ds^2} = \frac{d}{dr}\left(E^2 - (1 - \frac{2m}{r})(\lambda + \frac{J^2}{r^2})\right)$$

$$= -2\frac{m\lambda r^2 + 3mJ^2 - J^2 r}{r^4}\,. \tag{3.9.32}$$

Hence

$$m\lambda r^2 = J^2(r - 3m)\,. \tag{3.9.33}$$

Since $m\lambda \geq 0$ for causal geodesics the left-hand side is nonnegative, which together with $J^2 > 0$ requires that 1) $r \geq 3m$ with 2) $r = 3m$ if and only if $\lambda = 0$. We conclude that

> circular timelike geodesics exist if and only if $r > 3m$,

and that for such geodesics we have

$$J^2 = \frac{mr^2}{r - 3m}\,. \tag{3.9.34}$$

We also reach the surprising conclusion that

> there exist null geodesics circling at $r = 3m$.

For such geodesics we have, using (3.9.31) with $\lambda = 0$,

$$E^2 = \frac{(r - 2m)J^2}{r^3} = \frac{J^2}{3^3 m^2}\,. \tag{3.9.35}$$

Recall that the affine parameter on a null geodesic is only defined up to an affine transformations. In particular we can use t as an affine parameter. Simple algebra shows then that the null geodesics spiraling on the timelike cylinder $\{r = 3m\}$ are described by the equations

$$s \mapsto \gamma_\pm(s) = (t = s, \; r = 3m, \; \theta = \pi/2, \; \varphi = \pm 3^{-\frac{3}{2}} m^{-1} s). \tag{3.9.36}$$

Exercise 3.9.3 Alice circles planet X freely for a long time on a circular orbit of radius R, while her twin Bob remains motionless on the surface of the planet, at radius r_0. For $r \geq r_0$ the geometry of the gravitational field of the planet X is described by the Schwarzschild metric with mass $0 < m < r_0/2$. Derive a necessary and sufficient condition on R which guarantees that, on meeting Bob again, Alice will have the same age as Bob. You should assume that the time of travel back and forth from radius R to radius r_0 can be neglected compared to the time that Alice spent on the circular orbit.

We return to timelike geodesics, $\lambda = 1$. A somewhat miraculous calculation gives the simple formula

$$\left(\frac{d\varphi}{dt} \right)^2 = \left(\frac{\frac{d\varphi}{ds}}{\frac{dt}{ds}} \right)^2 = \frac{J^2}{r^4} \times \frac{\left(1 - \frac{2m}{r}\right)^2}{E^2}$$

$$= \frac{m}{r^3}. \tag{3.9.37}$$

To see this one inserts (3.9.34) into (3.9.31) with $\lambda = 1$ to obtain

$$E^2 = \left(1 + \frac{J^2}{r^2} \right) \left(1 - \frac{2m}{r}\right) = \frac{r}{r - 3m} \left(1 - \frac{2m}{r}\right)^2. \tag{3.9.38}$$

Dividing by J^2 as given by (3.9.34) and taking a square root one finds

$$E = \frac{|J|}{\sqrt{mr}} \left(1 - \frac{2m}{r}\right). \tag{3.9.39}$$

Inserting (3.9.39) into the rightermost expression in the first line of (3.9.37) leads to the desired formula.

Equation (3.9.37) leads to a simple expression for the *coordinate* angular-frequency Ω:

$$\varphi = \varphi_0 + \Omega(t - t_0), \quad \Omega := \frac{m^{\frac{1}{2}}}{r^{\frac{3}{2}}}. \tag{3.9.40}$$

Rather surprisingly, this is formally identical to the Newtonian formula (3.8.16).

For further reference we note that along timelike circular geodesics we have

$$t = t_0 + \frac{rE}{r - 2m}(s - s_0) = t_0 \pm \frac{J}{\sqrt{mr}}(s - s_0) \quad \Longleftrightarrow \quad \frac{dt}{ds} = \pm \frac{J}{\sqrt{mr}}. \tag{3.9.41}$$

Summarizing:

1. There are no circular timelike geodesics for $r \le 3m$.
2. For any given $r > 3m$ there exists an *equatorial future-directed timelike geodesic with constant r*, with

$$t - t_0 = \frac{|J|}{\sqrt{mr}}(s - s_0), \quad \varphi - \varphi_0 = \frac{J}{r^2}(s - s_0) = \pm \frac{m^{\frac{1}{2}}}{r^{\frac{3}{2}}}(t - t_0), \qquad (3.9.42)$$

where $|J|$ is defined by (3.9.34), and s is the proper time along the geodesic. The geodesic is defined uniquely by r up to the sign of J.

A key question is that of stability of circular geodesics. Recall the second order equation (3.9.15) for $u = m/r$:

$$\frac{d^2 u}{d\varphi^2} = -u + 3u^2 + \lambda \frac{m^2}{J^2}. \qquad (3.9.43)$$

We set $u = u_0 + \delta u$, where $du_0/d\varphi = 0$, and where δu is assumed to be small. Linearizing (3.9.43) at u_0 one finds

$$\frac{d^2 \delta u}{d\varphi^2} \approx -\delta u + 6u_0 \delta u = (6u_0 - 1)\delta u.$$

When $6u_0 > 1$ (equivalently, $r_0 < 6m$) the solutions are linear combinations of $e^{\pm \sqrt{6u_0 - 1}\varphi}$ and thus *linearization-unstable*: small perturbations can grow without bound in time. (Here one should keep in mind that the linearized equation ceases to provide a good approximation of the problem at hand when small perturbations grow large, at which time one needs to consider the effect of the nonlinearities.)

On the other hand, for $6u_0 < 1$ (equivalently, $r_0 > 6m$) the solutions are linear combinations of $\sin(\sqrt{|6u_0 - 1|}\varphi)$ and $\cos(\sqrt{|6u_0 - 1|}\varphi)$, and thus *linearization-stable*: small perturbations remain small. (One then expects that this feature will remain true for solutions of the full nonlinear equation, but a proof of this would require separate considerations.)

We conclude that circular timelike geodesics are linearization-stable if and only if $r > 6m$.

The orbits $r = 6m$ are called *Innermost stable circular orbits* (ISCOs) in the astrophysical literature.

3.9.5 Weak-Field Light Bending

For null geodesics (3.9.43) reads

$$\frac{d^2 u}{d\varphi^2} = -u + 3u^2. \qquad (3.9.44)$$

For u very small or, equivalently, for r large as compared to m, an excellent approximation is obtained by neglecting the quadratic term, leading to

$$u_0 = \alpha \cos(\varphi - \varphi_0),$$

for some (small) constant $\alpha \neq 0$ (otherwise $r = \infty$). By a redefinition of φ we can always achieve $\varphi_0 = 0$. Equivalently,

$$\alpha r \cos \varphi = m,$$

which is the equation for the straight line $x = m/\alpha$ in the (x, y) plane.

We can calculate the leading order correction to this by writing $u = \alpha \cos \varphi + v(\alpha, \varphi)$, where v is small: $v = O(\alpha^2)$. Inserting into (3.9.44) and neglecting terms which are $O(\alpha^3)$ one obtains

$$v'' + v = 3\alpha^2 \cos^2 \varphi.$$

This is easily integrated to give

$$v = A \cos \varphi + B \sin \varphi + \alpha^2 (1 + \sin^2 \varphi).$$

We choose A and B so that at $\varphi = 0$ the initial data for the orbit coincide with those for the unperturbed one,

$$u(0) = \alpha = \frac{m}{d}, \quad u'(0) = 0,$$

where d is the distance of the closest approach of the unperturbed orbit to the origin $r = 0$. This gives $A = -\alpha^2$, $B = 0$, and

$$u = \underbrace{(\alpha - \alpha^2) \cos \varphi + \alpha^2 (1 + \sin^2 \varphi)}_{=:u_1} + O(\alpha^3). \tag{3.9.45}$$

Representative plots of the approximate solution can be found in ◻ Fig. 3.9

Let us check that d is the distance of the closest approach of the orbit to the origin for α small: indeed,

$$u' \approx -(\alpha - \alpha^2) \sin \varphi + 2\alpha^2 \sin \varphi \cos \varphi$$

$$= \left(-(\alpha - \alpha^2) + 2\alpha^2 \cos \varphi \right) \sin \varphi$$

$$= \alpha \left(-1 + \underbrace{\alpha + 2\alpha \cos \varphi}_{\ll 1} \right) \sin \varphi$$

$$\neq 0 \quad \text{for } \varphi \in (-\pi, \pi) \setminus \{0\}.$$

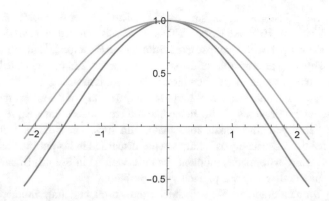

▫ Fig. 3.9 Sample plots of the function $f_\alpha := u_1/(m\alpha)$, where u_1 is given by (3.9.45). Here $f_{0.2} \geq f_{0.1} \geq f_0$, with f_0 corresponding to a null geodesic in flat spacetime. The zeros of the function f_α determine the asymptotic value, as the photon recedes to infinity, of the slope of the projection of the trajectory to the equatorial plane

Thus, up to α^3 corrections, u has only one extremum, at $\varphi = 0$, which is a maximum, and so r has a minimum at $\varphi = 0$.

Now, the "Newtonian" orbit u_0 was a straight line. By definition of u, we have $r \to \infty$ if and only if $u \to 0$; for u_0 this corresponds to $\varphi \to \pm\pi/2$. Note that

$$u_1 = \alpha \cos\varphi + \alpha^2 \underbrace{(1 - \cos\varphi + \sin^2\varphi)}_{\geq 0} \geq \alpha \cos\varphi, \qquad (3.9.46)$$

which shows that the corrected orbit u_1 will reach zero at angles $\varphi_\alpha = \pm(\pi/2 + \gamma_\alpha)$, slightly larger in modulus than $\pi/2$:

$$0 = \frac{u_1(\varphi_\alpha)}{\alpha} = (1 - \alpha)\cos\varphi_\alpha + \alpha(1 + \sin^2\varphi_\alpha)$$

$$= -(1 - \alpha)\sin\gamma_\alpha + \alpha(1 + \cos^2\gamma_\alpha).$$

Approximating $\sin\gamma_\alpha$ by γ_α, and $\cos\gamma_\alpha$ by one, one obtains

$$\gamma_\alpha = 2\alpha + O(\alpha^2).$$

The total bending of the orbit is 4α, giving the final SI formula for the angle deflection

$$\frac{4mG}{dc^2},$$

recall that d is the distance of closest approach to the center.

For a light-ray just grazing the surface of the Sun, so that $m = M_\odot$, $d = r_\odot$, one obtains a deflection of 10^{-5} radians or $2''$. This effect was claimed to have been observed by Eddington during the 1919 eclipse expedition to the Principe Island, off the coast of Africa, by comparing photographs of the star field near the Sun during an eclipse with a photograph of the same star field when the Sun was not interfering.

While there is no controversy nowadays about the reality of the effect, there appears to be one about Eddington's claim. Indeed, some researchers have expressed doubts about the reliability of the conclusions that could have been drawn from the data available at the time, especially as Eddington has discarded in his publication the results of his second team, who made simultaneous observations in Sobral, Brazil. There the effect was smaller, allegedly due to a telescope problem.

See http://jila.colorado.edu/~ajsh/insidebh/schw.html and http://homepage.univie. ac.at/Franz.Embacher/Rel/artLichtablenkung/start.html for informative animations.

3.9.6 The Shadow of a Black Hole

Let us continue with some remarks on strong-field light-deflection. In ◨ Fig. 3.10 one can see how the curvature of spacetime due to the central galaxy deforms the image of a much more distant one. The curving of geodesics by the geometry results in the existence of deformed, possibly multiple, images of a single object under certain circumstances. Because of the characteristic arc-shaped form, the resulting images are called *Einstein arcs*. If spacetime were flat, we would not see the distant galaxy at all, because the light-rays would propagate along straight lines, and the image would have been hidden by the star.

◨ **Fig. 3.10** An "Einstein horseshoe" around the Luminous Ring Galaxy LRG 3-757. Image Credit: ESA/Hubble & NASA, reproduced with permission. The blue Einstein arc is formed from multiple images of a distant galaxy located behind the central one

◻ Fig. 3.11 Einstein arcs in the Galaxy Cluster Abbel 2218, from the STScI Public Archive [47]

◻ Fig. 3.12 Gravitational lensing by the nearby galaxy ESO 325-G004 by ESO, ESA/Hubble, NASA, from https://www.spacetelescope.org/news/heic1812/, reproduced with permission. This is one of the nearest galactic gravitational lenses observed to date, with lens redshift $z = 0.035$, which is near enough to resolve stellar dynamics, providing a test of general relativity at galactic scale [24]

◻ Figure 3.10 provides an example of a general phenomenon referred to as *gravitational lensing*, an extensive treatment of the subject can be found in [38, 65, 77]. One of the important effects that arise here is a magnification of the image, allowing one to see objects which otherwise would have been much too faint for observations.

By now several Einstein arcs have been seen by astrophysicists. ◻ Figure 3.11 provides another example of such observations made by the Hubble space telescope. See the web site https://www.cfa.harvard.edu/castles/ for a whole catalogue. ◻ Figure 3.12, from [24], shows the image of a gravitational lens which is near enough to allow for tests of general relativity.

The question arises, how large is the region hidden from view when a black hole provides the lens. Keeping in mind that the event horizon is not an object that can be

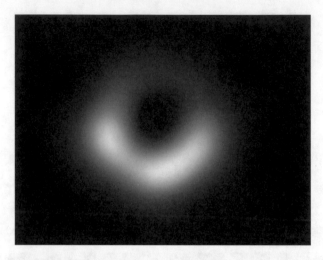

Fig. 3.13 The shadow of the supermassive black hole M87*, from https://www.eso.org/public/images/eso1907a. The mass of the black hole is estimated to be $M = (6.5 \pm 0.2|_{\text{stat}} \pm 0.7|_{\text{sys}}) \times 10^9\,M_\odot$ [33]. Credit: EHT Collaboration

seen as such, one can think of "viewing the shadow of the event horizon" instead. This shadow can be seen in the accretion disc around the supermassive black hole M87*, located in the galaxy Messier 87, as observed by the Event Horizon Telescope [33], see ▪ Fig. 3.13.

One is tempted to think that, in the Schwarzschild geometry, the hidden region corresponds to the $r = 2m$ black hole region. This is not the case, and the aim of what follows is to clarify this issue.

For this, let us return to (3.9.13) with $\lambda = 0$:

$$\left(\frac{du}{d\varphi}\right)^2 + u^2(1 - 2u) = \frac{m^2 E^2}{J^2} =: \mathscr{E}\,. \tag{3.9.47}$$

It is convenient to think of this equation as the *law of conservation of "energy $\mathscr{E}$"* for a particle of mass $m_0 = 2$, with position described by u, evolving in a time denoted by φ. The particle has therefore kinetic energy $(du/d\varphi)^2$, total energy $\mathscr{E}$, and moves in a potential which we denote by V_p,

$$V_p(u) = u^2(1 - 2u)\,, \tag{3.9.48}$$

plotted in ▪ Fig. 3.14. Note that $\mathscr{E}$ is determined by the ratio E/J, with small J's corresponding to large "energies" $\mathscr{E}$.

The unique maximum of V_p is easy to find:

$$V_p' = 2u - 6u^2 = 0 \iff u = \frac{1}{3}\,.$$

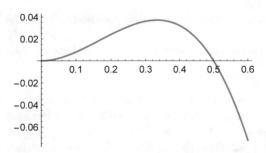

□ Fig. 3.14 The effective potential of a photon in the Schwarzschild geometry as a function of $u = m/r$

The corresponding null geodesics have already been seen in (3.9.36), and these are the photons circling around the timelike cylinder $\{r = 3m\}$. Let us call $\mathscr{E}_c$ the corresponding value of the potential:

$$\mathscr{E}_c = V_p|_{u=1/3} = 3^{-3}.$$

In what follows one needs to keep in mind that $u \to 0$ is equivalent to $r \to \infty$; the event horizon $r = 2m$ corresponds to $u = 1/2$; and $u \to \infty$ is equivalent to $r \to 0$. Hence the region $u > 1/2$ corresponds to the interior of the black hole, and $u < 1/2$ to its exterior.

A trajectory will be called "initially ingoing" when $dr/d\varphi < 0$ initially; this corresponds to $du/d\varphi > 0$ initially. "Initially outgoing" trajectories will start with $dr/d\varphi > 0$; equivalently, with u which is initially decreasing. Recall also that (3.9.47) makes sense only for photons with $J \neq 0$.

Elementary considerations based on the conservation equation (3.9.47) show that:

1.　Initially ingoing trajectories with critical energy $\mathscr{E} = \mathscr{E}_c$ starting at large r will take an infinite φ-time to reach the critical point $u = 1/3$. This is the same as saying that the photon orbits will rotate an infinite number of times around the spacetime cylinder $\{r = 3m\}$ as the φ-time progresses, asymptoting to this cylinder when φ tends to infinity.

2.　Initially ingoing trajectories with $\mathscr{E} < \mathscr{E}_c$ starting at large r will be reflected on the potential at a turning point $u = u_{\max} < 1/3$, at which r attains its minimum $r_{\min} = m/u_{\max}$, determined by the equation

$$\mathscr{E} = V_c(u_{\max}).$$

Hence, such trajectories will never cross the spacetime cylinder $\{r = 3m\}$. So, if a light-ray is emitted in the region $r > 3m$ towards the black hole, and is seen by observers which are arbitrary far away from the black hole, then this light-ray will never penetrate the region $\{r < 3m\}$.

Equivalently, suppose that you shine a torchlight towards a black hole, and you observe the result on a screen very far away from the black hole. The photons reaching the screen will never cross the "photosphere" $r = 3m$ before hitting the screen.

3. Initially ingoing trajectories with $\mathscr{E} > \mathscr{E}_c$ will always end up in the singularity $r = 0$.

Point 3 implies that if you are very far away in the region $\{r > 3m\}$, and you can observe a photon which has been emitted towards the star from the same region, then the photon must have $\mathscr{E} < \mathscr{E}_c$.

We conclude that the *black region of* ☐ *Fig. 3.14 corresponds to the region* $\{r < 3m\}$ *in spacetime.*

We also see that the *angular extent* of the directions from which all photons coming from far-away sources can reach an observer at $r > 3m$ is determined by the angular extent of all photons at r for which $\mathscr{E} < \mathscr{E}_c$.

To determine the corresponding angles, consider an equatorial photon with trajectory γ parameterized by φ:

$$\gamma(\varphi) = (t(\varphi), r(\varphi), \theta = \pi/2, \varphi) .$$

Its tangent vector is

$$\frac{d\gamma}{d\varphi} = \left(\frac{dt}{d\varphi}, \frac{dr}{d\varphi}, \frac{d\theta}{d\varphi} = 0, \frac{d\varphi}{d\varphi} = 1 \right) .$$

Recall that a stationary observer detecting the photon will have a world-line with unit-timelike tangent u given by

$$u = \frac{1}{\sqrt{1 - \frac{2m}{r}}} \partial_t .$$

The *space directions* (not to be confused with *spacelike* directions) of the observer are defined as directions orthogonal to u, hence vectors X satisfying

$$g(u, X) = 0 \quad \Longleftrightarrow \quad X = X^r \partial_r + X^\theta \partial_\theta + X^\varphi \partial_\varphi .$$

Equivalently, for a stationary observer in Schwarzschild spacetime, the space-part of a vector is obtained by dropping the time-component of the vector.

We wish to calculate the angle between the space-direction of the photon, say k, and the ingoing radial direction $\ell := -\partial_r$. Since the metric on the space orthogonal to u is Riemannian this angle, say θ, can be calculated from the usual formula

$$g(\ell, k) = \sqrt{g(\ell, \ell)} \sqrt{g(k, k)} \cos\theta . \tag{3.9.49}$$

We have:

$$k = \frac{dr}{d\varphi}\partial_r + \partial_\varphi \,,$$

$$g(\ell, k) = -g(\partial_r, k) = -\frac{\frac{dr}{d\varphi}}{1 - \frac{2m}{r}} \,,$$

$$g(\ell, \ell) = g(\partial_r, \partial_r) = \frac{1}{1 - \frac{2m}{r}} \,,$$

$$g(k, k) = \frac{\left(\frac{dr}{d\varphi}\right)^2}{1 - \frac{2m}{r}} + r^2 \,.$$

Hence

$$\cos^2\theta = \frac{(g(\ell, k))^2}{g(\ell, \ell), g(k, k)} = \frac{\left(\frac{\frac{dr}{d\varphi}}{1-\frac{2m}{r}}\right)^2}{\frac{1}{1-\frac{2m}{r}} \times \left(\frac{\left(\frac{dr}{d\varphi}\right)^2}{1-\frac{2m}{r}} + r^2\right)} = \frac{\left(\frac{dr}{d\varphi}\right)^2}{\left(\frac{dr}{d\varphi}\right)^2 + r^2(1 - \frac{2m}{r})} \,.$$

$$(3.9.50)$$

Since $dr/d\varphi = -mu^{-2}du/d\varphi$, we can rewrite this as

$$\cos^2\theta = \frac{\left(\frac{du}{d\varphi}\right)^2}{\left(\frac{du}{d\varphi}\right)^2 + u^2(1 - 2u)} = \frac{\mathscr{E} - u^2(1 - 2u)}{\mathscr{E}} \,. \tag{3.9.51}$$

Equivalently,

$$\sin^2\theta = \frac{u^2(1 - 2u)}{\mathscr{E}} = \frac{m^2(r - 2m)}{r^3\mathscr{E}} \,. \tag{3.9.52}$$

The aperture of the shadow is obtained when $\mathscr{E} = \mathscr{E}_c \equiv 3^{-3}$:

$$\boxed{\sin^2\theta = \frac{27m^2(r - 2m)}{r^3}} \,, \tag{3.9.53}$$

a formula first derived by Synge [83].

In order to make contact with "what we think we see," suppose that we are looking at a sphere of radius d which is a distance r away from us in Minkowski space-time. For $d \ll r$ the angle between the rays reaching us from the center of the sphere and those from its boundary equals

$$\theta \approx \tan\theta = d/r \,.$$

From this and from (3.9.53) with $m \ll r$ we find

$$\theta \approx \frac{\sqrt{27}\, m}{r} \approx \frac{5.2\, m}{r} = \frac{2.6\, r_s}{r} \,,$$

where $r_s := 2m$ is the Schwarzschild radius of the black hole. So, our Minkowskian praxis will lead us to think that we are seeing a sphere with a radius approximately equal to $2.6\, r_s$.

3.9.7 Perihelion/Periastron Precession

Given a Keplerian orbit around a star, the *periastron* is the point at which the orbit is closest to the star. When the star is our Sun, this point is usually called the *perihelion*.

One of the earliest tests of general relativity is provided by the precession of the orbit of Mercury. The aim of this section is to derive this effect.

More precisely, we want to calculate the deviation of the geodesics in the Schwarzschild metric from the Keplerian orbits, at distances large compared to m. Here the following should be kept in mind: for the orbit of Earth (so that $r = r_\oplus$ and $J = J_\oplus/M_\oplus$) around the Sun (so that $m = M_\odot$) we have (recall that we are using units where $c = G = 1$)

$$\frac{M_\odot}{r_\oplus} = u \sim 10^{-8}\,, \qquad \frac{J_\oplus^2}{M_\odot^2 r_\oplus^2} = J_\oplus^2 M_\odot^{-2} M_\oplus^{-2} u^2 \sim 10^{-8}\,, \qquad \frac{M_\odot^2 M_\oplus^2}{J_\oplus^2} \sim 10^{-8}\,.$$

(One should remember that, in the geodesic equation, m is the mass of the central body, while J is the angular momentum per unit mass of the orbiting one.)

The numbers for Mars and Mercury are of the same order. See http://www.zipcon.net/~swhite/docs/astronomy/Angular_Momentum.html for a useful summary of the relevant data for all planets in the solar system.

Recall the Newtonian equation of motion (3.8.7), p. 73, for $u = m/r$,

$$\frac{d^2 u}{d\varphi^2} = -u + \mathring{u}\,, \quad \text{where} \quad \boxed{\mathring{u} := \frac{G m_0^2 m^2}{J_N^2}}\,, \tag{3.9.54}$$

and its solution (3.8.14),

$$u = \mathring{u}(e \cos \varphi + 1) = \frac{G m_0^2 m^2}{J_N^2}(e \cos \varphi + 1)\,, \tag{3.9.55}$$

chosen so that $du/d\varphi = 0$ at $\varphi = 0$. In the Schwarzschild metric, the corresponding equation of motion (3.9.15) for u with $\lambda = 1$ and $G = 1 = m_0$ reads

$$\frac{d^2 u}{d\varphi^2} = -u + 3u^2 + \frac{m^2}{J^2}\,. \tag{3.9.56}$$

We assume that $\mathring{u}$ is small. We write $u = u_0 + v$, where $u_0 = O(\mathring{u})$ is the Newtonian solution and $v = O(\mathring{u}^2)$, and insert in (3.9.56). One finds, neglecting terms which are $O(\mathring{u}^3)$,

$$\frac{d^2v}{d\varphi^2} + v = 3\frac{m^4}{J^4}(1 + e\cos\varphi)^2$$

$$= 3\frac{m^4}{J^4}(1 + 2e\cos\varphi + \underbrace{e^2\cos^2\varphi}_{e^2(1+\cos(2\varphi))/2}). \tag{3.9.57}$$

By standard results the solution takes the form

$$v(\varphi) = A + B\cos\varphi + C\sin\varphi + D\cos(2\varphi) + E\varphi\sin\varphi + F\varphi\cos\varphi, \tag{3.9.58}$$

with constants A, B, C, D, E, and F which need to be determined from the equation together with the initial conditions. The term $E\varphi\sin\varphi + F\varphi\cos\varphi$, called *secular*, arises from an effect referred to as *resonance*: the source term of the equation contains functions which have exactly the same period as the solutions of the homogeneous equation. We will see in our "Solution 2" below how this term leads to a precession of the perihelion.

However, it is convenient to proceed differently because (3.9.58) is obviously wrong for large φ, equivalently, for long times: Indeed, solutions of (3.9.54) have a conserved energy (compare (3.8.6), p. 73)

$$E = \frac{1}{2}\left(\frac{du}{d\varphi}\right)^2 + \underbrace{\frac{u^2}{2} - u^3 - \frac{m^2}{J^2}u}_{V(u)}. \tag{3.9.59}$$

This shows that (3.9.58) can be thought of as the equation of motion of a particle of mass equal to one moving in the potential V. For $0 < m/J < 1/4$ and for all u sufficiently small these solutions are periodic in φ, as can be seen by inspection of the right plot in ◘ Fig. 3.15. Closed orbits require solutions which are 2π-periodic when parameterized by the polar angle φ, and *the precession of the perihelion results from the fact that the period is different from 2π*. Our aim is to approximately calculate this period.

Remark 3.9.4 A comment on periodicity might be in order. For this, consider a solution $u(\varphi)$ of (3.9.58) with $\frac{m}{J} < \frac{1}{5}$, with negative total energy E_0, with $u(0) < 0.1$ (cf. the right ◘ Fig. 3.15), and with $u'(0) > 0$. It follows from the graph of V that there exist two unique solutions $u_\pm$, satisfying $0 < u_- < u_+ < 0.1$, of the equation

$$E_0 = V(u_\pm).$$

Then the solution $u(\varphi)$ attains a maximum u_+ at a first time $\varphi_+ > 0$, followed by a minimum at a first time $\varphi_- > \varphi_+$, and a next maximum at $u(\varphi_+ + T)$ for some $T > 0$. Uniqueness of solutions of ODEs shows that all subsequent maxima will be attained at $u(\varphi_+ + nT)$, $n \in \mathbb{N}$, and that in fact we must have $u(\varphi + T) = u(\varphi)$. This establishes periodicity of the solutions.

We can obtain an explicit formula for the period as follows: We rewrite (3.9.59) in the form

$$\frac{d\varphi}{du} = \pm \frac{1}{\sqrt{2(E_0 - V(u))}} \,.$$

Integrating these two equations from u_- to u_+ (one equation for each choice of sign △) shows that the "φ-time" needed for the solution to go from u_+ to u_- is the same as that needed to go from u_- to u_+. Hence this time is equal half of the period T, leading to

$$T = 2 \int_{u_-}^{u_+} \frac{du}{\sqrt{2(E_0 - V(u))}} \,. \tag{3.9.60}$$

One can derive the precession of the perihelion using this equation, but a simpler argument is provided by what follows.

In view of what has been said so far, instead of analyzing (3.9.58) we will use the *Poincaré-Lindstedt (PL) method* to obtain a solution where the secular term $E\varphi \sin \varphi + F\varphi \cos \varphi$ of (3.9.58) is absorbed in a change of period; as a bonus we will obtain a formula for the new period. The PL method gives an equivalent result as the standard perturbation calculation outlined above for small times, but provides a better approximation to the solution of the original equation for large times. As significantly, it respects the periodicity property of the solution of the original equation.

The idea is to set up a perturbative scheme which includes explicitly the information about the change of the period. For this we start by setting

$$\varepsilon := \frac{m^2}{J^2} \,,$$

and introduce a new "time variable" ϕ by letting

$$\varphi = (1 + \varepsilon\delta)\phi \,, \tag{3.9.61}$$

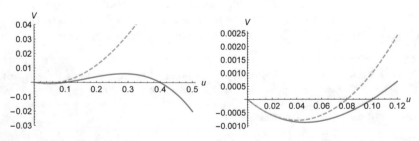

■ **Fig. 3.15** The analysis of geodesics in the Schwarzschild metric can be reduced to the analysis of one-dimensional motion $u = u(\varphi)$ in a potential $V(u)$ given by (3.9.59), with $u = m/r$. The plot is typical for $0 < m/J < 1/4$, here $m/J = 1/5$. The left plot shows the overall form of the potential, with the dashed plot corresponding to the Newtonian case. The right one is a zoom into the region of small u. All negative-energy orbits meeting the region covered by the second plot remain there and are periodic in φ with a period different from 2π, except for the constant orbit at the minimum of the potential

with $\delta \in \mathbb{R}$ to be chosen so that the solution will be 2π-periodic in ϕ. Keeping in mind that ε is assumed to be very small, we have

$$\frac{d}{d\varphi} = \frac{1}{1+\varepsilon\delta}\frac{d}{d\phi} \approx (1-\varepsilon\delta)\frac{d}{d\phi}, \quad \frac{d^2}{d\varphi^2} \approx (1-\varepsilon\delta)^2\frac{d^2}{d\phi^2} \approx (1-2\varepsilon\delta)\frac{d^2}{d\phi^2}.$$

We also rescale u_0 and v to obtain functions U_0 and V which we expect to be of order one:

$$u_0 = \varepsilon U_0, \quad v = \varepsilon^2 V, \tag{3.9.62}$$

thus

$$u = u_0 + v = \varepsilon U_0 + \varepsilon^2 V.$$

After dividing by ε, Eq. (3.9.56) becomes

$$(1 - 2\varepsilon\delta)\frac{d^2(U_0 + \varepsilon V)}{d\phi^2} \approx -(U_0 + \varepsilon V) + 3\varepsilon(U_0 + \varepsilon V)^2 + 1. \tag{3.9.63}$$

The terms independent of ε give

$$\frac{d^2 U_0}{d\phi^2} = -U_0 + 1.$$

Choosing the solution U_0 which attains a maximum at $\phi = 0$ (compare with (3.9.54), but we emphasize that $\varphi \neq \phi$),

$$U_0 = e\cos\phi + 1, \tag{3.9.64}$$

(3.9.63) becomes

$$\frac{d^2 V}{d\phi^2} \approx -V + 3U_0^2 + 2\delta\frac{d^2 U_0}{d\phi^2} \quad \Longleftrightarrow \quad \frac{d^2 V}{d\phi^2} \approx -V + 3(e\cos\phi + 1)^2 - 2\delta e\cos\phi.$$

$$\tag{3.9.65}$$

Hence

$$\frac{d^2 V}{d\phi^2} + V \approx -3(e^2\cos^2\phi + 1) + 2(3 - \delta)e\cos\phi. \tag{3.9.66}$$

The resonant term $2(\delta - 3)e\cos\phi$ will disappear if we choose

$$\delta = 3, \tag{3.9.67}$$

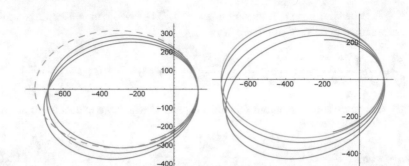

☐ Fig. 3.16 Left figure: approximate relativistic orbits calculated using (3.9.69) with $m/J = 1/15$, $e = .7$. The dotted curve is the Newtonian orbit tangent to the general relativistic one at $\varphi = 0$. The growth of the major axis apparent from the plot, and the existence of two points through which all orbits pass repeatedly, are artifacts of the approximate solution (3.9.69). Right figure: approximate orbits obtained by replacing in (3.9.69) the "secular term" $\varphi \sin \varphi$ by the modified period of a Keplerian approximation to the Schwarzschildian orbit, as derived by the Poincaré–Lindstedt method

leading to a solution V, and hence u, periodic in ϕ. One can easily find an explicit formula for V (see the right ☐ Fig. 3.16) but this is irrelevant for what follows.

Choosing V so that the minimum distance to the origin is achieved at $\phi = 0$, the next minimum approach will take place at $\phi = 2\pi$, i.e.

$$\varphi = (1 + \varepsilon\delta)2\pi = \left(1 + \frac{3m^2}{J^2}\right)2\pi \; .$$

This corresponds to an angle $\varphi = 2\pi + \gamma$ with, in SI units,

$$\boxed{\gamma \approx 6\pi \frac{m^2 G^2}{J^2 c^2} \; .}$$

(3.9.68)

(Recall that $J \neq 0$ is the orbital angular momentum per unit mass of the orbiting object, while m is the mass of the central object.) This is the *perihelion advance* as predicted by general relativity.

SOLUTION 2: We solve (3.9.57) with the boundary conditions $v(0) = v'(0) = 0$, so that the solution coincides to first order with the Newtonian one at the angle of minimum approach. This leads to (see the left ☐ Fig. 3.16)

$$v(\varphi) = \frac{m^4}{J^4}\left[-(3 + e^2)\cos\varphi + 3\left(1 + \frac{e^2}{2}\right) - \frac{e^2}{2}\cos 2\varphi + 3e\varphi \sin\varphi\right] \; .$$

(3.9.69)

Thus,

$$u(\varphi) \approx \frac{m^2}{J^2}(1 + e\cos\varphi)$$

$$+ \frac{m^4}{J^4}\left[-(3 + e^2)\cos\varphi + 3\left(1 + \frac{e^2}{2}\right) - \frac{e^2}{2}\cos 2\varphi + 3e\varphi\sin\varphi\right]. \tag{3.9.70}$$

As already mentioned, the *perihelion* is the point of closest approach to the center, hence a maximum of u. At $\varphi = 0$ we have $\partial_\varphi u = 0$, so this is indeed an extremum, and it is clear from the Newtonian solution (3.9.54) that this is a maximum of u for m/J small enough. The next maximum will be at $\varphi = 2\pi + \gamma$, with γ small:

$$0 = \partial_\varphi u \approx -\frac{m^2}{J^2}e\sin\varphi + \frac{m^4}{J^4}\Big[\underbrace{(3 + e^2)\sin\varphi + e^2\sin 2\varphi + 3e\sin\varphi}_{\approx 0} + 3e\varphi\cos\varphi\Big]$$

$$\approx -\frac{m^2}{J^2}e\gamma + 6\frac{m^4 e\pi}{J^4}.$$

This leads again to (3.9.68). □

For Mercury the effect is sometimes called the *perihermion* advance, and equals about

40″ per century .

A more detailed calculation gives the observed value of 43″ per century. This value was known to astronomers at the beginning of the twentieth century, one suggested explanation being the existence of another planet between Mercury and the Sun.

For the Taylor–Hulse pulsar (cf. ◼ Fig. 4.1, p. 120) the advance is around $4°$ per year. This has to be corrected to account for the gravitational waves emitted by the system; the observed corrections agree extremely well with the theory, and provide an indirect proof of existence of gravitational waves.

Incidentally The effect above should not be confused with the *precession of Earth's equinoxes*, observed already by the Greek astronomer Hipparchus around year 130 BC when comparing his observations with Babylonian records. This precession consists of a cyclic wobbling in the orientation of Earth's axis of rotation. Currently, this annual motion is about 50.3 s of arc per year or 1 degree every 71.6 years. The process is slow and takes 25,772 years for a full precession to occur. The cause of this was already discussed by Isaac Newton, who correctly attributed it to the fact that Earth is a spheroid and not a perfect sphere: Indeed, the Sun's gravitational pull is stronger on the portion of Earth that is tilted towards it, thus creating a torque effect on the planet. If Earth were a perfect sphere, there would be no precession.

For more on this, see www-istp.gsfc.nasa.gov/stargaze/Sprecess.htm or www.universetoday.com/77640/precession-of-the-equinoxes.

3.10 Gyroscope Precession

3.10.1 General Considerations

Consider an observer in Minkowski spacetime moving on a geodesic $\gamma(\tau)$. At each proper time τ of the observer, the collection of points which can be reached by a geodesic starting at $\gamma(\tau)$ in a direction orthogonal to $u \equiv \dot{\gamma} := d\gamma/d\tau$ forms a Euclidean space, with a preferred set of coordinates x^i in which the Euclidean metric takes the usual explicitly flat form. Requiring that the observer is located at the origin of the coordinate system, at each τ the manifestly Euclidean coordinates x^i are uniquely defined up to a rotation, and at different times a different rotation can be chosen to define the preferred coordinates. The question then arises, how to synchronize those coordinates along γ in an optimal way, to define the notion of *nonrotating space-coordinates*.

The obvious way to do that proceeds as follows: given the geodesic γ, consider a global manifestly Minkowskian coordinate system x^μ in which $\gamma(\tau) = (\tau, \vec{0})$. Then the resulting coordinates x^i can be used as manifestly flat coordinates on each slice $t = \tau$. One way of justifying that these are the correct nonrotating coordinates stems from Newtonian intuition: in a Newtonian spacetime consider a gyroscope rotating along a space direction $\vec{s}$ without any forces acting on it. In a Newtonian inertial frame the direction of rotation of the gyroscope will not change. It should then also be true that in a special relativistic inertial frame the direction of the gyroscope will not change. But the special relativistic inertial frames are precisely defined by manifestly Minkowskian coordinates, so in those coordinates the direction of the gyroscope should not change. This gives then precise contents to the notion of nonrotating coordinates.

Note that the direction four-vector $s = (0, \vec{s})$ of such a gyroscope satisfies, along γ, the set of equations

$$\boxed{\frac{Ds}{d\tau} = 0\,, \quad g(u, s) = 0\,,} \tag{3.10.1}$$

with $g = \eta$, the Minkowski metric. One can then invoke the correspondence principle to argue that (3.10.1) is the right general-relativistic equation to describe gyroscopes or, equivalently, nonrotating frames.

An argument to support this proposal has been put forward by Synge, using light-rays. Suppose that a geodesic observer in Minkowski spacetime sends photons which are reflected by a mirror to return to the observer. In the manifestly Minkowskian coordinates the direction of the returning photon will coincide with that of the emitted one. Had one, at different moments of time, used space-coordinates which are rotated with respect to the manifestly Minkowskian ones, the returning photon would be seen as coming from different directions. So the nonrotating space coordinates can be defined by the condition that the direction of return of the photon coincides with the direction of emission.

Such an experiment can be carried out in any spacetime. A field of unit spacelike vectors, say e_1, defined along a timelike geodesic γ, with e_1 orthogonal to $e_0 := \dot{\gamma}$, will correspondingly be said to be nonrotating if, after hitting a mirror orthogonally,

the photons emitted in the direction of e_1 will return in the direction $-e_1$.   (3.10.2)

Now, in a general spacetime, it is not clear at all that (3.10.2) will *ever* happen for mirrors situated at finite distance, because of the light-bending. However, since the metric is nearly Minkowskian at very small scales, one could think that this should be approximately true for sufficiently nearby mirrors. Assuming, for example, that the spacetime is exactly Minkowskian near the world-line of the observer, it should be clear that e_1 should satisfy (3.10.1) with $s = e_1$ if e_1 is nonrotating in the sense just defined.

Incidentally   There might be some interest to *derive* (3.10.1) for e_1, using Synge's proposal, *assuming that there exists along γ a direction e_1 such that the returning photons remain on the surface spanned by the outgoing ones*; this proceeds as follows:[2]

Let S denote the two-dimensional submanifold of $\mathcal{M}$ obtained by shooting null geodesics with initial direction $\ell(0) := e_0 + e_1$ from all points on γ. Imposing affine parameterization, this defines on S a vector field ℓ tangent to those geodesics, solution of the equation

$$\nabla_\ell \ell = 0 \, .$$

On S we can define an orthonormal (ON) frame, still denoted by e_0 and e_1, by parallel-propagating e_0 and e_1 along the integral curves of ℓ:

$$\nabla_\ell e_0 = 0 = \nabla_\ell e_1 \, .$$

Uniqueness of solutions of ODEs implies

$$\ell = e_0 + e_1 \, .$$

By assumption, the returning geodesics lie in the same surface S. Hence the vector field

$$m := e_0 - e_1$$

is geodesic:

$$\nabla_m m = \lambda m \, ,$$

for some function λ on S. Now, the left-hand side of the last equation can be rewritten as

$$\nabla_m m = \nabla_{e_0 - e_1} m = \nabla_{2e_0 - e_0 - e_1} m = 2\nabla_{e_0} m - \underbrace{\nabla_\ell m}_{=0} = 2\nabla_{e_0} m$$

$$= 2\nabla_{e_0} e_0 - 2\nabla_{e_0} e_1 \, .$$

On the central geodesic $\nabla_{e_0} e_0$ vanishes, and we conclude that there we have

$$\nabla_{e_0} e_1 = -\frac{1}{2}\lambda(e_0 - e_1) \, .$$

---

[2]I am grateful to R. Beig for pointing out this derivation.

But e_1 has unit length, hence on γ

$$0 = e_0(g(e_1, e_1)) = 2g(\nabla_{e_0}e_1, e_1) = -\lambda g(e_0 - e_1, e_1) = \lambda.$$

We conclude that $\nabla_{e_0}e_1 = 0$ along γ. In other words, e_1 is parallelly propagated along γ, as claimed.

3.10.2 The Parallel Transport Equation

Consider a timelike geodesic $\gamma(\tau)$ with tangent vector $u \equiv \dot{\gamma} := d\gamma/d\tau$. As argued above, a gyroscope along γ can be described by a vector s orthogonal to u which is *parallel transported along* γ:

$$\boxed{g(s, u) = 0, \quad \frac{Ds}{d\tau} = 0.} \tag{3.10.3}$$

The vector s is pointing along the axis of rotation of the gyroscope, and its length can be thought of as describing the rotation rate. Here we are mostly interested in the direction of rotation of the gyroscope, so the length of s is actually irrelevant for the current purposes.

We start with some general remarks about the parallel transport equation. Let X and Y be two vectors which are parallelly propagated along *any curve* $\gamma(\tau)$, not necessarily geodesic. Thus

$$\frac{DX}{d\tau} = 0 = \frac{DY}{d\tau}, \quad \text{with} \quad \frac{DX^\alpha}{d\tau} := \frac{dX^\alpha}{d\tau} + \Gamma^\alpha{}_{\mu\nu}X^\mu\frac{d\gamma^\nu}{d\tau}. \tag{3.10.4}$$

Then

$$\frac{d}{d\tau}(g(X, Y)) = g(\frac{DX}{d\tau}, Y) + g(X, \frac{DY}{d\tau}) = 0.$$

Thus, the angle between parallelly-propagated vectors, and their lengths, are preserved along the curve. In particular if $X \perp Y$ at one point, then it will be so along the whole curve. Similarly if $g(X, X) = \pm 1$ at one point, it will be so along the whole curve.

Suppose that we have an ON basis of the tangent space, say $e_a, a = 0, \ldots, n$ at some point of γ, then parallelly transporting the e_a's we will obtain an ON basis of $T\mathcal{M}$ at any point lying on the image of γ.

If γ is an affinely parameterized geodesic, then the tangent vector u is parallel along γ. So, for a gyroscope, the first condition $g(s, u) = 0$ in (3.10.3) is consistent with the evolution equations for s and u. Equivalently, if $g(s, u)$ vanishes at one point of the geodesic, it will vanish everywhere.

3.10.3 Geodetic Precession on Circular Geodesics

Returning to the Schwarzschild metric, to illustrate the gyroscope-precession effect we will consider the parallel transport equation on affinely parameterized timelike circular geodesics lying in the $\{z = 0\}$ plane:

$$\gamma(\tau) = \left(t(\tau), r, \frac{\pi}{2}, \varphi(\tau)\right).$$

We have

$$0 = g(s, u) = -\left(1 - \frac{2m}{r}\right) s^t u^t + r^2 s^\varphi u^\varphi,$$

and since $u^t \neq 0$ for timelike geodesics, we can calculate s^t if s^φ is known:

$$s^t = \frac{r^2 u^\varphi}{\left(1 - \frac{2m}{r}\right) u^t} s^\varphi. \tag{3.10.5}$$

Explicitly, the equation for s reads

$$\frac{ds^\alpha}{d\tau} = -\Gamma^\alpha{}_{\mu\nu} s^\mu u^\nu$$

$$= -\Gamma^\alpha{}_{\mu t} s^\mu u^t - \Gamma^\alpha{}_{\mu\varphi} s^\mu u^\varphi. \tag{3.10.6}$$

The required Christoffel symbols of the metric could be read from the variational principle for geodesics, with the Lagrange function

$$\mathscr{L} = \frac{1}{2}\left(-\left(1 - \frac{2m}{r}\right)\dot{t}^2 + \frac{\dot{r}^2}{\left(1 - \frac{2m}{r}\right)} + r^2\dot{\theta}^2 + r^2 \sin^2\theta\,\dot{\varphi}^2\right),$$

where a dot denotes a τ-derivative; see (3.9.1)–(3.9.4). However, for the current problem it is faster to determine the required Christoffels by hand, as follows: We only need those Christoffels for which one of the lower indices is t or φ. Now,

$$\Gamma^\beta{}_{\mu t}\partial_\beta = \frac{1}{2}g^{\beta\alpha}\left(\partial_\mu g_{t\alpha} + \underbrace{\partial_t g_{\mu\alpha}}_{0} - \partial_\alpha g_{\mu t}\right)\partial_\beta$$

$$= \frac{1}{2}g^{tt}\partial_\mu g_{tt}\,\partial_t - \frac{1}{2}g^{rr}\partial_r g_{\mu t}\,\partial_r.$$

It follows that the nonvanishing Christoffel symbols with one lower index equal to t are

$$\Gamma^t{}_{rt} = \Gamma^t{}_{tr} = \frac{1}{2}g^{tt}\partial_r g_{tt}, \qquad \Gamma^r{}_{tt} = -\frac{1}{2}g^{rr}\partial_r g_{tt}. \tag{3.10.7}$$

Similarly

$$\Gamma^{\beta}{}_{\mu\varphi}\partial_{\beta} = \frac{1}{2}g^{\beta\alpha}\left(\partial_{\mu}g_{\varphi\alpha} + \underbrace{\partial_{\varphi}g_{\mu\alpha}}_{0} - \partial_{\alpha}g_{\mu\varphi}\right)\partial_{\beta}$$

$$= \frac{1}{2}g^{\varphi\varphi}\partial_{\mu}g_{\varphi\varphi}\partial_{\varphi} - \frac{1}{2}g^{rr}\partial_{r}g_{\mu\varphi}\partial_{r}.$$

Here we have used the fact that

$$\partial_{\theta}g_{\varphi\varphi} = 2r^2\sin\theta\cos\theta$$

vanishes at $\theta = \pi/2$. Hence the nonvanishing Christoffel symbols with one lower index equal to φ are

$$\Gamma^{\varphi}{}_{r\varphi} = \Gamma^{\varphi}{}_{\varphi r} = \frac{1}{2}g^{\varphi\varphi}\partial_{r}g_{\varphi\varphi}, \quad \Gamma^{r}{}_{\varphi\varphi} = -\frac{1}{2}g^{rr}\partial_{r}g_{\varphi\varphi}. \tag{3.10.8}$$

The above allows us immediately to conclude that

$$\frac{ds^{\theta}}{d\tau} = 0,$$

and s^{θ} is constant.

We continue with

$$\frac{ds^{r}}{d\tau} = -\Gamma^{r}{}_{\mu t}s^{\mu}u^{t} - \Gamma^{r}{}_{\mu\varphi}s^{\mu}u^{\varphi}$$

$$= -\Gamma^{r}{}_{tt}s^{t}u^{t} - \Gamma^{r}{}_{\varphi\varphi}s^{\varphi}u^{\varphi}$$

$$= \frac{1}{2}g^{rr}\left(\partial_{r}g_{tt}s^{t}u^{t} + \partial_{r}g_{\varphi\varphi}s^{\varphi}u^{\varphi}\right)$$

$$= \frac{1}{2}\left(1 - \frac{2m}{r}\right)\left(-\frac{2m}{r^2}\frac{r^2}{\left(1 - \frac{2m}{r}\right)} + 2r\right)s^{\varphi}u^{\varphi}$$

$$= (r - 3m)s^{\varphi}u^{\varphi}, \tag{3.10.9}$$

where in the before-last step we have used (3.10.5). Since

$$u^{\varphi} = \frac{d\varphi}{d\tau} = \frac{dt}{d\tau}\frac{d\varphi}{dt} = u^{t}\Omega,$$

where (see (3.9.40))

$$\Omega = \sqrt{\frac{m}{r^3}}$$

is the coordinate angular velocity of the orbit, we obtain the desired equation for s^r,

$$\frac{ds^r}{dt} = \frac{d\tau}{dt}\frac{ds^r}{d\tau} = \frac{1}{u^t}\frac{ds^r}{d\tau} = (r - 3m)s^\varphi\frac{u^\varphi}{u^t}$$

$$= (r - 3m)\Omega s^\varphi . \tag{3.10.10}$$

We continue with the s^φ equation:

$$\frac{ds^\varphi}{d\tau} = -\Gamma^\varphi{}_{\mu t}s^\mu u^t - \Gamma^\varphi{}_{\mu\varphi}s^\mu u^\varphi = -\frac{1}{2}g^{\varphi\varphi}\partial_r g_{\varphi\varphi}s^r u^\varphi$$

$$= -\frac{1}{r}s^r u^\varphi . \tag{3.10.11}$$

Hence

$$\frac{ds^\varphi}{dt} = \frac{d\tau}{dt}\frac{1}{u^t}\frac{ds^\varphi}{d\tau} = -\frac{1}{r}s^r\frac{u^\varphi}{u^t}$$

$$= -\frac{\Omega}{r}s^r . \tag{3.10.12}$$

Summarizing,

$$\boxed{\frac{ds^r}{dt} = (r - 3m)\Omega s^\varphi , \quad \frac{ds^\varphi}{dt} = -\frac{\Omega}{r}s^r .} \tag{3.10.13}$$

This set of equations can be solved as follows: Differentiating the first equation and using the second one we find

$$\frac{d^2 s^r}{dt^2} = -\frac{(r - 3m)\Omega^2}{r}s^r , \tag{3.10.14}$$

with an identical equation for s^φ. This is a harmonic oscillator with frequency

$$\omega = \sqrt{1 - \frac{3m}{r}}\,\Omega \approx \left(1 - \frac{3m}{2r}\right)\Omega , \tag{3.10.15}$$

where the second, approximate, equality holds for small m/r.

To get some insight into this formula, consider a vector $(\mathring{s}^i)$ in $\mathbb{R}^3$ with constant entries in the usual Cartesian coordinates on $\mathbb{R}^3$. When transformed to (r, θ, φ) coordinates we have

$$\mathring{s}^r = \mathring{s}(r) = \frac{\mathring{s}^i x^i}{r} .$$

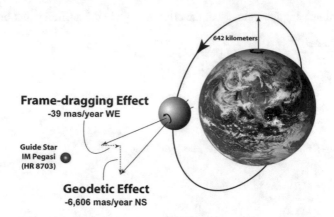

Fig. 3.17 The results of Gravity Probe B, reprinted with permission from [35]. © 2019 by the American Physical Society

On the circular equatorial orbit $\vec{x}(t) = (r\cos(\Omega t), r\sin(\Omega t), 0)$ this reads

$$\mathring{s}^r = \mathring{s}(r) = \mathring{s}^x \cos(\Omega t) + \mathring{s}^y \sin(\Omega t).$$

Thus $\mathring{s}^r$ is periodic with frequency Ω. Similarly, using $d\varphi = (xdy - ydx)/r^2$,

$$\mathring{s}^\varphi = \mathring{s}(\varphi) = \frac{\mathring{s}^y \cos(\Omega t) - \mathring{s}^x \sin(\Omega t)}{r}.$$

Hence, both the r and φ components of a vector with constant Euclidean components rotate with frequency Ω along the orbit.

Returning to (3.10.15), we conclude that the (r, φ) components of the spin vector of a gyroscope moving along a timelike circular geodesic in the Schwarzschild metric will rotate with frequency *smaller* than the frequency of the orbit. This effect is known under the name of *geodetic precession*. The geodetic effect was first predicted by Willem de Sitter in 1916, who provided relativistic corrections to Earth–Moon system's motion.

For a satellite orbiting around Earth, the effect is of 6 arcsec/year. It was verified to a precision of better than 0.5% by the Gravity Probe B (GP-B), a satellite-based mission which launched on 20 April 2004 on a Delta II rocket, and has been taking data for 17 months, see ◻ Fig. 3.17. The analysis of the GP-B experimental results was plagued by Newtonian sources of error, due to "patch potentials" on the gyroscope rotor and housing surfaces. See [34] for details.

Incidentally The above frequency-comparison argument can be tied to the parallel transport equation in Minkowski spacetime along a circular spiral $\mathring{\gamma}(t) = (t, r, \pi/2, \Omega t)$:[3]

$$\frac{d\mathring{s}^{\alpha}}{dt} + \mathring{\Gamma}^{\alpha}{}_{\beta\gamma} \mathring{s}^{\beta} \frac{d\mathring{x}^{\gamma}}{dt} = 0,$$

where the Christoffel symbols $\mathring{\Gamma}^{\alpha}{}_{\beta\gamma}$ are now those of the Minkowski metric in spherical coordinates. The curve $\mathring{\gamma}$ is *not* a geodesic, but the parallel transport equation still makes sense along the spacetime spiral. However, there is no reason for the conservation condition $g(u, s) = 0$ to be satisfied; recall that this equation previously followed from parallel-transport when the curve was a geodesic, and that the equation has been used to derive (3.10.5).

Nevertheless, those Christoffel symbols that have already been calculated provide the Minkowskian ones by setting $m = 0$ there. Further, it is easily checked that *all* $\mathring{\Gamma}^{t}{}_{\alpha\beta}$'s vanish, leading to

$$\frac{d\mathring{s}^{t}}{dt} = 0,\tag{3.10.16}$$

instead of (3.10.5). (Alternatively, note that s^{t} in (t, r, θ, φ) coordinates equals s^{t} in the manifestly flat (t, x, y, z) coordinates, and (3.10.16) immediately follows.) (So a spin vector $\mathring{s}$ which equals $(0, \vec{\mathring{s}})$ initially will remain orthogonal to ∂_t when Minkowski-parallel-transported along $\mathring{\gamma}$, but this is irrelevant for our further calculations). Since (3.10.5) has been assumed in the derivation of (3.10.13), we cannot simply assume that the last equation holds with $m = 0$ in Minkowski, and so we need to revisit the whole calculation.

As such, (3.10.11) remains unchanged up to trivial renaming:

$$\frac{d\mathring{s}^{\varphi}}{dt} = -\frac{1}{r} \mathring{s}^{r} \frac{d\mathring{\varphi}}{dt} = \frac{\Omega}{r} \mathring{s}^{r}.\tag{3.10.17}$$

The $\mathring{s}^{r}$-equation can be obtained from (3.10.18) by setting $m = 0$ there, since the term involving $\partial_r g_{tt}$ drops out then:

$$\frac{d\mathring{s}^{r}}{dt} = r\mathring{s}^{\varphi} \frac{d\mathring{\varphi}}{dt} = r\Omega\mathring{s}^{\varphi}.$$

We conclude that, in spite of all warnings, parallel transport leads now to the Eq. (3.10.13) with $m = 0$ there, and hence to periodic solutions with frequency $\mathring{\omega} = \Omega$.

We could have guessed the result by noting that the Minkowskian vector fields $\mathring{e}_i := \partial_i$ are parallel:

$$\mathring{\nabla}_{\mu}\mathring{e}_i = 0.$$

In particular they satisfy the Minkowski-parallel-transport equation,

$$\mathring{\gamma}^{\mu}\mathring{\nabla}_{\mu}\mathring{e}_i = 0,$$

---

[3]Note that a gyroscope in Minkowski spacetime will *not* be parallel transported around the spiral, it will instead undergo Fermi–Walker transport, see ▶ Sect. 3.10.4. The current calculation is done for the sake of comparison only.

along *any* curve γ. This implies that a vector field which is parallelly transported along γ will have constant entries in the basis $\mathring{e}_i$. Since the vectors $\mathring{e}_1$ and $\mathring{e}_2$ rotate with angular velocity Ω along the spacetime spiral with respect to the basis ∂_r and ∂_φ, a parallelly transported vector will have Ω-periodic components in the basis ∂_r and ∂_φ.

3.10.4 Fermi–Walker Transport and Thomas Precession

Given a proper-time-parameterized world-line $\gamma^\mu(\tau)$, with unit tangent $u = \dot{\gamma}$ and acceleration four-vector $a := Du/d\tau$, we would like to describe the motion of a gyroscope subject to no external forces other than those that keep it on the world-line *without torque*. Keeping in mind that a gyroscope is described by a vector s satisfying $g(s, u) = 0$, the parallel transport equation is not adequate since then

$$\frac{d(g(u,s))}{d\tau} = g(\frac{Du}{d\tau}, s) + g(u, \frac{Ds}{d\tau}) = g(a, s),$$

which is not zero in general. The simplest modification to the parallel transport equation is to write

$$\frac{Ds}{d\tau} \sim u.$$

Denoting by ϕ the proportionality factor, one then has

$$\frac{d(g(u,s))}{d\tau} = g(a, s) + g(u, \underbrace{\frac{Ds}{d\tau}}_{\phi u})$$

$$= g(a, s) + \phi \underbrace{g(u, u)}_{=-1}.$$

This will vanish if we choose $\phi = g(a, s)$. So a possible equation reads

$$\boxed{\frac{Ds}{d\tau} = g(s, a)u.} \tag{3.10.18}$$

The equation leads to an effect known as the *Thomas precession*. It should be stressed that this is independent of the geodetic precession, as it gives a nonzero correction to parallel transport even in Minkowski spacetime. Interesting animations illustrating the effect can be found on http://www.gbeyerle.de/twr, see also [18].

Incidentally There appears to be in the literature some confusion about the exact meaning of "Thomas precession." Some authors tie it to the fact that the composition of two boost transformations with velocities u and v is a Lorentz transformation which is the composition of a boost and a rotation. The rotation part of this composition is then called Thomas precession. One can invoke that last property to derive a relativistic correction to the spin-orbit coupling in nonrelativistic quantum mechanics. (Surprisingly, the resulting effect corrects by a desired factor 1/2 the calculations of spin-orbit interactions in nuclear physics.)

Equation (3.10.18) is a special case of the *Fermi–Walker transport* (FW-transport) of any vector s along a world-line γ, not necessarily orthogonal to the four-velocity. The Fermi–Walker transport equation reads

$$\boxed{\frac{Ds^{\mu}}{d\tau} = g(s, a)u^{\mu} - g(s, u)a^{\mu}.}$$

(3.10.19)

This equation has interesting features. First,

$$\frac{Du^{\mu}}{d\tau} = \underbrace{g(u, a)}_{=0}u^{\mu} - g(u, u)a^{\mu} = a^{\mu},$$

(3.10.20)

which shows that the four-velocity is FW-transported along γ. Next, consider two vectors, say X and Y, which are FW-transported along γ. Then

$$\frac{d(g(X, Y))}{d\tau} = g(\frac{DX}{d\tau}, Y) + g(X, \frac{DY}{d\tau})$$

$$= g(g(X, a)u - g(X, u)a, Y) + g(X, g(Y, a)u - g(Y, u)a)$$

$$= 0.$$

Hence, Fermi–Walker transport preserves scalar products. In particular a vector initially orthogonal to u will remain orthogonal when FW transported.

Finally, Fermi–Walker transport reduces to parallel transport for geodesics.

It is argued in the literature that the Fermi–Walker equation provides the right equation for describing the motion of gyroscopes along general worldlines. However, a gyroscope is a complicated object, and its behavior under relativistic accelerations and velocities is by no means obvious.

Incidentally It is tempting to raise the question, what is most general *linear* transport law along a curve which preserves scalar products. Linearity implies that the equation has to take the form

$$\frac{Ds^{\mu}}{d\tau} = A^{\mu}{}_{\nu}s^{\nu}$$

(3.10.21)

for some matrix $A^{\mu}{}_{\nu}$. The requirement that the length of s is preserved gives

$$0 = \frac{d(s_{\alpha}s^{\alpha})}{d\tau} = 2A_{\mu\nu}s^{\mu}s^{\nu}.$$

Hence $A_{\mu\nu}s^\mu s^\nu$ must vanish for all vectors s^μ. Differentiating this equation with respect to s^α and s^β we obtain

$$0 = \frac{\partial^2(2A_{\mu\nu}s^\mu s^\nu)}{\partial s^\alpha \partial s^\beta} = 2(A_{\alpha\beta} + A_{\beta\alpha}).$$

So $A_{\alpha\beta}$ must be antisymmetric. Invoking the polarization formula, or by a direct calculation, one finds that any antisymmetric matrix works: indeed, if X and Y satisfy (3.10.21), then

$$\frac{d(g_{\alpha\beta}X^\alpha Y^\beta)}{d\tau} = g_{\alpha\beta}(A^\alpha{}_\mu X^\mu Y^\beta + X^\alpha A^\beta{}_\mu Y^\mu) = (A_{\beta\alpha} + A_{\alpha\beta})X^\alpha Y^\beta = 0.$$

Note that the Fermi–Walker transport equation is of this form, with

$$A_{\mu\nu} = u_\mu a_\nu - a_\mu u_\nu.$$

One can thus think of the FW equation as the simplest evolution equation which preserves scalar products.

3.10.5 The Lense–Thirring Effect

In ▶ Sect. 4.1 we will derive the following leading-order behavior of the metric for weak and slowly-varying gravitational fields:

$$g_{00} \approx -1 + \frac{2m}{r}, \qquad g_{ij} \approx \left(1 + \frac{2m}{r}\right)\delta_{ij}.$$

This is consistent with the Schwarzschild metric in *isotropic coordinates*

$$g = -\left(\frac{1 - \frac{m}{2r}}{1 + \frac{m}{2r}}\right)^2 dt^2 + \left(1 + \frac{m}{2r}\right)^4(dx^2 + dy^2 + dz^2).$$

So, the Schwarzschild calculations give a good approximation for the motion of gyroscopes in all weak gravitational fields at large distances.

It turns out that a *rotating source* will give a further contribution to the gravitational field of the form

$$g_{0i} \approx -2\epsilon_{ijk}\frac{x^j J^k}{r^3},$$

where J^k is the angular-momentum vector of the source, and we will show this in ▶ Sect. 4.7. Such terms are second-order corrections in the post-Newtonian approximation, and they decay faster in r than the mass corrections in any case.

The inclusion of such terms in the gyroscope equation gives rise to a new effect, first discussed by Lense and Thirring, known as *frame dragging*, or as the *Lense–Thirring effect*.

We will not derive the correct equation for the effect, as the calculation is lengthy, complicated, and not especially enlightening. Instead, we illustrate the effect on a toy model, where the post-Newtonian metric above is replaced by

$$g = -c^2 dt^2 + dx^2 + dy^2 + dz^2 - \frac{4GJ}{c^3 r^3}(c\,dt)\Big(\frac{xdy - ydx}{r}\Big).$$

(3.10.22)

We will consider a gyroscope moving along the curve

$$\gamma(\tau) = (x^\mu(\tau)) = \big(t(\tau), 0, 0, z(\tau)\big).$$

In order to crosscheck that this is a geodesic,

$$\frac{d^2 x^\mu}{d\tau^2} = -\Gamma^\mu_{\alpha\beta}\frac{dx^\alpha}{d\tau}\frac{dx^\beta}{d\tau},$$

we need those Christoffels where one of the α and β equals t or z.

Now along γ we have $g_{\mu\nu} = \eta_{\mu\nu}$ and further

$$\Gamma^\mu_{z\alpha}\big|_\gamma = \frac{1}{2}\eta^{\mu\nu}(\partial_z g_{v\alpha} + \partial_\alpha \underbrace{g_{vz}}_{0 \text{ or } 1} - \partial_v g_{z\alpha}) = 0.$$

Indeed,

$$\partial_z g_{ty}\big|_\gamma = \partial_z\Big(-\frac{4GJx}{c^2 r^3}\Big)\Big|_\gamma = x\partial_z\Big(-\frac{4GJ}{c^2 r^3}\Big)\Big|_\gamma = 0,$$

similarly for $\partial_z g_{tx}\big|_\gamma$, while the remaining ∂_z derivatives of $g_{\alpha\nu}$ obviously vanish. Further,

$$\Gamma^\mu_{0\alpha}\big|_\gamma = \frac{1}{2}\eta^{\mu\nu}(\partial_0 g_{v\alpha} + \partial_\alpha g_{v0} - \partial_v g_{0\alpha})$$

$$= \frac{1}{2}\eta^{\mu\nu}(\partial_\alpha g_{v0} - \partial_v g_{0\alpha}),$$

(3.10.23)

$$\Gamma^\mu_{00}\big|_\gamma = \frac{1}{2}\eta^{\mu\nu}(\partial_0 g_{v0} - \partial_v g_{00}) = 0,$$

(3.10.24)

and so this is a affinely parameterized geodesic, as claimed, when

$$(x^\mu(\tau)) = \big(u^t, 0, 0, u^z\big)\tau,$$

with u^t and u^z being constants satisfying $u^t = \sqrt{1 + (u^z)^2}$.

Along γ the orthogonality relation $g(s, u) = 0$ simply reads

$$u^t s^t = u^z s^z \qquad \Longleftrightarrow \qquad s^t = \frac{u^z}{u^t}s^z.$$

(3.10.25)

The z-component of the parallel-transport equation reads

$$0 = \frac{ds^z}{d\tau} + \Gamma^z_{\alpha\beta}u^\alpha s^\beta = \frac{ds^z}{d\tau} + \Gamma^z_{0\beta}u^t s^\beta = \frac{ds^z}{d\tau},$$

and s^z is constant. Note that if $s^z(0) = 0$, then both s^z and s^t vanish for all τ.

We choose for simplicity $s = (0, s^x, s^y, 0)$. One then has

$$0 = \frac{ds^x}{d\tau} + \Gamma^x_{\alpha\beta}u^\alpha s^\beta = \frac{ds^x}{d\tau} + \Gamma^x_{0x}u^t s^x + \Gamma^x_{0y}u^t s^y. \qquad (3.10.26)$$

Recall (3.10.23)

$$\Gamma^\mu_{0\alpha}|_\gamma = \frac{1}{2}\eta^{\mu\nu}(\partial_\alpha g_{\nu 0} - \partial_\nu g_{0\alpha}), \qquad (3.10.27)$$

thus

$$\Gamma^x_{0x}|_\gamma = \frac{1}{2}\eta^{xx}(\partial_x g_{x0} - \partial_x g_{0x}) = 0 = \Gamma^y_{0y}|_\gamma,$$

$$\Gamma^x_{0y}|_\gamma = \frac{1}{2}\eta^{xx}(\partial_y g_{x0} - \partial_x g_{0y}),$$

$$\partial_x g_{0y}\big|_\gamma = \partial_x\left(-\frac{4GJx}{c^2r^3}\right)\bigg|_\gamma = -\frac{4GJ}{c^2r^3}\bigg|_\gamma = -\partial_x g_{0y}\big|_\gamma,$$

$$\Gamma^x_{0y}|_\gamma = \frac{4GJ}{c^2r^3}\bigg|_\gamma.$$

Coming back to (3.10.26)

$$0 = \frac{ds^x}{d\tau} + \Gamma^x_{0x}u^t s^x + \Gamma^x_{0y}u^t s^y$$

$$= \frac{ds^x}{d\tau} + \frac{4GJ}{c^2r^3}u^t s^y.$$

By a similar calculation, or by symmetry, one finds

$$0 = \frac{ds^y}{d\tau} - \frac{4GJ}{c^2r^3}u^t s^x.$$

For large distances and small radial velocities, so that $r \approx$ const, $u^t \approx 1$, we view

$$\omega := \frac{4GJ}{c^2r^3}$$

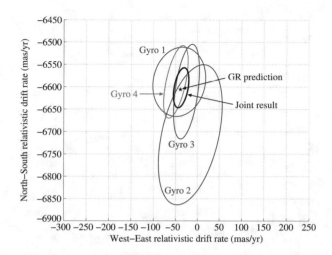

▣ **Fig. 3.18**   The results of Gravity Probe B, gyroscope by gyroscope, reprinted with permission from [35]. © 2019 by the American Physical Society

as a constant, and so

$$\frac{ds^x}{d\tau} = -\omega s^y, \qquad \frac{ds^y}{d\tau} = \omega s^x,$$

hence

$$\frac{d^2 s^x}{d\tau^2} = -\omega^2 s^x, \qquad \frac{d^2 s^y}{d\tau^2} = -\omega^2 s^y,$$

and (s^x, s^y) is precessing with angular velocity ω proportional to the angular momentum parameter J.

The experimental verification of the real frame-dragging formula has been carried out by the Gravity Probe B, see ▣ Fig. 3.18. The formula was confirmed up to a statistical uncertainty of 14% and a systematic uncertainty of 10%. See http://einstein.stanford.edu/highlights/status1.html#frame-draggingr for more information.

Weak Fields and Gravitational Waves

Piotr T. Chruściel

© Springer Nature Switzerland AG 2019
P. T. Chruściel, *Elements of General Relativity*, Compact Textbooks in Mathematics,
https://doi.org/10.1007/978-3-030-28416-9_4

The aim of this chapter is to present some essential ingredients of the proof of the *Einstein quadrupole formula* for the rate of loss of energy by a gravitating system into gravitational waves. On the way towards this we will also verify that Einstein equations reduce to Newton's in an appropriate limit.

As such, one of the major predictions of general relativity is the existence of gravitational waves. By now we have both direct and indirect evidence of their existence.

The indirect evidence comes from the observation of binary pulsars, whose time-evolution agrees to very impressive accuracy with the general-relativistic predictions, including Einstein's famous quadrupole formula for emission of energy by gravitational waves: Indeed, the general-relativistic calculations have found experimental confirmation in the observations by Taylor and Hulse of the millisecond pulsar PSR 1913+16, rewarded by a Nobel prize in 1993, cf. ◘ Fig. 4.1.[1] The result has been confirmed since in other binary systems, see [13, 55] and references therein.

The direct evidence comes from laser interferometric detectors. As of today, there exist four such detectors specifically built for the purpose: two detectors in the USA, forming the *Laser Interferometric Gravitational Observatory* (LIGO, see ◘ Fig. 4.2), the GEO600 detector in Germany, and the Virgo detector in Italy. It should be said that GEO600 is thought to be a toy for testing the technology rather than a real detector. Two further instruments are being built, one in India and one in Japan.

At the time of writing of this work there have been 11 signals interpreted as direct detections of gravitational waves [7]. The first detection took place in September 2015 [2], with a second one in December 2015 [1], and a third one in January 2017 [5]. Each of these events is a detection by itself, but they come with a twist: One can think of them as observations, by means of gravitational waves, of black-hole mergers.

While there is widespread consensus that the waves have been detected by now, some scientific skepticism is in order. The observations require the extraction of

---

[1] An excellent elementary description of the PSR 1913+16 pulsar and of the Taylor–Hulse observations can be found at http://astrosun.tn.cornell.edu/courses/astro201/psr1913.htm.

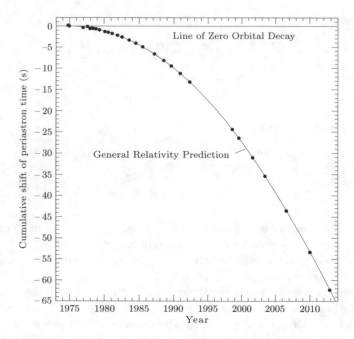

◘ Fig. 4.1 Orbital decay of PSR B1913+16, from [87]. The data points indicate the observed shift of the periastron, while the parabola is the shift predicted by general relativity when emission of energy by gravitational waves is taken into account. © AAS. Reproduced with permission

◘ Fig. 4.2 The LIGO detectors in Washington (left) and Louisiana (right), reproduced with permission courtesy Caltech/MIT/LIGO Laboratory

an absurdly small signal from overwhelmingly noisy data using sophisticated data analysis techniques. While scientists working on the problem have made many efforts to ensure the validity of the claims, there always remains the possibility of instrumental, interpretational, or data analysis errors; see, e.g., [26]. In any case there is a strong case for several direct observations of gravitational waves by now, and we can hope that evidence will keep growing stronger.

The first event, christened GW150914 (for "Gravitational Wave observed on September 14, 2015"), is thought to have been created by two black-holes with respective masses $36^{+5}_{-4}M_\odot$ and $29^{+4}_{-4}M_\odot$, merging into a final black hole with mass $62^{+4}_{-4}M_\odot$. An astounding $3^{+0.5}_{-0.5}M_\odot c^2$ amount of energy has been released within a fraction of a second into gravitational waves. The signal observed can be seen in the left ◻ Fig. 4.3.

The second event GW151226, illustrated by the right ◻ Fig. 4.3, is interpreted as representing the merger of two black holes of respective masses $14^{+8.3}_{-3.7}M_\odot$ and $7.5^{+2.3}_{-2.3}M_\odot$, leading to a final black holes of mass $21^{+5.9}_{-1.9}M_\odot$. Inspection of ◻ Fig. 4.3 reveals that the GW151226 signal is nowhere as striking as GW150914, with a maximal amplitude smaller than the residual noise. In fact, the signal there resembles much more what one expected before the detections than GW150914. Nevertheless, the estimated probability of a false detection for GW151226 is smaller than the convincingly small number 10^{-7}.

The third event GW170104, with wave forms displayed in ◻ Fig. 4.4, detects the merger of two black holes of respective masses $31^{+8.6}_{-6}M_\odot$ and $20^{+5}_{-6}M_\odot$. The signal is somewhat reminiscent of that of GW150914, compare ◻ Fig. 4.3.

In addition to direct detection of gravitational waves, the LIGO events gave the first evidence of existence of black hole binaries, and of black holes with masses in the $10M_\odot - 100M_\odot$ range.

Undoubtedly, one of the most spectacular detections was that on August 17, 2017. The waveform GW170817 observed corresponds to a catastrophic merger of two neutron stars, resulting in a kilonova. The event has been seen in several electromagnetic bands, see ◻ Fig. 4.5, p. 124, opening the era of "multimessenger astronomy." It turns out that no signal was seen in the VIRGO detector, which suggested a source located in VIRGO's dead angle. This location was subsequently confirmed by optical observations, see ◻ Fig. 4.6.

No significant neutrino flux has been observed around the time of the event, which is consistent with model predictions of short gamma ray bursts (GRB) observed at large off-axis angle, or of GRBs with (relatively) low luminosity [9].

All LIGO detections until December 2018 are summarized in Table 4.1, p. 126, with the corresponding gravitational wave signals shown in ◻ Fig. 4.7, p. 127; see also ◻ Fig. 4.8, p. 128.

The relativity community still awaits a first detection of continuous (as opposed to the catastrophic GW170817) gravitational waves emitted by neutron star-neutron star binaries, which are estimated to occur (see [11] and references therein) at a rate of between 0.04 to 400 events per year and 0.2 to 300 events per year for neutron star-black hole binaries. This lack of detection of continuous signals can be used to put constraints on the existing source candidates [3, 8]. As another example, in the case of two recent nearby gamma ray bursts (GRBs), the non-detection of a gravitational wave signal allows one to exclude a merger as the GRB central engine [11].

4.1 Weak Gravitational Fields

4.1.1 Small Perturbations of Minkowski Spacetime

Consider $\mathbb{R}^{n+1}$ with a metric which, in the natural coordinates on $\mathbb{R}^{n+1}$, takes the form

$$g_{\mu\nu} = \eta_{\mu\nu} + h_{\mu\nu}, \tag{4.1.1}$$

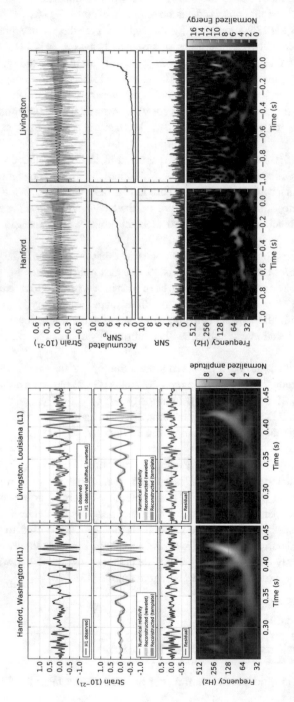

☐ **Fig. 4.3** GW150914 (left) and GW151226 (right) as observed in the Hanford and Livingston detectors by the LIGO Scientific and Virgo Collaborations, reprinted with permission from [1, 2]. The bottom rows show the evolution of frequency of the signal in time. Left figure: The top row is the signal observed, after filtering out the low-frequency and high-frequency noise. In the top-right corner of the left figure, the Hanford signal has been inverted when superposing with the Livingston signal because of the opposite orientation of the detector arms. Right figure: The top row is the signal observed, after filtering out the low-frequency and high-frequency noise, superposed with the black curves corresponding to the best-fit general-relativistic template. The second row shows the accumulated-in-time signal-to-noise ratio (SNR), superposed with the best-match template waveform and computing the integrated SNR at each point in time

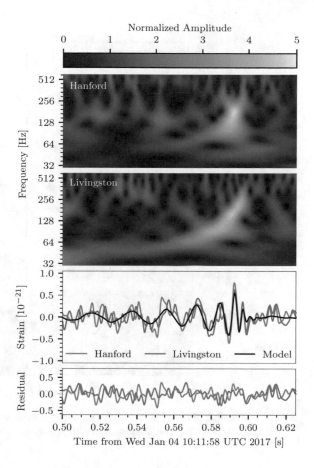

◻ Fig. 4.4 GW170104 as observed in the Hanford and Livingston detectors by the LIGO and Virgo Scientific Collaboration, reprinted with permission from [5]. The top two rows show the evolution in time of the frequency spectrum of the signal, after filtering out the low- and high-frequency noise. The third row is a superposition of the filtered signals together with the black curve corresponding to the best-fit general-relativistic template. The last row shows residuals from the best fit

and suppose that there exists a small constant ϵ such that we have

$$|h_{\mu\nu}|, \ |\partial_\sigma h_{\mu\nu}|, \ |\partial_\sigma \partial_\rho h_{\mu\nu}| \leq \epsilon. \tag{4.1.2}$$

We wish to calculate the explicit form of the Einstein equations for such metrics, disregarding terms which are quadratic or higher in ϵ.

It is then easy to check that

$$g^{\mu\nu} = \eta^{\mu\nu} - h^{\mu\nu} + O(\epsilon^2). \tag{4.1.3}$$

Incidentally Note that (4.1.1)–(4.1.3) only make sense if ϵ, and hence both the metric components and all coordinate functions, are unitless. One can think that all coordinates have been rescaled with suitable scale factors relevant for the physical problem at hand; otherwise explicit scale factors should be introduced in equations such as (4.1.2).

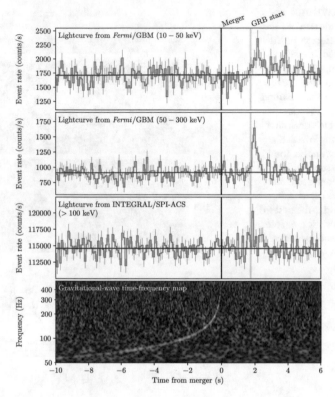

◘ Fig. 4.5 A merger of two neutron stars, as observed in three γ-ray bands and as a gravitational wave in July 2017, from [4]. © AAS. Reproduced with permission

Throughout this section we use the metric η to raise and lower indices, *e.g.*

$$h^{\alpha}{}_{\beta} := \eta^{\alpha\mu}h_{\mu\beta}, \qquad h^{\alpha\beta} := \eta^{\alpha\mu}\eta^{\beta\nu}h_{\mu\nu} = \eta^{\beta\nu}h^{\alpha}{}_{\nu}.$$

Next,

$$\begin{aligned}
\Gamma^{\alpha}{}_{\beta\gamma} &= \frac{1}{2}g^{\alpha\sigma}\{\partial_{\beta}g_{\sigma\gamma} + \partial_{\gamma}g_{\sigma\beta} - \partial_{\sigma}g_{\beta\gamma}\} \\
&= \frac{1}{2}g^{\alpha\sigma}\{\partial_{\beta}h_{\sigma\gamma} + \partial_{\gamma}h_{\sigma\beta} - \partial_{\sigma}h_{\beta\gamma}\} \\
&= \frac{1}{2}\eta^{\alpha\sigma}\{\partial_{\beta}h_{\sigma\gamma} + \partial_{\gamma}h_{\sigma\beta} - \partial_{\sigma}h_{\beta\gamma}\} + O(\epsilon^2) \\
&= \frac{1}{2}\{\partial_{\beta}h^{\alpha}{}_{\gamma} + \partial_{\gamma}h^{\alpha}{}_{\beta} - \partial^{\alpha}h_{\beta\gamma}\} + O(\epsilon^2) = O(\epsilon).
\end{aligned} \tag{4.1.4}$$

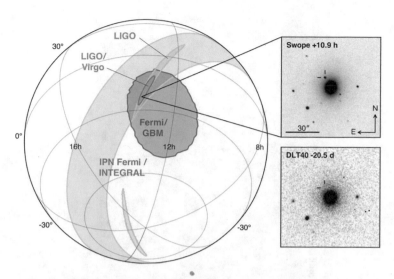

◘ Fig. 4.6 The localization of the neutron-star merger at the origin of GW170817 using gravitational waves, gamma-rays, and visible light, from [6]. © AAS. Reproduced with permission

The Ricci tensor is easily found from (1.4.2)

$$R_{\beta\delta} = \partial_\alpha \Gamma^\alpha{}_{\beta\delta} - \partial_\delta \Gamma^\alpha{}_{\beta\alpha} + O(\epsilon^2)$$

$$= \frac{1}{2}\left[\partial_\alpha\{\partial_\beta h^\alpha{}_\delta + \partial_\delta h^\alpha{}_\beta - \partial^\alpha h_{\beta\delta}\} - \partial_\delta\{\partial_\beta h^\alpha{}_\alpha + \partial_\alpha h^\alpha{}_\beta - \partial^\alpha h_{\beta\alpha}\}\right] + O(\epsilon^2)$$

$$= \frac{1}{2}\left[\partial_\alpha\{\partial_\beta h^\alpha{}_\delta + \partial_\delta h^\alpha{}_\beta - \partial^\alpha h_{\beta\delta}\} - \partial_\delta\partial_\beta h^\alpha{}_\alpha\right] + O(\epsilon^2). \qquad (4.1.5)$$

4.1.2 Coordinate Conditions and Wave Coordinates

The expression (4.1.5) for the Ricci tensor is still more complicated than desired: we will be interested in solving, e.g., the vacuum equation $R_{\mu\nu} = 0$, and it is far from clear how that can be done using (4.1.5). It turns out that one can obtain considerable simplifications if one imposes a set of *coordinate conditions*. Recall that a tensor field g is represented by matrices $g_{\mu\nu}$ in many different ways, depending upon the coordinate system chosen: if a point p has coordinates y^μ in a coordinate system $\{y^\mu\}$, and coordinates x^α in a second coordinate system $\{x^\alpha\}$, then we have

$$g = g_{\mu\nu}(y^\sigma)dy^\mu dy^\nu = g_{\mu\nu}(y^\sigma(x^\alpha))\left(\frac{\partial y^\mu}{\partial x^\beta}dx^\beta\right)\left(\frac{\partial y^\nu}{\partial x^\gamma}dx^\gamma\right)$$

$$= g_{\mu\nu}(y^\sigma(x^\alpha))\frac{\partial y^\mu}{\partial x^\beta}\frac{\partial y^\nu}{\partial x^\gamma}dx^\beta dx^\gamma$$

$$= g_{\beta\gamma}(x^\alpha)dx^\beta dx^\gamma,$$

4

Table 4.1 The physical parameters associated with the eleven gravitational wave detections observed until December 2018, from [7]

Event	$m_1/M_\odot$	$m_2/M_\odot$	$\mathcal{M}/M_\odot$	χ_{eff}	$M_f/M_\odot$	a_f	$E_{\text{rad}}/(M_\odot c^2)$	$\ell_{\text{peak}}/\text{erg s}^{-1}$	d_L/Mpc	z	$\Delta\Omega/\text{deg}^2$
GW150914	$35.6^{+4.8}_{-3.0}$	$30.6^{+3.0}_{-4.4}$	$28.6^{+1.6}_{-1.5}$	$-0.01^{+0.12}_{-0.13}$	$63.1^{+3.3}_{-3.0}$	$0.69^{+0.05}_{-0.04}$	$3.1^{+0.4}_{-0.4}$	$3.6^{+0.4}_{-0.4}\times10^{56}$	430^{+150}_{-170}	$0.09^{+0.03}_{-0.03}$	180
GW151012	$23.3^{+14.0}_{-5.5}$	$13.6^{+4.1}_{-4.8}$	$15.2^{+2.0}_{-1.1}$	$0.04^{+0.28}_{-0.19}$	$35.7^{+9.9}_{-3.8}$	$0.67^{+0.13}_{-0.11}$	$1.5^{+0.5}_{-0.5}$	$3.2^{+0.8}_{-1.7}\times10^{56}$	1060^{+540}_{-480}	$0.21^{+0.09}_{-0.09}$	1555
GW151226	$13.7^{+8.8}_{-3.2}$	$7.7^{+2.2}_{-2.6}$	$8.9^{+0.3}_{-0.3}$	$0.18^{+0.20}_{-0.12}$	$20.5^{+6.4}_{-1.5}$	$0.74^{+0.07}_{-0.05}$	$1.0^{+0.1}_{-0.2}$	$3.4^{+0.7}_{-1.7}\times10^{56}$	960^{+430}_{-410}	$0.09^{+0.04}_{-0.04}$	1033
GW170104	$31.0^{+7.2}_{-5.6}$	$20.1^{+4.9}_{-4.5}$	$21.5^{+2.1}_{-1.7}$	$-0.04^{+0.17}_{-0.20}$	$49.1^{+5.2}_{-3.9}$	$0.66^{+0.08}_{-0.10}$	$2.2^{+0.5}_{-0.5}$	$3.3^{+0.6}_{-0.9}\times10^{56}$	960^{+430}_{-410}	$0.19^{+0.07}_{-0.08}$	924
GW170608	$10.9^{+5.3}_{-1.7}$	$7.6^{+1.3}_{-2.1}$	$7.6^{+0.2}_{-0.2}$	$0.03^{+0.19}_{-0.07}$	$17.8^{+3.2}_{-0.7}$	$0.69^{+0.04}_{-0.04}$	$0.9^{+0.05}_{-0.1}$	$3.5^{+0.4}_{-1.3}\times10^{56}$	320^{+120}_{-110}	$0.07^{+0.02}_{-0.02}$	396
GW170729	$50.6^{+16.6}_{-10.2}$	$34.3^{+9.1}_{-10.1}$	$35.7^{+6.5}_{-4.7}$	$0.36^{+0.21}_{-0.25}$	$80.3^{+14.6}_{-10.2}$	$0.18^{+0.21}_{-0.13}$	$4.8^{+1.7}_{-1.7}$	$4.2^{+0.9}_{-1.5}\times10^{56}$	2750^{+1350}_{-1320}	$0.48^{+0.19}_{-0.20}$	1033
GW170809	$35.2^{+8.3}_{-6.0}$	$23.8^{+5.2}_{-5.1}$	$25.0^{+2.1}_{-1.6}$	$0.07^{+0.16}_{-0.16}$	$56.4^{+5.2}_{-3.7}$	$0.70^{+0.08}_{-0.09}$	$2.7^{+0.6}_{-0.6}$	$3.5^{+0.6}_{-0.9}\times10^{56}$	990^{+320}_{-380}	$0.20^{+0.05}_{-0.07}$	340
GW170814	$30.7^{+5.7}_{-3.0}$	$25.3^{+2.9}_{-4.1}$	$24.2^{+1.4}_{-1.1}$	$0.07^{+0.12}_{-0.11}$	$53.4^{+3.2}_{-2.4}$	$0.72^{+0.07}_{-0.05}$	$2.7^{+0.4}_{-0.3}$	$3.7^{+0.4}_{-0.5}\times10^{56}$	580^{+160}_{-210}	$0.12^{+0.03}_{-0.04}$	87
GW170817	$1.46^{+0.12}_{-0.10}$	$1.27^{+0.09}_{-0.09}$	$1.186^{+0.001}_{-0.001}$	$0.00^{+0.02}_{-0.01}$	≤ 2.8	≤ 0.89	≤ 0.04	$\geq 0.1\times10^{56}$	40^{+10}_{-10}	$0.01^{+0.00}_{-0.00}$	16
GW170818	$35.5^{+7.5}_{-4.7}$	$26.8^{+5.2}_{-5.2}$	$26.7^{+2.1}_{-1.7}$	$-0.09^{+0.18}_{-0.21}$	$59.8^{+4.8}_{-3.8}$	$0.67^{+0.07}_{-0.08}$	$2.7^{+0.5}_{-0.5}$	$3.4^{+0.5}_{-0.7}\times10^{56}$	1020^{+430}_{-360}	$0.20^{+0.07}_{-0.07}$	39
GW170823	$39.6^{+10.0}_{-6.6}$	$29.4^{+6.3}_{-7.1}$	$29.3^{+4.2}_{-3.2}$	$0.08^{+0.20}_{-0.22}$	$65.6^{+9.4}_{-6.6}$	$0.71^{+0.08}_{-0.10}$	$3.3^{+0.9}_{-0.8}$	$3.6^{+0.6}_{-0.9}\times10^{56}$	1850^{+840}_{-840}	$0.34^{+0.13}_{-0.14}$	1651

One notices that the values of the parameter a_f, which is the ratio spin-to-mass of the final object, cluster around 0.7

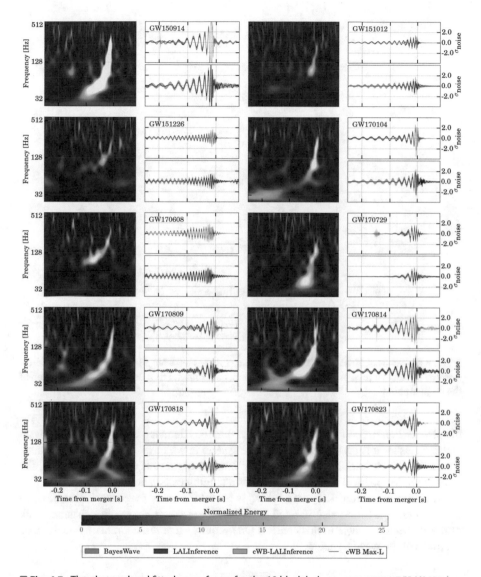

Fig. 4.7 The observed and fitted wave-forms for the 10 black hole mergers seen up to December 2018 by the LIGO Scientific and Virgo Collaborations. Reproduced with permission from [7]

so that

$$g_{\beta\gamma}(x^\alpha) = g_{\mu\nu}(y^\sigma(x^\alpha))\frac{\partial y^\mu}{\partial x^\beta}\frac{\partial y^\nu}{\partial x^\gamma}.$$ (4.1.6)

One can make use of this transformation law to obtain a form of the matrix $g_{\mu\nu}$ which is convenient for the problem at hand. (This property is referred to as *"gauge freedom"* in

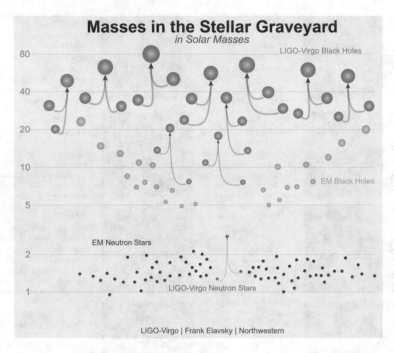

□ Fig. 4.8 Masses of "dead stars," from https://ciera.northwestern.edu/gallery, © Frank Elavsky, Northwestern IT, LIGO-Virgo, reproduced with permission. I am grateful to T. Bulik for a color-inversion of the original figure

the physics literature.) For example, to show that the Riemann tensor of the Minkowski metric vanishes it is best to use a coordinate system in which all the $g_{\mu\nu}$'s are constants, hence the $\Gamma^\alpha{}_{\beta\gamma}$'s vanish, which obviously implies the result. On the other hand, the same result in spherical coordinates requires a lengthy calculation.

A choice of coordinate system, which is useful for many purposes, is that of *wave coordinates*, sometimes also referred to as *harmonic coordinates*; physicists also talk of the *de Donder gauge* in this context: One requires that the coordinate functions be solutions of the wave equation:

$$\Box_g x^\alpha = 0 \, , \tag{4.1.7}$$

where $\Box_g$ is the wave operator associated with the metric g; in (any) local coordinates:

$$\Box_g f := g^{\mu\nu} \nabla_\mu \nabla_\nu f \, . \tag{4.1.8}$$

For further purposes it is convenient to rewrite (4.1.8) as

$$\Box_g f = \frac{1}{\sqrt{|\det g_{\mu\nu}|}} \partial_\rho \left(\sqrt{|\det g_{\mu\nu}|} g^{\rho\sigma} \partial_\sigma f \right) \, . \tag{4.1.9}$$

Incidentally In order to show that this formula is indeed correct, we calculate

$$\Box_g f = g^{\mu\nu}\left(\partial_\mu\partial_\nu f - \Gamma^\gamma{}_{\mu\nu}\partial_\gamma f\right)$$

$$= g^{\mu\nu}\left(\partial_\mu\partial_\nu f - \frac{1}{2}g^{\gamma\sigma}(\partial_\mu g_{\sigma\nu} + \partial_\nu g_{\sigma\mu} - \partial_\sigma g_{\mu\nu})\partial_\gamma f\right)$$

$$= g^{\mu\nu}\left(\partial_\mu\partial_\nu f - \frac{1}{2}g^{\gamma\sigma}(2\partial_\mu g_{\sigma\nu} - \partial_\sigma g_{\mu\nu})\partial_\gamma f\right)$$

$$= g^{\sigma\gamma}\partial_\sigma\partial_\gamma f - \left(\underbrace{g^{\mu\nu}g^{\gamma\sigma}\partial_\mu g_{\sigma\nu}}_{=:a} - \frac{1}{2}g^{\gamma\sigma}\underbrace{g^{\mu\nu}\partial_\sigma g_{\mu\nu}}_{=:b}\right)\partial_\gamma f . \qquad (4.1.10)$$

Differentiating the identity

$$g^{\gamma\sigma}g_{\sigma\nu} = \delta^\gamma_\nu$$

we obtain

$$g^{\gamma\sigma}\partial_\mu g_{\sigma\nu} = -g_{\sigma\nu}\partial_\mu g^{\gamma\sigma} . \qquad (4.1.11)$$

It follows that the function a of (4.1.10) equals

$$a = -g^{\mu\nu}g_{\sigma\nu}\partial_\mu g^{\gamma\sigma} = -\partial_\sigma g^{\gamma\sigma} . \qquad (4.1.12)$$

To continue, let us work out

$$\frac{\partial\sqrt{|\det g|}}{\partial g^{\mu\nu}} ,$$

this proceeds as follows: Fix an index μ, then the determinant of the matrix $g_{\alpha\beta}$ can be calculated by expanding in the μ'th column:

$$\det g_{\alpha\beta} = \sum_\nu g_{\mu\nu}\Delta^{\mu\nu}$$

(no summation over μ), where $\Delta^{\mu\nu}$ is the matrix of co-factors. Since $\Delta^{\mu\nu}$ does not involve the $g_{\mu\nu}$ entry of the matrix g, we have

$$\frac{\partial(\det g_{\alpha\beta})}{\partial g_{\mu\nu}} = \Delta^{\mu\nu} .$$

Now, the matrix of co-factors is related to the matrix $g^{\mu\nu}$, inverse to $g_{\mu\nu}$, by the formula:

$$\Delta^{\mu\nu} = (\det g_{\alpha\beta})g^{\mu\nu} ,$$

so that

$$\frac{\partial (\det g_{\alpha\beta})}{\partial g_{\mu\nu}} = (\det g_{\alpha\beta}) g^{\mu\nu}. \tag{4.1.13}$$

It then follows that

$$\frac{\partial \sqrt{|\det g_{\alpha\beta}|}}{\partial g_{\mu\nu}} = \frac{1}{2} \sqrt{|\det g_{\alpha\beta}|} g^{\mu\nu}. \tag{4.1.14}$$

The identity

$$g^{\alpha\beta} g_{\beta\gamma} = \delta^\alpha_\gamma$$

leads to the equation

$$\frac{\partial}{\partial g^{\mu\nu}} = -g_{\alpha\mu} g_{\beta\nu} \frac{\partial}{\partial g_{\alpha\beta}}. \tag{4.1.15}$$

Further, (4.1.14) shows that the function b of (4.1.10) equals

$$b = g^{\mu\nu} \partial_\sigma g_{\mu\nu} = \frac{2}{\sqrt{|\det g_{\alpha\beta}|}} \partial_\sigma \left(\sqrt{|\det g_{\alpha\beta}|} \right). \tag{4.1.16}$$

Inserting all this into (4.1.10) we obtain

$$\begin{aligned}
\Box_g f &= g^{\sigma\gamma} \partial_\sigma \partial_\gamma f + \left(\partial_\sigma g^{\gamma\sigma} + g^{\gamma\sigma} \frac{1}{\sqrt{|\det g_{\alpha\beta}|}} \partial_\sigma \left(\sqrt{|\det g_{\alpha\beta}|} \right) \right) \partial_\gamma f \\
&= \frac{1}{\sqrt{|\det g_{\mu\nu}|}} \partial_\rho \left(\sqrt{|\det g_{\mu\nu}|} g^{\rho\sigma} \partial_\sigma f \right),
\end{aligned} \tag{4.1.17}$$

as claimed.

(Local) solutions of (4.1.7) are easily constructed as follows: one chooses any spacelike hypersurface $\mathscr{S} \subset \mathscr{M}$—by definition, this means that the metric γ induced on $\mathscr{S}$ from g is Riemannian; the explicit formula for γ reads

$$\forall\, X, Y \in T\mathscr{S} \quad \gamma(X, Y) := g(X, Y).$$

As g is Lorentzian, Eq. (4.1.7) is a second order hyperbolic equation. The standard theory of hyperbolic PDEs (*cf., e.g.,* [85]) asserts that for any functions $k, h : \mathscr{S} \to \mathbb{R}$ and any vector field X defined along $\mathscr{S}$ and *transverse* to $\mathscr{S}$ there exists a neighborhood of $\mathscr{S}$ and a unique solution of the wave equation $\Box_g f = 0$ satisfying

$$f|_\mathscr{S} = k, \qquad X(f)|_\mathscr{S} = h.$$

So, in order to construct wave coordinates around a point p one chooses any spacelike $\mathscr{S}$ passing through p, together with a coordinate patch $\mathscr{U}$ on $\mathscr{S}$ with coordinates $\{y^i\}$.

Replacing $\mathscr{S}$ by $\mathscr{U}$ one can without loss of generality assume that $\mathscr{S} = \mathscr{U}$. Then one solves the Cauchy problem with initial data given, e.g., by

$$x^0|_{\mathscr{S}} = 0, \qquad n(x^0)|_{\mathscr{S}} = 1, \tag{4.1.18a}$$

$$x^i|_{\mathscr{S}} = y^i, \qquad n(x^i)|_{\mathscr{S}} = 0, \tag{4.1.18b}$$

where n is the field of unit-normals to $\mathscr{S}$. Let $\mathscr{O} \subset \mathscr{M}$ denote the neighborhood of $\mathscr{S}$ on which the solution exists: (4.1.18) shows that for any coordinate system y^α around p the matrix $\partial x^\mu / \partial y^\alpha$ will be nondegenerate on $\mathscr{S}$. It follows that, passing to a subset of $\mathscr{O}$ if necessary, the x^μ's will form a coordinate system on $\mathscr{O}$.

4.1.3 Linearized Einstein Equations in Wave Coordinates

Let us return to metrics of the form (4.1.1) satisfying (4.1.2). As explained in the previous section we can choose a coordinate system in which the coordinate functions will satisfy the wave equation (4.1.7). We wish to show that the expression (4.1.5) for the Ricci tensor simplifies considerably when wave coordinates are chosen. Indeed, it then follows from (4.1.9) that we have

$$
\begin{aligned}
0 &= \Box_g x^\alpha \\
&= \frac{1}{\sqrt{|\det g_{\mu\nu}|}} \partial_\rho \left(\sqrt{|\det g_{\mu\nu}|} g^{\rho\sigma} \underbrace{\partial_\sigma x^\alpha}_{\delta^\alpha_\sigma} \right) \\
&= \frac{1}{\sqrt{|\det g_{\mu\nu}|}} \partial_\rho \left(\sqrt{|\det g_{\mu\nu}|} g^{\rho\alpha} \right).
\end{aligned}
\tag{4.1.19}
$$

We need to calculate this expression up terms of order ϵ^2. In order to do this, we first use (4.1.14) to obtain

$$
\begin{aligned}
\sqrt{|\det g_{\mu\nu}|} &= \sqrt{|\det g_{\mu\nu}|}\Big|_{g=\eta} + \left.\frac{\partial \sqrt{|\det g_{\mu\nu}|}}{\partial g_{\alpha\beta}}\right|_{g=\eta} h_{\alpha\beta} + O(\epsilon^2) \\
&= 1 + \frac{1}{2}\eta^{\alpha\beta} h_{\alpha\beta} + O(\epsilon^2) \\
&= 1 + \frac{1}{2} h^\alpha{}_\alpha + O(\epsilon^2).
\end{aligned}
$$

This, together with (4.1.3) leads to

$$
\begin{aligned}
0 &= \partial_\rho \left(\sqrt{|\det g_{\mu\nu}|} g^{\rho\alpha} \right) \\
&= \partial_\rho \left((1 + \frac{1}{2} h^\beta{}_\beta)(\eta^{\rho\alpha} - h^{\rho\alpha}) \right) + O(\epsilon^2) \\
&= \frac{1}{2}\partial^\alpha h^\beta{}_\beta - \partial_\rho h^{\rho\alpha} + O(\epsilon^2).
\end{aligned}
\tag{4.1.20}
$$

Equivalently,

$$\partial_\rho h^\rho{}_\alpha = \frac{1}{2}\partial_\alpha h^\beta{}_\beta + O(\epsilon^2).$$

(4.1.21)

This allows us to rewrite (4.1.5) as

$$R_{\beta\delta} = \frac{1}{2}\left[\partial_\alpha\{\partial_\beta h^\alpha{}_\delta + \partial_\delta h^\alpha{}_\beta - \partial^\alpha h_{\beta\delta}\} - \partial_\delta\partial_\beta h^\alpha{}_\alpha\right] + O(\epsilon^2)$$

$$= \frac{1}{2}\left[\partial_\beta \underbrace{\partial_\alpha h^\alpha{}_\delta}_{=-\partial_\delta h^\alpha{}_\alpha/2} + \partial_\delta \underbrace{\partial_\alpha h^\alpha{}_\beta}_{=-\partial_\beta h^\alpha{}_\alpha/2} - \partial_\alpha\partial^\alpha h_{\beta\delta} - \partial_\delta\partial_\beta h^\alpha{}_\alpha\right] + O(\epsilon^2)$$

$$= -\frac{1}{2}\partial_\alpha\partial^\alpha h_{\beta\delta} + O(\epsilon^2)$$

$$= -\frac{1}{2}\Box_\eta h_{\beta\delta} + O(\epsilon^2).$$

(4.1.22)

It follows that—up to higher order terms and a constant multiplicative factor—the Ricci tensor is the Minkowski wave operator acting on h:

$$\boxed{R_{\alpha\beta} = -\tfrac{1}{2}\Box_\eta h_{\alpha\beta}} + O(\epsilon^2).$$

(4.1.23)

4.1.4 First-Order Post-Newtonian Expansion

Einstein's theory is supposed to be a theory of gravitation. We already have one such theory, due to Newton, which works pretty well in several situations. It would thus be desirable if Einstein's theory contained Newton's theory in some limit. This is indeed the case and can be easily established using the calculations done so far.

There are a few conditions which should obviously hold when trying to recover Newton's theory: since that last theory is a linear one, and Einstein's is not, the gravitational field should be sufficiently weak in order that the nonlinearities do not matter. This is taken care of by the parameter ϵ in (4.1.1). Next, the wave operator arising in Eq. (4.1.23) leads to radiation phenomena when systems with bodies with large relative velocities are considered. On the other hand, Newton's equation

$$\Delta_\delta \phi = 4\pi G\mu$$

(4.1.24)

does not exhibit any wave behavior. In (4.1.24) μ is the matter density, ϕ is the Newtonian potential, Δ_δ is the Euclidean Laplace operator, and G is Newton's constant. This suggests that a regime in which approximate agreement with (4.1.24) can be obtained is one where time derivatives are smaller than what has been assumed in (4.1.2):

$$\partial_t h_{\mu\nu} = O(\epsilon^2), \quad \partial_t\partial_\alpha h_{\mu\nu} = O(\epsilon^2).$$

(4.1.25)

Let us write the Einstein equations as

$$R_{\mu\nu} - \frac{1}{2}Rg_{\mu\nu} = \lambda T_{\mu\nu}\,,$$

(4.1.26)

as one of the goals of our calculation will be to determine the value of the constant λ needed to recover Newton's theory.

We consider a spacetime containing a body made of dust with small energy-density,

$$T_{\mu\nu} = \rho u_\mu u_\nu\,, \quad \rho = O(\epsilon)\,.$$

(4.1.27)

The body is assumed to be moving slowly,

$$u_\mu dx^\mu = u_0 dt + O(\epsilon) \quad \Longleftrightarrow \quad u_i = O(\epsilon)\,.$$

(4.1.28)

We further assume Einstein's equation describing the equivalence of *active gravitational mass density*—by definition, this is the function μ which appears in (4.1.24)—and of the energy density:

$$\rho \equiv \mu c^2 = \mu$$

(4.1.29)

(recall that most of the time we are using units in which the speed of light c equals one). We note that in the calculations below it would suffice that (4.1.29) holds up to terms $O(\epsilon^2)$.

To continue, we rewrite (4.1.26) in a more convenient form. For this we start by taking the trace of (4.1.26) to obtain

$$g^{\mu\nu}(R_{\mu\nu} - \frac{1}{2}Rg_{\mu\nu}) = R - 2R = \lambda g^{\mu\nu} T_{\mu\nu} =: \lambda T\,,$$

(4.1.30)

so that $T = -R$ and (4.1.26) becomes

$$R_{\mu\nu} = \lambda\left(T_{\mu\nu} - \frac{1}{2}T g_{\mu\nu}\right)\,.$$

(4.1.31)

Under (4.1.27)–(4.1.28) we have $T = -\rho$ and, in a coordinate system adapted to u_μ,

$$T_{\mu\nu} - \frac{1}{2}T g_{\mu\nu} = \begin{pmatrix} \rho & 0 & 0 & 0 \\ 0 & 0 & 0 & 0 \\ 0 & 0 & 0 & 0 \\ 0 & 0 & 0 & 0 \end{pmatrix} + \frac{1}{2}\begin{pmatrix} -\rho & 0 & 0 & 0 \\ 0 & \rho & 0 & 0 \\ 0 & 0 & \rho & 0 \\ 0 & 0 & 0 & \rho \end{pmatrix}$$

$$= \frac{\rho}{2}\delta_{\mu\nu}\,,$$

(4.1.32)

where $\delta_{\mu\nu}$ stands for the usual Kronecker delta.

Replacing the left-hand side of (4.1.31) by the right-hand side of (4.1.23), using (4.1.32) and systematically ignoring the error terms we obtain the key equation

$$\boxed{-\tfrac{1}{2}\Box_\eta h_{\alpha\beta} = \tfrac{\lambda\rho}{2}\delta_{\mu\nu}}.$$

(4.1.33)

The wave operator $-\partial_t^2 + \Delta_\delta$ reduces to the flat Laplacian Δ_δ by (4.1.25), hence in that approximation we obtain

$$\Delta_\delta h_{\alpha\beta} = -\lambda\rho\delta_{\mu\nu}.$$

(4.1.34)

If we consider a bounded body, so that ρ is compactly supported, then the Newtonian potential of ρ is defined as the unique solution of the equation

$$\Delta\phi = 4\pi G\rho$$

(4.1.35)

which approaches zero at large distances. Hence the unique solution of (4.1.34) which tends to zero at infinity is

$$h_{\alpha\beta} = -\frac{\lambda}{4\pi G}\phi\delta_{\alpha\beta}.$$

(4.1.36)

In summary, assuming that all error terms are dominated by the leading one we have obtained:

$$g_{00} = -1-\frac{\lambda\phi}{4\pi G} + O\left(\left(\frac{\lambda\phi}{4\pi G}\right)^2\right), \quad g_{0i} = O\left(\left(\frac{\lambda\phi}{4\pi G}\right)^2\right),$$

$$g_{ij} = \left(1-\frac{\lambda\phi}{4\pi G}\right)\delta_{ij} + O\left(\left(\frac{\lambda\phi}{4\pi G}\right)^2\right).$$

(4.1.37)

When the configuration of the system is bounded in space and has finite total mass M we have

$$\phi = -\frac{GM}{r} + O(r^{-2})$$

at large distances in the vacuum region, which leads to the following asymptotic form of the gravitational field

$$g_{00} = -1+\frac{\lambda M}{4\pi r} + O\left(\left(\frac{M}{r}\right)^2\right), \quad g_{0i} = O\left(\left(\frac{M}{r}\right)^2\right),$$

$$g_{ij} = \left(1+\frac{\lambda M}{4\pi r}\right)\delta_{ij} + O\left(\left(\frac{M}{r}\right)^2\right).$$

(4.1.38)

At this stage one should redo the whole calculation using (4.1.37) or (4.1.38) as a starting point, to make sure that the final result is consistent with the remaining calculations and hypotheses, as turns out to be the case when $\lambda M/r$ is much smaller than one.

> **Incidentally** We note that for gravitational fields which are static and vacuum at large distances one can obtain complete asymptotic expansions in the spirit above by a recursive use of the Einstein equations, the reader is referred to [14, 86] for details. Compare ▶ Sect. 4.7, p. 162 below.

4.1.5 Newton's Equations of Motion, and Why $8\pi G$ Is $8\pi G$

In ▶ Sect. 2.3.2 we have shown that the integral curves $x^\mu(s)$ of the vector field u^μ appearing in (4.1.28) are affinely parameterized timelike geodesics:

$$\frac{dx^\alpha}{ds} = u^\alpha, \qquad \frac{d^2 x^\mu}{ds^2} = -\Gamma^\mu{}_{\alpha\beta} \frac{dx^\alpha}{ds} \frac{dx^\beta}{ds}. \qquad (4.1.39)$$

We wish to use this fact to derive the equations of motion of the dust particles considered in the last section. In order to do that we calculate

$$\frac{dx^\alpha}{ds} = u^\alpha = g^{\alpha\beta} u_\beta = \eta^{\alpha\beta} u_\beta + O(\epsilon) = \eta^{\alpha 0} u_0 + O(\epsilon),$$

leading to

$$\frac{dx^0}{ds} = 1 + O(\epsilon), \qquad \frac{dx^i}{ds} = O(\epsilon).$$

This implies that

$$\Gamma^\mu{}_{\alpha\beta} \frac{dx^\alpha}{ds} \frac{dx^\beta}{ds} = \Gamma^\mu{}_{00} \frac{dx^0}{ds} \frac{dx^0}{ds} + O(\epsilon) = \Gamma^\mu{}_{00} + O(\epsilon).$$

In order to calculate the space-acceleration

$$\frac{d^2 x^i}{dt^2} = \frac{d^2 x^i}{ds^2} + O(\epsilon^2)$$

of a geodesic it remains to calculate $\Gamma^i{}_{00}$:

$$\Gamma^i{}_{00} = \frac{1}{2} g^{i\upsilon} \left(2\partial_0 g_{\sigma 0} - \partial_\sigma g_{00}\right) = -\frac{1}{2} \partial_i h_{00} + O(\epsilon).$$

From (4.1.38) we thus obtain

$$\Gamma^i{}_{00} = -\frac{\lambda}{8\pi G} \partial_i \phi.$$

It follows that the geodesic equation implies

$$\frac{d^2 x^i}{dt^2} = \frac{\lambda}{8\pi G} \partial_i \phi, \qquad (4.1.40)$$

which is identical with Newton's equations of motion,

$$\frac{d^2 x^i}{ds^2} = \partial_i \phi \,,$$

(4.1.41)

if and only if

$$\boxed{\lambda = 8\pi G \,.}$$

(4.1.42)

We have thus shown that the choice (4.1.42) of the constant appearing in (4.1.26) leads to a theory which reproduces Newton's theory of gravitation, in the limit of weak fields, for slowly moving low density bodies made of dust.

It should be borne in mind that all our arguments above have been carried through at a somewhat heuristic level. The problem is that we have considered metrics defined on a neighborhood of the hypersurface $\{t = 0\}$ in $\mathbb{R}^4$. As already pointed out in the gravitational radiation context, hypotheses (4.1.2) on non-compact subsets of spacetime need justification. A rigorous treatment would require careful estimates, to show that the terms which we have neglected can indeed be neglected. This can be done in some situations [62, 63, 90].

Incidentally We close this section by noting the elegant framework of Ehlers [30] which geometrizes the Newtonian limit of Einstein's theory. Ehlers' approach has been used by Heilig [42] to construct general relativistic axially-symmetric stationary star models, using the implicit function theorem in a neighborhood of the corresponding Newtonian solutions of Lichtenstein. See also Rendall [10, 12, 72].

4.2   Linearized Plane Waves

We consider the simplest possible solutions of the vacuum linearized Einstein equations,

$$\Box_\eta h_{\alpha\beta} = 0$$

(4.2.1)

(compare (4.1.23)), namely

$$h_{\alpha\beta} = \Re(A_{\alpha\beta} e^{ik_\mu x^\mu}) \,,$$

(4.2.2)

where $\Re$ denotes the real part, for some constant real vector k_μ and a matrix of (possibly complex) numbers $A_{\alpha\beta}$. The wave equation (4.2.1) requires that k^μ be null in the Minkowski metric,

$$k^\mu k_\mu = 0 \,.$$

(4.2.3)

Here and elsewhere in this section, in a manner similar to that in ▶ Sect. 4.1.1, all indices are raised and lowered with the Minkowski metric η.

The "transversality condition"

$$\partial_\mu \bar{h}^\mu{}_\nu = 0, \text{ where } \bar{h}^\mu{}_\nu = h^\mu{}_\nu - \frac{1}{2} h^\alpha{}_\alpha \delta^\mu_\nu \tag{4.2.4}$$

(compare (4.1.21)), will be satisfied if

$$\underbrace{(A_{\alpha\beta} - \frac{1}{2} A^\mu{}_\mu \eta_{\alpha\beta})}_{=: \bar{A}_{\alpha\beta}} k^\beta = 0. \tag{4.2.5}$$

It turns out that there remains some gauge-freedom in (4.2.2), which we will explore now. Our aim is to show that one can always achieve the gauge conditions (4.2.18) below.

For this, consider a coordinate-transformation of the form

$$x^\mu \mapsto \hat{x}^\mu = x^\mu + \underbrace{\Im(f^\mu e^{ik_\alpha x^\alpha})}_{=: \zeta^\mu}, \tag{4.2.6}$$

where $\Im$ denotes the imaginary part, for some possibly complex constant vector f^μ. Similarly to our hypotheses on $h_{\mu\nu}$, the vector f^μ is assumed to be small, in the sense that there exists $0 < \epsilon \ll 1$ such that

$$|A_{\mu\nu}|, \; |f^\mu| \le \epsilon.$$

This will allow us to neglect all terms quadratic or higher in ϵ in the calculations. Writing $\hat{g}_{\mu\nu}$ for the components of the metric g in the coordinate system $\hat{x}^\mu$, we have

$$\hat{g}_{\alpha\beta} d\hat{x}^\alpha d\hat{x}^\beta = (\eta_{\alpha\beta} + \hat{h}_{\alpha\beta}) d\hat{x}^\alpha d\hat{x}^\beta = (\eta_{\alpha\beta} + \hat{h}_{\alpha\beta}) \frac{\partial \hat{x}^\alpha}{\partial x^\mu} \frac{\partial \hat{x}^\beta}{\partial x^\nu} dx^\mu dx^\nu$$

$$= (\eta_{\alpha\beta} + \hat{h}_{\alpha\beta}) \left(\delta^\alpha_\mu + \frac{\partial \zeta^\alpha}{\partial x^\mu} \right) \left(\delta^\beta_\nu + \frac{\partial \zeta^\beta}{\partial x^\nu} \right) dx^\mu dx^\nu$$

$$= \left(\eta_{\mu\nu} + \underbrace{\hat{h}_{\mu\nu} + \frac{\partial \zeta_\nu}{\partial x^\mu} + \frac{\partial \zeta_\mu}{\partial x^\nu} + O(\epsilon^2)}_{=: h_{\mu\nu}} \right) dx^\mu dx^\nu.$$

Neglecting higher-order terms in ϵ, we have established the transformation law

$$\hat{h}_{\mu\nu} = h_{\mu\nu} - \partial_\mu \zeta_\nu - \partial_\nu \zeta_\mu. \tag{4.2.7}$$

When ζ_μ is given by (4.2.6), we find

$$\partial_\mu \zeta_\nu + \partial_\nu \zeta_\mu = \Re\left((k_\mu f_\nu + k_\nu f_\mu) e^{ik_\alpha x^\alpha} \right), \tag{4.2.8}$$

which for metrics of the form (4.2.2) gives

$$\hat{h}_{\alpha\beta} = \Re(\hat{A}_{\alpha\beta} e^{ik_\mu x^\mu}), \quad \text{where } \hat{A}_{\alpha\beta} = A_{\alpha\beta} - k_\alpha f_\beta - k_\beta f_\alpha. \tag{4.2.9}$$

Clearly $\hat{h}_{\alpha\beta}$ still solves the homogeneous wave equation. Next, we have

$$\overline{\hat{A}}_{\alpha\beta} := \hat{A}_{\alpha\beta} - \frac{1}{2} \hat{A}^\gamma{}_\gamma \eta_{\alpha\beta}$$

$$= \bar{A}_{\alpha\beta} - k_\alpha f_\beta - k_\beta f_\alpha + k^\gamma f_\gamma \eta_{\alpha\beta}, \tag{4.2.10}$$

which easily implies that the transversality condition (4.2.5) continues to be satisfied by $\hat{A}_{\alpha\beta}$:

$$(\hat{A}_{\alpha\beta} - \frac{1}{2} \hat{A}^\mu{}_\mu \eta_{\alpha\beta}) k^\beta = 0. \tag{4.2.11}$$

We can choose f_μ so that $\eta^{\alpha\beta} \hat{A}_{\alpha\beta} = 0$:

$$\eta^{\alpha\beta} \hat{A}_{\alpha\beta} = \eta^{\alpha\beta}(A_{\alpha\beta} - k_\alpha f_\beta - k_\beta f_\alpha) = 0 \iff k^\alpha f_\alpha = \frac{1}{2} \eta^{\alpha\beta} A_{\alpha\beta}. \tag{4.2.12}$$

We choose any f_α satisfying (4.2.12), and make the gauge transformation just described. To avoid an explosion of notations we call $h_{\mu\nu}$ the new field, and $A_{\mu\nu}$ the resulting matrix, which now satisfies

$$\eta^{\alpha\beta} A_{\alpha\beta} = 0, \quad A_{\alpha\beta} k^\beta = 0. \tag{4.2.13}$$

Note that further coordinate transformations of the form (4.2.6) will preserve traceless-ness provided that $k^\alpha f_\alpha = 0$.

Consider, now, any Minkowskian observer with (unit) four-velocity vector u^μ. We claim that we can choose f_μ, orthogonal to k_μ, such that

$$\hat{A}_{\alpha\beta} u^\alpha = 0. \tag{4.2.14}$$

Indeed, choose a coordinate system in which $u^\mu \partial_\mu = \partial_0$, then

$$\hat{A}_{00} = 0 \iff A_{00} - 2k_0 f_0 = 0 \iff f_0 = \frac{1}{2k_0} A_{00}, \tag{4.2.15}$$

which determines f_0. The equation $\hat{A}_{0i} = 0$ determines f_i:

$$\hat{A}_{0i} = A_{0i} - k_0 f_i - k_i f_0 = 0$$

$$\iff k_0 f_i = A_{0i} - k_i f_0 = A_{0i} - \frac{1}{2k_0} A_{00} k_i. \tag{4.2.16}$$

It only remains to check that the vector f_μ so determined is orthogonal to k_μ. For this, we use (4.2.14) contracted with k^β to obtain

$$0 = \hat{A}_{\alpha\beta} u^\alpha k^\beta \equiv (A_{\alpha\beta} - k_\alpha f_\beta - k_\beta f_\alpha) u^\alpha k^\beta$$
$$= \underbrace{A_{\alpha\beta} k^\beta}_{0} u^\alpha - \underbrace{k_\alpha u^\alpha}_{\neq 0} k^\beta f_\beta - \underbrace{k^\beta k_\beta}_{0} u^\alpha f_\alpha, \tag{4.2.17}$$

whence $k_\beta f^\beta = 0$, as desired.

Performing the gauge transformation, and removing hats, we conclude that plane gravitational waves (4.2.2) can be brought to the form

$$\boxed{h_{\alpha\beta} = \Re(A_{\alpha\beta} e^{ik_\mu x^\mu}), \text{ with } \eta^{\alpha\beta} A_{\alpha\beta} = 0, \ A_{\alpha\beta} k^\beta = 0, \ A_{\alpha\beta} u^\beta = 0.} \tag{4.2.18}$$

For definiteness, we choose a coordinate system such that

$$u^\mu \partial_\mu = \partial_0, \quad k^\mu \partial_\mu = \partial_0 + \partial_3.$$

The matrix $A_{\mu\nu}$ must then be of the form

$$(A_{\mu\nu}) = \begin{pmatrix} 0 & 0 & 0 & 0 \\ 0 & a_+ & a_\times & 0 \\ 0 & a_\times & -a_+ & 0 \\ 0 & 0 & 0 & 0 \end{pmatrix}, \tag{4.2.19}$$

with $a_+, a_\times \in \mathbb{C}$. Because there remain two free parameters, we say that gravitational waves have two polarizations.

We also say that gravitational waves are *transversal waves*, because all the action is taking place in directions orthogonal to the direction of propagation of the wave.

Exercise 4.2.1 So far we have been considering the linearized Einstein equations near Minkowski spacetime. However, consider the metric

$$g_{\mu\nu} = \eta_{\mu\nu} + f k_\mu k_\nu,$$

where f is a function on $\mathbb{R}^4$ such that $\eta^{\mu\nu} k_\mu \partial_\nu f = 0$, and k_μ is a vector which is covariantly constant for the Minkowski metric $\eta_{\mu\nu}$.
1. Show that $g^{\mu\nu}$ takes the form $\eta^{\mu\nu} - f k^\mu k^\nu$, where $k^\mu = \eta^{\mu\nu} k_\nu$.
2. Show that the Christoffel coefficients of $g_{\mu\nu}$ are of the form

$$\Gamma^\mu{}_{\nu\gamma} = \tfrac{1}{2}\eta^{\mu\alpha}\left(k_\alpha k_\nu \partial_\gamma f + k_\alpha k_\gamma \partial_\nu f - k_\nu k_\gamma \partial_\alpha f\right).$$

3. Show that

$$R_{\mu\nu} = -\frac{1}{2} k_\mu k_\nu \eta^{\alpha\beta} \partial_\alpha \partial_\beta f. \tag{4.2.20}$$

[Hint: Check that $\Gamma^\alpha{}_{\beta\alpha}$ vanishes, and that any contraction of k^ν with the Christoffel symbols vanishes.]

4. Conclude that Einstein's vacuum field equations have solutions as above with $f = \Re(\alpha e^{ik_\mu x^\mu})$, with $\alpha \in \mathbb{C}$. This provides thus an *exact* solution of the vacuum Einstein equations with properties strikingly similar to the linearized waves we have discussed above.

Let us analyze some properties of spacetimes with a metric

$$g_{\mu\nu} = \eta_{\mu\nu} + h_{\mu\nu},$$

with $h_{\mu\nu}$ small, given by (4.2.18)–(4.2.19). Consider an observer with four-velocity $u^\mu \partial_\mu = u^0 \partial_t$. Since $h_{0\mu} = 0$, the normalization condition $g(u, u) = -1$ gives $u^0 = 1$. Next, we have

$$\Gamma^\lambda_{\mu\nu} u^\mu u^\nu = \Gamma^\lambda_{00} = \frac{1}{2} g^{\lambda\sigma} (2\partial_0 g_{0\sigma} - \partial_\sigma g_{00}) = 0,$$

which implies that the curves $t \mapsto (t, \vec{x})$, with $\vec{x}$-fixed (so that $\frac{dx^\mu}{dt} \partial_\mu = \partial_t$) are affinely parameterized geodesics:

$$\frac{d^2 x^\lambda}{dt^2} + \Gamma^\lambda_{\mu\nu} \frac{dx^\mu}{dt} \frac{dx^\nu}{dt} = \Gamma^\lambda_{00} = 0.$$

(This could also have been inferred by general considerations from $g(u, u) = -1$ together with $u = \nabla t$.)

Consider, then two freely falling point objects with space coordinates differing by a vector

$$\vec{\ell}_\alpha = \ell(\cos\alpha, \sin\alpha, 0),$$

thus $\vec{\ell}_\alpha$ is located in the plane perpendicular to the direction ∂_z of propagation of the wave. Let $(x^a) = (x, y)$, and at any moment of time consider the space-metric

$$\gamma := g_{ab} dx^a dx^b = (\delta_{ab} + h_{ab}) dx^a dx^b$$

on the planes $z = \text{const}$. We have $\partial_a g_{bc} = 0$, which shows that γ is flat (even though the spacetime metric $g_{\mu\nu} dx^\mu dx^\nu$, or the space metric $g_{ij} dx^i dx^j$, are not). The straight lines on those planes are therefore geodesics. Writing $h_{11} = -h_{22} = h_+$, and $h_{12} = h_\times$, the physical distance between the objects within the plane equals thus

$$\sqrt{g_{ab} \ell^a_\alpha \ell^b_\alpha} = \ell \sqrt{(\delta_{ab} + h_{ab}) \frac{\ell^a_\alpha}{\ell} \frac{\ell^b_\alpha}{\ell}} \approx \ell \left(1 + \frac{1}{2} h_{ab} \frac{\ell^a_\alpha}{\ell} \frac{\ell^b_\alpha}{\ell}\right)$$

$$= \ell \left(1 + \frac{1}{2} h_+ (\cos^2\alpha - \sin^2\alpha) + h_\times \cos\alpha \sin\alpha\right)$$

$$= \ell \left(1 + \frac{1}{2} (h_+ \cos(2\alpha) + h_\times \sin(2\alpha))\right). \tag{4.2.21}$$

□ Fig. 4.9 Time evolution of a circle perpendicular to the direction of a +-polarized gravitational wave (left figure) and ×-polarized wave (right figure)

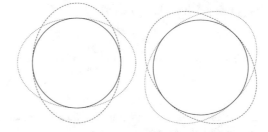

Consider the circle obtained by making α vary from 0 to 2π. If the cross polarization is turned-off, $h_\times = 0$, and if h_+ varies periodically, the coordinate circle is actually an ellipse, whose minor and major axes coincide with the coordinate axes, and whose lengths vary periodically in time. See □ Fig. 4.9. If the plus polarization is turned-off, $h_+ = 0$, the coordinate circle is again an ellipse, whose minor and major axes are rotated by 45° with respect to the coordinate axes, and whose lengths vary periodically in time as $h_\times$ does.

Incidentally Interesting nontrivial examples of linearized gravitational waves can be found in [19].

4.3 Remarks on Submanifolds, Integration, and Stokes' Theorem

In order to address the question of emission of energy by gravitational waves we need to understand integral conservation laws. While we will mostly be concerned with this issue in Minkowski spacetime, it is useful to outline the general picture, as even in Minkowski spacetime we will need to integrate on general submanifolds.

Our starting observation is that the integral of a *scalar function* f with respect to the coordinate measure

$$d^n x := dx^1 \cdots dx^n$$

is *not* a coordinate-independent operation. This is due to the fact that, under a change of variables $x \mapsto \bar{x}(x)$, one has

$$\int_{\mathbb{R}^n} \bar{f}(\bar{x}) d^n \bar{x} = \int_{\mathbb{R}^n} \underbrace{\bar{f}(\bar{x}(x))}_{f(x)} |J_{x \mapsto \bar{x}}(x)| d^n x , \qquad (4.3.1)$$

where the *Jacobian* $J_{x \mapsto \bar{x}}$ is the determinant of the Jacobi matrix,

$$J_{x \mapsto \bar{x}} = \left| \frac{\partial(\bar{x}^1, \ldots, \bar{x}^n)}{\partial(x^1, \ldots, x^n)} \right| .$$

Supposing that we have a metric

$$g = g_{ij}(x)dx^i dx^j = g_{ij}(x)\frac{\partial x^i}{\partial \bar{x}^k}(\bar{x}(x))\frac{\partial x^j}{\partial \bar{x}^\ell}(\bar{x}(x))\,d\bar{x}^k d\bar{x}^\ell = \bar{g}_{k\ell}(\bar{x}(x))d\bar{x}^k d\bar{x}^\ell$$

$$(4.3.2)$$

at our disposal, the integral can then be made manifestly coordinate-invariant by introducing the *metric measure*

$$d\mu_g := \sqrt{\det g_{ij}}d^n x \,. \tag{4.3.3}$$

Indeed, if we use

$$x\,(\bar{x}(x)) = x \implies \frac{\partial x^k}{\partial \bar{x}^\ell}(\bar{x}(x))\frac{\partial \bar{x}^\ell}{\partial x^i}(x) = \delta_i^k \implies J_{\bar{x}\mapsto x}\,(\bar{x}(x))\,J_{x\mapsto \bar{x}}(x) = 1\,,$$

where "1" stands for the identity matrix, we find from (4.3.2) that

$$\sqrt{\det \bar{g}_{ij}\,(\bar{x}(x))} = \sqrt{\det g_{ij}(x)}|J_{\bar{x}\mapsto x}\,(\bar{x}(x))| = \frac{\sqrt{\det g_{ij}(x)}}{|J_{x\mapsto \bar{x}}(x)|}\,,$$

hence

$$d\mu_g \equiv \sqrt{\det g_{ij}(x)}d^n x = \sqrt{\det \bar{g}_{ij}\,(x(\bar{x}))}|J_{x\mapsto \bar{x}}\,(x)\,|d^n x \,. \tag{4.3.4}$$

This shows that

$$\int_{\mathbb{R}^n} f(x)\sqrt{\det g_{ij}}d^n x = \int_{\mathbb{R}^n} f(x)\sqrt{\det \bar{g}_{ij}}|J_{x\mapsto \bar{x}}\,(x)\,|d^n x \,.$$

Comparing with (4.3.1), we see that

$$\int_{\mathbb{R}^n} f(x)d\mu_g = \int_{\mathbb{R}^n} f\,(x(\bar{x}))\,\sqrt{\det \bar{g}_{ij}}d^n \bar{x} = \int_{\mathbb{R}^n} \bar{f}(\bar{x})d\mu_{\bar{g}}\,.$$

A similar formula holds for subsets of $\mathbb{R}^n$. We conclude that the metric measure $d\mu_g$ is the right thing to use when integrating scalars over a manifold.

Now, when defining conserved charges we have been integrating on submanifolds. The first naive thought would be to use the spacetime metric determinant as above for that, e.g., in spacetime dimension $n + 1$,

$$\int_{\{x^0=0\}} f = \int_{\mathbb{R}^n} f(0, x^1, \ldots, x^n)\sqrt{\det g_{\mu\nu}}dx^1 \ldots dx^n \,.$$

This does not work because if, e.g., we take g to be the Minkowski metric on $\mathbb{R}^n$, and replace x^0 by $\bar{x}^0$ using $x^0 = 2\bar{x}^0$, then the only thing that will change in the last integral is the determinant $\sqrt{\det g_{\mu\nu}}$, giving a different value for the answer.

So, to proceed, it is useful to make first a short excursion into hypersurfaces, induced metrics, and measures.

4.3.1 Hypersurfaces

A subset $\mathscr{S} \subset \mathscr{M}$ is called a *hypersurface* if near every point $p \in \mathscr{S}$ there exists a coordinate system $\{x^1, \ldots, x^n\}$ on a neighborhood $\mathscr{U}$ of p in $\mathscr{M}$ and a constant C such that

$$\mathscr{S} \cap \mathscr{U} = \{x^1 = C\}.$$

For example, any hyperplane $\{x^1 = \text{const}\}$ in $\mathbb{R}^n$ is a hypersurface. Similarly, a sphere $\{r = R\}$ in $\mathbb{R}^n$ is a hypersurface if $R > 0$.

Further examples include graphs,

$$x^1 = f(x^2, \ldots, x^{n-1}),$$

which is seen by introducing new coordinates $(\bar{x}^i) = (x^1 - f, x^2, \ldots x^n)$.

A standard result in analysis asserts that if φ is a differentiable function on an open set Ω such that $d\varphi$ is nowhere zero on $\Omega \cap \{\varphi = c\}$ for some constant c, then

$$\Omega \cap \{\varphi = c\}$$

forms a hypersurface in Ω.

A vector $X \in T_p\mathscr{M}$, $p \in \mathscr{S}$, is said to be *tangent to* $\mathscr{S}$ if there exists a differentiable curve γ with image lying on $\mathscr{S}$, with $\gamma(0) = p$, such that $X = \dot{\gamma}(0)$. One denotes by $T\mathscr{S}$ the set of such vectors. Clearly, the bundle $T\mathscr{S}$ of all vectors tangent to $\mathscr{S}$, defined when $\mathscr{S}$ is viewed as a manifold on its own, is naturally diffeomorphic with the bundle $T\mathscr{S} \subset T\mathscr{M}$ just defined.

As an example, suppose that $\mathscr{S} = \{x^1 = C\}$ for some constant C, then $T\mathscr{S}$ is the collection of vectors defined along $\mathscr{S}$ for which $X^1 = 0$.

As another example, suppose that

$$\mathscr{S} = \{x^0 = f(x^i)\} \tag{4.3.5}$$

for some differentiable function f. Then a curve γ lies on $\mathscr{S}$ if and only if

$$\gamma^0 = f(\gamma^1, \ldots, \gamma^n),$$

and so its tangent satisfies

$$\dot{\gamma}^0 = \partial_1 f \dot{\gamma}^1 + \ldots + \partial_n f \dot{\gamma}^n.$$

We conclude that $X = X^0 \partial_0 + X^1 \partial_1 + \ldots + X^n \partial_n$ is tangent to $\mathscr{S}$ if and only if

$$X^0 = X^1 \partial_1 f + \ldots + X^n \partial_n f = X^i \partial_i f \quad \Longleftrightarrow \quad X = X^i \partial_i f \partial_0 + X^i \partial_i . \quad (4.3.6)$$

Equivalently, the vectors

$$\partial_i f \partial_0 + \partial_i$$

form a basis of the tangent space $T\mathscr{S}$.

Finally, if

$$\mathscr{S} = \Omega \cap \{\varphi = c\}, \quad\quad\quad\quad (4.3.7)$$

then for any curve lying on $\mathscr{S}$ we have

$$\varphi(\gamma(s)) = c \quad \Longleftrightarrow \quad \dot{\gamma}^\mu \partial_\mu \varphi = 0 \text{ and } \varphi(\gamma(0)) = c .$$

Hence, a vector $X \in T_p\mathscr{M}$ is tangent to $\mathscr{S}$ if and only if $\varphi(p) = c$ and

$$X^\mu \partial_\mu \varphi = 0 \quad \Longleftrightarrow \quad X(\varphi) = 0 \quad \Longleftrightarrow \quad d\varphi(X) = 0 . \quad (4.3.8)$$

A one-form α is said to *annihilate* $T\mathscr{S}$ if

$$\forall X \in T\mathscr{S} \quad\quad \alpha(X) = 0 . \quad\quad\quad\quad (4.3.9)$$

The set of such one-forms is called *the annihilator* of $T\mathscr{S}$, and denoted as $(T\mathscr{S})^o$. By elementary algebra, $(T\mathscr{S})^o$ is a one-dimensional subset of $T^*\mathscr{M}$. So, (4.3.8) can be rephrased as the statement that $d\varphi$ annihilates $T\mathscr{S}$.

A vector $Y \in T_p\mathscr{M}$ is said to be *normal* to $\mathscr{S}$ if Y is orthogonal to every vector $X \in T_p\mathscr{S}$, where $T_p\mathscr{S}$ is viewed as a subset of $T_p\mathscr{M}$. Equivalently, the one-form $g(Y, \cdot)$ annihilates $T_p\mathscr{S}$. If N has unit length, $g(N, N) \in \{-1, +1\}$, then N is said to be the *unit normal*. Thus, for a unit normal,

$$\forall X \in T\mathscr{S} \quad\quad g(X, N) = 0, \quad g(N, N) = \epsilon \in \{\pm 1\} . \quad (4.3.10)$$

In Riemannian geometry only the plus sign is possible, and a unit normal vector always exists. This might not be the case in Lorentzian geometry: Indeed, consider the hypersurface

$$\mathscr{S} = \{t = x\} \subset \mathbb{R}^{1,1} \quad\quad\quad\quad (4.3.11)$$

in two-dimensional Minkowski spacetime. A curve lying on $\mathscr{S}$ satisfies $\gamma^0(s) = \gamma^1(s)$, hence X is tangent to $\mathscr{S}$ if and only if $X^0 = X^1$. Let Y be orthogonal to $X \neq 0$, then

$$0 = \eta(X, Y) = X^0(-Y^0 + Y^1),$$

whence

$$Y^0 = Y^1,$$ (4.3.12)

which has zero Lorentzian length. So no vector orthogonal to the hypersurface $\mathscr{S}$ of (4.3.11) can have length one or minus one.

Since vectors of the form (4.3.12) are tangent to $\mathscr{S}$, we also reach the surprising conclusion that vectors normal to $\mathscr{S}$ coincide with vectors tangent to $\mathscr{S}$ in this case.

Suppose that the direction normal to $\mathscr{S}$ is timelike or spacelike. Then the *metric h induced by g on $\mathscr{S}$* is defined as

$$\forall\, X, Y \in T\mathscr{S} \quad h(X, Y) = g(X, Y).$$ (4.3.13)

In other words, $h(X, Y)$ coincides with $g(X, Y)$ when both expressions are defined, but we are only allowed to consider vectors tangent to $\mathscr{S}$ when using h.

Some comments are in order: If g is Riemannian, then normals to $\mathscr{S}$ are spacelike, and (4.3.13) defines a Riemannian metric on $\mathscr{S}$. For Lorentzian g's, it is easy to see that h is Riemannian if and only if vectors orthogonal to $\mathscr{S}$ are timelike, and then $\mathscr{S}$ is called *spacelike*. Similarly, h is Lorentzian if and only if vectors orthogonal to $\mathscr{S}$ are spacelike, and then $\mathscr{S}$ is called *timelike*. When the normal direction to $\mathscr{S}$ is null, then (4.3.13) defines a symmetric tensor on $\mathscr{S}$ with signature $(0, +, \cdots, +)$, which is degenerate and therefore not a metric; such hypersurfaces are called *null*, or *degenerate*.

If $\mathscr{S}$ is *not* degenerate, it comes equipped with a Riemannian or Lorentzian metric h. This metric defines a measure $d\mu_h$, using (4.3.3) with g there replaced by h, which can be used to integrate over $\mathscr{S}$.

We are ready now to formulate the Stokes theorem for open bounded sets: Let Ω be a bounded open set with piecewise differentiable boundary and assume that there exists a well-defined field of *exterior-pointing conormals* $N = N_\mu dx^\mu$ to Ω. Then for any differentiable vector field X it holds that

$$\int_\Omega \nabla_\alpha X^\alpha d\mu_g = \int_{\partial\Omega} X^\mu N_\mu dS.$$ (4.3.14)

If $\partial\Omega$ is nondegenerate, N_μ can be normalized to have unit length, and then dS is the measure $d\mu_h$ associated with the metric h induced on $\partial\Omega$ by g.

The definition of dS for null hypersurfaces is more complicated. Since we will not need it in these lectures, it will not be discussed any further. Let us simply emphasize that (4.3.14) remains valid for a suitable measure dS on null components of the boundary. This measure is not uniquely defined by the geometry of the problem, but the product $N_\mu dS$ is.

Remark 4.3.1 The reader might wonder how (4.3.14) fits with the usual version of the divergence theorem

$$\int_\Omega \partial_\alpha X^\alpha \, d^n x = \int_{\partial\Omega} X^\mu dS_\mu,$$ (4.3.15)

as used in advanced calculus for sets Ω which can be covered by a single coordinate chart. For this we note the identity

$$\nabla_\mu X^\mu = \frac{1}{\sqrt{|\det g|}} \partial_\mu \left(\sqrt{|\det g|} X^\mu \right),$$ (4.3.16)

which gives

$$\int_\Omega \nabla_\alpha X^\alpha d\mu_g = \int_\Omega \frac{1}{\sqrt{|\det g|}} \partial_\alpha \left(\sqrt{|\det g|} X^\alpha \right) \sqrt{|\det g|} d^n x$$

$$= \int_\Omega \partial_\alpha \left(\sqrt{|\det g|} X^\alpha \right) d^n x.$$ (4.3.17)

This should make clear the relation between (4.3.15) and (4.3.14).

4.4 Energy-Emission by Solutions of Wave Equations

We wish to obtain a formula for the amount of energy radiated away by a linearized gravitational wave. Now, in the vacuum region and in an appropriate gauge, such gravitational fields satisfy the massless wave equation. For this, as a first step, we will attempt to understand how energy is radiated away by a scalar field ϕ satisfying a sourceless wave equation in Minkowski spacetime:

$$\Box_\eta \phi = 0.$$

We will use the energy-momentum tensor

$$T_{\mu\nu} = \nabla_\mu \phi \nabla_\nu \phi - \frac{1}{2} \nabla^\alpha \phi \nabla_\alpha \phi \, \eta_{\mu\nu}$$

$$= \partial_\mu \phi \partial_\nu \phi - \frac{1}{2} \eta^{\alpha\beta} \partial_\alpha \phi \partial_\beta \phi \, \eta_{\mu\nu}.$$ (4.4.1)

Strictly speaking, we should include an overall multiplicative constant in the definition of $T_{\mu\nu}$. We shall ignore this constant in this section, keeping in mind that the exact value of this constant will be important in specific physical models when comparing with experiments.

The tensor $T_{\mu\nu}$ is symmetric, and satisfies the conservation equation

$$\nabla_\nu T^\nu{}_\mu = 0.$$ (4.4.2)

It holds that

$$T_{00} = \frac{1}{2} (\dot\phi^2 + |D\phi|^2) \geq 0,$$ (4.4.3)

where $D\phi$ denotes the space-gradient of ϕ.

Exercise 4.4.1

Show that $T_{\mu\nu}$ satisfies the *dominant energy condition:* $T_{\mu\nu}X^{\mu}Y^{\nu} \geq 0$ for all causal future directed vectors X, Y. ∎

Throughout the remainder of this section we will be working in Minkowskian coordinates, in which the Minkowski metric η equals $\mathrm{diag}(-1, +1, \ldots, +1)$. Then (4.4.1) reads

$$\partial_{\nu}T^{\nu}{}_{\mu} = 0. \tag{4.4.4}$$

Given a hypersurface $\mathscr{S}$ in Minkowski spacetime, the *total energy-momentum vector* of $\mathscr{S}$ is defined as

$$p_{\mu}(\mathscr{S}) = -\int_{\mathscr{S}} T_{\mu}{}^{\nu}n_{\nu}\,d\mu_{\mathscr{S}}, \tag{4.4.5}$$

where n^{μ} is a field of conormals to $\mathscr{S}$. For simplicity we will assume that $\mathscr{S}$ is nowhere characteristic, so that $d\mu_{\mathscr{S}}$ is the (well-defined) natural measure on $\mathscr{S}$. If n^{μ} is timelike, which is the case of main interest, one chooses n^{μ} to be future-directed. (Note the minus sign above, which is related to the current signature $(-, +, \ldots, +)$, and is chosen so that $p^0 \geq 0$, compare (4.4.3).)

We emphasize that these integrals might change in an uncontrollable way *when using coordinates other than the manifestly Minkowskian ones*, because of the free tensor index on p_{μ}. So, even though $T_{\mu\nu}$ is a tensor field, $p_{\mu}(\mathscr{S})$ is *not* a vector field but rather a collection of numbers. Here one should keep in mind that a vector is always attached to a point, but where should $p_{\mu}(\mathscr{S})$ be attached?

Further note that if we stick to manifestly Minkowskian coordinates and make a nontrivial boost, then $\mathscr{S}$ will not be the hypersurface $\{x^0 = \mathrm{const}\}$ anymore if it was before the boost.

Stokes' theorem provides the key to understand the properties of p_{μ}: Indeed, as already mentioned, this theorem asserts that for any *bounded* set Ω with piecewise differentiable boundary for each μ we have

$$\int_{\partial\Omega} T_{\mu}{}^{\nu}N_{\nu} = \int_{\Omega} \partial_{\nu}T_{\mu}{}^{\nu},$$

where N_{μ} is the field of outer-pointing conormals to $\partial\Omega$. Since the right-hand side vanishes in manifestly Minkowskian coordinates, we conclude that in such coordinates it holds

$$\boxed{\int_{\partial\Omega} T_{\mu}{}^{\nu}N_{\nu} = 0}. \tag{4.4.6}$$

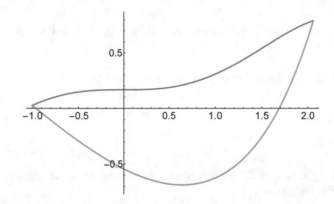

■ **Fig. 4.10** Two hypersurfaces in Minkowski spacetime with timelike normals and common boundary

As an application, in situations depicted in ■ Fig. 4.10 we obtain:

Proposition 4.4.2

Let $\mathscr{S}_1$ and $\mathscr{S}_2$ be two spacelike hypersurfaces in Minkowski spacetime with future-pointing timelike normals n^μ such that

$$\mathscr{S}_1 \cup \mathscr{S}_2 = \partial\Omega \tag{4.4.7}$$

for some bounded set Ω. Then

$$p_\mu(\mathscr{S}_1) = p_\mu(\mathscr{S}_2). \tag{4.4.8}$$

Proof

Assume, first, that $\mathscr{S}_2$ lies to the future of $\mathscr{S}_1$. Then $N^\mu = n^\mu$ along $\mathscr{S}_2$ and $N^\mu = -n^\mu$ along $\mathscr{S}_1$, and (4.4.6) gives

$$0 = \int_{\partial\Omega} T_\mu{}^\nu N_\nu = \int_{\mathscr{S}_2} T_\mu{}^\nu n_\nu - \int_{\mathscr{S}_1} T_\mu{}^\nu n_\nu = p_\mu(\mathscr{S}_1) - p_\mu(\mathscr{S}_2). \tag{4.4.9}$$

The case where $\mathscr{S}_2$ lies to the past of $\mathscr{S}_1$ is obtained by interchanging $\mathscr{S}_1$ and $\mathscr{S}_2$ in (4.4.9).   □

It is often the case that when two bounded hypersurfaces $\mathscr{S}_1$ and $\mathscr{S}_2$, with compact closures, share a common boundary,

$$\partial\mathscr{S}_1 = \partial\mathscr{S}_2,$$

then one can find a bounded set Ω such that (4.4.7) holds.

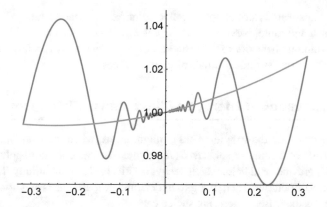

Fig. 4.11 Two hypersurfaces in Minkowski spacetime with common boundary which do not obviously bound a well-behaved set. The normal direction is timelike in both cases, which is not immediately apparent because of different scaling of the axes

Incidentally In cases where the existence of such a set Ω is not completely clear, as, e.g., in Fig. 4.11, we proceed as follows: Note, first, that a hypersurface in Minkowski spacetime with timelike normal is always a graph over a subset of $\{t = 0\} \approx \mathbb{R}^n$, hence there exist a bounded subset $\mathscr{U} \subset \mathbb{R}^n$ and functions $f_1, f_2 : \mathscr{U} \to \mathbb{R}$ such that

$$\mathscr{S}_a = \{t = f_a(\vec{x}), \ \vec{x} \in \mathscr{U}\}.$$

(Actually, one needs to justify that the set $\mathscr{U}$ is the same for $\mathscr{S}_1$ and $\mathscr{S}_2$; this follows from the fact that both hypersurfaces are bounded and their boundaries coincide.) We let $M :=$ $\min(\inf_{\mathscr{U}} f_1, \inf_{\mathscr{U}} f_2) - 1$, and consider the sets

$$\Omega_a = \{\vec{x} \in \mathscr{U}, \ M \le t \le f_a(\vec{x})\}.$$

Then

$$\partial\Omega_a = \mathscr{S}_a \cup \Sigma, \quad \text{where} \quad \Sigma = \{\vec{x} \in \partial\mathscr{U}, \ M \le t \le f_a(\vec{x})\} \cup \{\vec{x} \in \mathscr{U}, \ t = M\}.$$

Applying (4.4.6) on Ω_a we obtain, with the obvious choice of normals,

$$p_\mu(\mathscr{S}_a) + p_\mu(\Sigma) = 0, \ a = 1, 2, \quad \text{whence} \quad p_\mu(\mathscr{S}_1) = p_\mu(\mathscr{S}_2).$$

We conclude that:

Corollary 4.4.3
For connected bounded hypersurfaces with boundary, $p_\mu(\mathscr{S})$ depends only upon $\partial\mathscr{S}$.

One sometimes thinks of (4.4.8) as the statement that p_μ does not depend upon $\mathscr{S}$. However, care needs to be taken with such statements for hypersurfaces $\mathscr{S}$ extending to infinity. First, there are issues of convergence, to make sure that the integral defining p_μ is well defined and finite. Next, whether or not $p_\mu(\mathscr{S})$ depends upon $\mathscr{S}$ will depend upon the asymptotic behavior of $\mathscr{S}$. The aim of what follows is to study this question in more detail for two specific families of hypersurfaces.

4.4.1 Conservation of Energy on Minkowskian-Time Slices

We will consider only the simplest field configurations, where both $\phi|_{t=0}$ and $\partial_t\phi|_{t=0}$ are zero outside of a compact set, say $B(R)$. (Note that this does not apply directly to the linearized gravitational field, which decays relatively slowly at infinity. This is at the origin of many difficulties when trying to understand properly the issues arising, but we will not address those issues here.) In such a case

$$p_\mu(\{t = \tau\})$$

is well defined, and independent of τ. To see this, we start by noting the fundamental property of the wave equation that, for the solutions under consideration, for any τ the function $\phi|_{t=\tau}$ will be zero outside of $B(R + |\tau|)$. This is due to the fact that solutions of the wave equation propagate with the speed of light, cf. the left ◘ Fig. 4.12. Let then $\tau \in \mathbb{R}$ and choose $T \in \mathbb{R}^+$ such that $|\tau| \leq T$. We then have

$$p_\mu(\{t = \tau\}) \equiv -\int_{\{t=\tau\}} T_\mu{}^\nu n_\nu \, d^3x = -\int_{\{t=\tau, |\vec{x}|\leq R+T\}} T_\mu{}^\nu n_\nu \, d^3x \,.$$

Let

$$\Omega = \{\tau_1 \leq t \leq \tau_2, \ |\vec{x}| \leq R + T\}, \tag{4.4.10}$$

cf. the right ◘ Fig. 4.12. Using (4.4.6) we have

$$0 = \int_{\partial\Omega} T_\mu{}^\nu N_\nu$$

$$= \int_{\{t=\tau_2, |\vec{x}|\leq R+T\}} T_\mu{}^\nu N_\nu + \int_{\{t=\tau_1, |\vec{x}|\leq R+T\}} T_\mu{}^\nu N_\nu$$

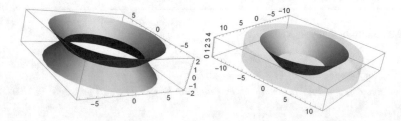

◘ **Fig. 4.12** Time runs along the vertical axis. Left figure: The solution of $\Box_\eta\phi = 0$ is zero outside the characteristic hypersurfaces emanating to the future and the past from the sphere $S(R)$ if the initial data for ϕ vanish outside the ball $B(R)$. Right figure: the shaded region is the set Ω of (4.4.10)

$$+ \underbrace{\int_{\{\tau_1 \le t \le \tau_2, \, |\vec{x}|=R+T\}} T_\mu{}^\nu N_\nu}_{0}$$

$$= \int_{\{t=\tau_2\}} T_\mu{}^\nu n_\nu - \int_{\{t=\tau_1\}} T_\mu{}^\nu n_\nu$$

$$= -p_\mu(\{t = \tau_2\}) + p_\mu(\{t = \tau_1\})$$

(recall that N_μ denotes the field of conormals while $n_\mu \in \{\pm N_\mu\}$ are the future directed ones on the spacelike parts of the boundary). Since T was arbitrary, we conclude that for all $\tau_1, \tau_2 \in \mathbb{R}$

$$p_\mu(\{t = \tau_2\}) = p_\mu(\{t = \tau_1\});$$

equivalently,

$$\boxed{\tfrac{d}{d\tau} p_\mu(\{t = \tau\}) = 0}. \tag{4.4.11}$$

4.4.2 Energy-Radiation on Hypersurfaces Asymptotic to Light-Cones

We pass next to the question of radiation of energy. Assuming again that both $\phi|_{t=0}$ and $\partial_t \phi|_{t=0}$ are zero outside of $B(R)$, one can show that the field ϕ satisfies the following property: there exists a function $f(u, \theta, \varphi)$ such that for $r \to \infty$ with $t - r$ being fixed it holds that

$$\phi(t, \vec{x}) = \frac{f(t - r, \theta, \varphi)}{r} + O(r^{-2}), \quad \partial_\mu \phi(t, \vec{x}) = \partial_\mu \left(\frac{f(t - r, \theta, \varphi)}{r} \right) + O(r^{-3}).$$

$$\tag{4.4.12}$$

Exercise 4.4.4 Check that for any differentiable function f the function

$$\phi(t, \vec{x}) = \frac{f(t - r)}{r}, \quad r > 0,$$

solves the wave equation in four-dimensional Minkowski spacetime (away from $r = 0$).

We consider therefore hypersurfaces $\mathscr{S}_u$ such that $t - r$ approaches u along $\mathscr{S}_u$ as r goes to infinity. Such hypersurfaces are called *hyperboloidal*, and they are asymptotic to the light-cones $t - r = $ const, see ◻ Fig. 4.13. We will see that the total energy on such hypersurfaces is not conserved, but is radiated away.

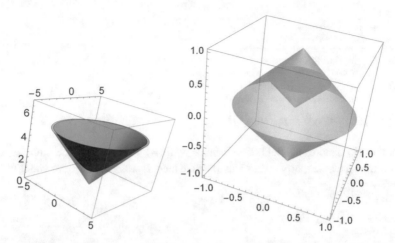

◘ Fig. 4.13 The hyperboloid $t = \sqrt{5 + x^2 + y^2}$ provides an example of a hyperboloidal hypersurface in Minkowski spacetime. The left figure shows the hyperboloid using manifestly flat coordinates, as well as the light-cone emanating from the origin. In the right figure the hyperboloid is shown in the conformally rescaled Minkowski spacetime, compare the right ◘ Fig. 3.4, p. 72

Incidentally In order to visualize such hypersurfaces globally it is convenient to transform first the whole Minkowski spacetime to a bounded set. This is best done using null coordinates as in (3.7.4), p. 71, and leads to a representation of Minkowski spacetime as in ◘ Fig. 3.4.

For simplicity we take the $\mathscr{S}_u$'s to be spherically symmetric graphs of the form

$$\mathscr{S}_u = \{t = \chi(r) + u\}, \quad \text{with} \quad \lim_{r \to \infty} (\chi(r) - r) = 0, \tag{4.4.13}$$

where $u \in \mathbb{R}$. It follows from arguments similar to those of the proof of Corollary 4.4.3 that the final result applies, however, to *any* family such that $t - r$ approaches u along $\mathscr{S}_u$ as r goes to infinity.

We choose $u_0 \in \mathbb{R}$ and some large $R > 0$, and for $u \geq u_0$ we consider (4.4.6) on the set

$$\Omega = \{\chi(r) + u_0 \leq t \leq \chi(r) + u, \ r \leq R\}.$$

Hence

$$0 = \int_{\partial\Omega} T_\mu{}^\nu N_\nu$$

$$= \int_{\mathscr{S}_u \cap \{r \leq R\}} T_\mu{}^\nu n_\nu - \int_{\mathscr{S}_{u_0} \cap \{r \leq R\}} T_\mu{}^\nu n_\nu$$

$$+ \int_{\{\chi(R) + u_0 \leq t \leq \chi(R) + u, \ |\vec{x}| = R\}} T_\mu{}^\nu N_\nu. \tag{4.4.14}$$

The last integral above equals

$$\int_{t=\chi(R)+u_0}^{\chi(R)+u} \int_{S(R)} T_\mu{}^i N_i d^2 S dt \,,$$

and so

$$\frac{d}{du} \int_{\mathscr{S}_u \cap \{r \le R\}} T_\mu{}^\nu n_\nu = \int_{S(R)} T_\mu{}^i N_i d^2 S \,. \tag{4.4.15}$$

Passing to the limit $R \to \infty$ we obtain (happily assuming that the derivative of the limit is the limit of the derivative...)

$$\frac{d}{du} p_\mu(\mathscr{S}_u) = - \lim_{R \to \infty} \int_{S(R)} T_\mu{}^i N_i d^2 S \,. \tag{4.4.16}$$

To calculate the right-hand side with $\mu = 0$, we use the fact that the relevant normal direction is $N_i = N^i = \frac{x^i}{r}$, thus

$$T_0{}^i n_i = \partial_t \phi \partial_i \phi \frac{x^i}{r} = \partial_t \phi \partial_r \phi \,.$$

Using (4.4.12) we obtain

$$T_0{}^i n_i = \partial_t \phi \partial_r \phi$$

$$= \partial_t \left(\frac{f(t-r,\theta,\varphi)}{r} + O(r^{-2}) \right) \partial_r \left(\frac{f(t-r,\theta,\varphi)}{r} + O(r^{-2}) \right)$$

$$= -\frac{\dot{f}^2}{r^2} + O(r^{-3}) \,, \text{ where } \dot{f} = \partial_u f \,.$$

Hence, keeping in mind that $p^0 = -p_0$,

$$\frac{d}{du} p^0(\mathscr{S}_u) = - \int_{S(1)} \dot{f}^2 d^2 S \,. \tag{4.4.17}$$

We see that the total energy $p^0(\mathscr{S}_u)$ contained in $\mathscr{S}_u$ decreases, with energy flux equal to $-\dot{f}^2$.

Remark 4.4.5 For those who are not at ease with spacetime normals it might be helpful to redo the whole calculation using only vector calculus in $\mathbb{R}^3$. This proceeds as follows:
Recall (4.4.8), which shows that for the hypersurfaces (4.4.13) we have

$$p_\mu(\mathscr{S}_u \cap \{|\vec{x}| \le R\}) = p_\mu(\{t = \chi(R) + u, |\vec{x}| \le R\}) \,, \tag{4.4.18}$$

as both those hypersurfaces are graphs over $B(R)$ and share the common boundary $\{t = \chi(R) + u, |\vec{x}| = R\}$. Hence, to prove (4.4.17), instead of working with the original family of hypersurfaces $\mathscr{S}_u \cap \{|\vec{x}| \leq R\}$ we can use the family of hypersurfaces $\{t = \chi(R) + u, |\vec{x}| \leq R\}$. We then consider the "spacetime cylinder"

$$\Omega_R = \{\chi(R) + u_0 \leq t \leq \chi(R) + u, |\vec{x}| \leq R\},$$

and the identity

$$0 = \int_{\Omega_R} \partial_\nu T_\mu{}^\nu d^4x = \int_{\Omega_R} (\partial_0 T_\mu{}^0 + \partial_i T_\mu{}^i) d^4x. \tag{4.4.19}$$

The $\partial_0 T_\mu{}^0$ terms integrate to

$$\int_{\Omega_R} \partial_0 T_\mu{}^0 d^4x = \int_{t=\chi(R)+u_0}^{\chi(R)+u} \int_{B(R)} \partial_0 T_\mu{}^0 d^3x \, dt$$

$$= \int_{B(R)} T_\mu{}^0(t = \chi(R) + u, \vec{x}) d^3x - \int_{B(R)} T_\mu{}^0(t = \chi(R) + u_0, \vec{x}) d^3x$$

$$= -\int_{B(R)} T_{\mu 0}(t = \chi(R) + u, \vec{x}) d^3x + \int_{B(R)} T_{\mu 0}(t = \chi(R) + u_0, \vec{x}) d^3x$$

$$= p_\mu(\{t = \chi(R) + u, |\vec{x}| \leq R\}) - p_\mu(\{t = \chi(R) + u_0, |\vec{x}| \leq R\})$$

$$= p_\mu(\mathscr{S}_u \cap \{|\vec{x}| \leq R\}) - p_\mu(\mathscr{S}_{u_0} \cap \{|\vec{x}| \leq R\});$$

in the before last-step we have used $n^\mu \partial_\mu = \partial_0$, hence $T_{\mu\nu} n^\nu = T_{\mu 0}$, while in the last step (4.4.18) was invoked. Note the change of sign arising from the fact that $T_{\mu 0} = -T_\mu{}^0$.

Using the divergence theorem on $B(R)$, the $\partial_i T_\mu{}^i$ term in (4.4.19) integrates to

$$\int_\Omega \partial_i T_\mu{}^i d^4x = \int_{t=\chi(R)+u_0}^{\chi(R)+u} \int_{B(R)} \partial_i T_\mu{}^i d^3x \, dt$$

$$= \int_{t=\chi(R)+u_0}^{\chi(R)+u} \int_{S(R)} T_\mu{}^i \frac{x^i}{r} d^2S.$$

Collecting all this, we obtain

$$p_\mu(\mathscr{S}_u \cap \{|\vec{x}| \leq R\}) = p_\mu(\mathscr{S}_{u_0} \cap \{|\vec{x}| \leq R\}) - \int_{t=\chi(R)+u_0}^{\chi(R)+u} \int_{S(R)} T_\mu{}^i \frac{x^i}{r} d^2S.$$

This reproduces (4.4.15).

Remark 4.4.6 For those with some knowledge of differential forms and who are confused with the signs involved (as this author often is), we recall that in (4.4.6) one integrates $T_\mu{}^\nu dS_\nu$ over $\partial\Omega$, where the three-forms dS_ν are defined as

$$dS_\nu = \sqrt{|\det g_{\alpha\beta}|} \partial_\nu \lrcorner dx^0 \wedge \cdots \wedge dx^n.$$

For the last integral in (4.4.14) it is convenient to go to spherical coordinates, since $dr = 0$ on the set integrated upon. Then the only form above which gives a nontrivial contribution to the integral is

$$dS_r = r^2 \sin\theta \partial_r \,\rfloor\, dt \wedge dr \wedge d\theta \wedge d\varphi = -r^2 \sin\theta\, dt \wedge d\theta \wedge d\varphi .$$

In this formalism, the integral over that part of the boundary where r is constant gives a contribution with a minus sign when the normal vector $n^i = x^i/r$ is used. However, in our signature, $n_\mu dx^\mu = -dt$ on slices $x^0 = $ const, which gives another negative sign in the formulae above.

4.5 The Quadrupole Formula

Recall, now, that the linearized gravitational field $\bar{h}_{\mu\nu}$ satisfies

$$\Box \bar{h}_{\mu\nu} = -\frac{16\pi G}{c^4} T_{\mu\nu} . \tag{4.5.1}$$

Every solution of (4.5.1) is the sum of the retarded solution,[2]

$$\bar{h}_{\mu\nu}(t, \vec{x}) = \frac{4G}{c^4} \int_{\mathbb{R}^3} \frac{T_{\mu\nu}(t - |\vec{x} - \vec{y}|, \vec{y})}{|\vec{x} - \vec{y}|} d^3 y , \tag{4.5.2}$$

and a solution of the homogeneous wave equation. It is usual to assume that the latter part is irrelevant for the problem at hand, and to take $\bar{h}_{\mu\nu}$ of the form (4.5.2).

Outside of the sources we have

$$\Box \bar{h}_{\mu\nu} = 0 .$$

So, one expects that a formula in the spirit of (4.4.16) should apply in this case.

Let us thus derive the leading-order behavior of $\bar{h}_{\mu\nu}$ for large r and t, with $u := t - r$ fixed. Assuming that the sources are confined to a spatially bounded region, to leading order we can approximate $|\vec{x} - \vec{y}|$ as $r := |\vec{x}|$. Then

$$t - |\vec{x} - \vec{y}| \approx t - r = u ,$$

and (4.5.2) becomes

$$\bar{h}_{\mu\nu}(t, \vec{x}) \approx \frac{\kappa}{|\vec{x}|} \int_{\mathbb{R}^3} T_{\mu\nu}(t - |\vec{x}|, \vec{y}) d^3 y , \quad \text{where } \kappa := \frac{4G}{c^4} . \tag{4.5.3}$$

---

[2]The readers expecting a different overall sign in the integral representation (4.5.2) of the solution should keep in mind our signature $(-, +, +, +)$ together with the definition $\Box = \eta^{\mu\nu}\partial_\mu\partial_\nu$, which gives a sign of our $\Box$-operator opposite to that in some standard textbooks.

For $\bar{h}_{0\mu}$ we obtain

$$\bar{h}_{0\mu}(t, \vec{x}) \approx \frac{\kappa}{|\vec{x}|} \int_{\mathbb{R}^3} T_{0\mu}(t - |\vec{x}|, \vec{y}) d^3 y = -\frac{\kappa p_\mu}{|\vec{x}|}, \tag{4.5.4}$$

where p^μ is the *total four-momentum of the slices* $t = $ const, also known as the Arnowitt-Deser-Misner (ADM) four-momentum. Since this p^μ is time-independent, it will give no contribution to energy emission.

To continue, we need the following:

Lemma 4.5.1
Suppose that $\partial_\mu T^{\mu\nu} = 0$, and that there exists $R > 0$ such that the sources are confined to a region $|\vec{x}| \le R$. Then

$$\boxed{\int_{\mathbb{R}^3} T_{ij}(t, \vec{y}) d^3 y = \frac{1}{2} \frac{d^2}{dt^2} \int_{\mathbb{R}^3} T_{00}(t, \vec{y}) y^i y^j d^3 y.}$$

Remark 4.5.2 For further use we note that if $T_{\mu\nu}$ is time-independent, then $\int_{\mathbb{R}^3} T_0{}^i y^j$ is antisymmetric under the exchange of the indices i and j, as follows from (4.5.5) below. □

Proof
The Minkowskian divergence identity $\partial_\mu T^\mu{}_\nu = 0$ can be rewritten as

$$\partial_t T_{0\nu} = \partial_i T^i{}_\nu.$$

We have, assuming as always that the integral of the time-derivative is the time-derivative of the integral,

$$\frac{d}{dt} \int_{\mathbb{R}^3} T_{00}(t, \vec{y}) y^i y^j d^3 y = \int_{B(R)} \frac{\partial T_{00}(t, \vec{y})}{\partial t} y^i y^j d^3 y = \int_{B(R)} \frac{\partial T_0{}^k(t, \vec{y})}{\partial y^k} y^i y^j d^3 y$$

$$= \int_{B(R)} \frac{\partial}{\partial y^k} \left(T_0{}^k(t, \vec{y}) y^i y^j \right) d^3 y - \int_{B(R)} T_0{}^k(t, \vec{y}) \frac{\partial}{\partial y^k} \left(y^i y^j \right) d^3 y$$

$$= \underbrace{\int_{S(R)} T_0{}^k(t, \vec{y}) y^i y^j n^k - \int_{B(R)} T_0{}^k(t, \vec{y}) \frac{\partial}{\partial y^k} \left(y^i y^j \right) d^3 y}_{0}$$

$$= -\int_{\mathbb{R}^3} (T_0{}^i(t, \vec{y}) y^j + T_0{}^j(t, \vec{y}) y^i) d^3 y. \tag{4.5.5}$$

Next,

$$\frac{d}{dt}\int_{\mathbb{R}^3} T_0{}^i(t,\vec{y})y^j d^3y = \int_{B(R)} \frac{\partial T_0{}^i(t,\vec{y})}{\partial t}y^j d^3y = \int_{B(R)} \frac{\partial T^{ki}(t,\vec{y})}{\partial y^k}y^j d^3y$$

$$= \int_{B(R)} \frac{\partial}{\partial y^k}\left(T^{ki}(t,\vec{y})y^j\right)d^3y - \int_{B(R)} T^{ki}(t,\vec{y})\frac{\partial y^j}{\partial y^k}d^3y$$

$$= \underbrace{\int_{S(R)} T^{ki}(t,\vec{y})y^j n^k}_{0} - \int_{B(R)} T^{ji}(t,\vec{y})d^3y$$

$$= -\int_{\mathbb{R}^3} T^{ij}d^3y, \tag{4.5.6}$$

and the result easily follows. □

Returning to (4.5.3), Lemma 4.5.1 gives

$$\bar{h}_{ij}(t,\vec{x}) \approx \frac{\kappa}{2|\vec{x}|}\frac{d^2}{dt^2}\int_{\mathbb{R}^3} T_{00}(t-|\vec{x}|,\vec{y})y^i y^j d^3y. \tag{4.5.7}$$

Summarizing, after performing a boost if necessary so that the total space-momentum vanishes,

$$\bar{h}_{00} \approx \frac{4GM}{c^4r}, \quad \bar{h}_{0i} \approx 0, \quad \bar{h}_{ij} \approx \frac{2G}{3c^2r}\ddot{q}_{ij}(t-r)$$

where M is the total mass and q_{ij} the quadrupole moments of the energy density:

$$M(\tau) = \int_{t=\tau} T_{00}d^3x = M(0) =: M, \quad q_{ij}(\tau) = \int_{t=\tau} T_{00}x^i x^j d^3x.$$

So, from (4.4.17) one expects that the energy flux will be proportional to some combination of the squares of third derivatives of q_{ij}. A more complete analysis, which requires the derivation of the energy-momentum tensor for weak gravitational fields, leads to the *Einstein quadrupole formula*, first derived by Einstein in 1917:

$$\boxed{-L := \frac{d}{dt}p^0(\mathscr{S}_t) = -\frac{G}{5c^5}\sum_{ij}\left(\frac{d^3 Q_{ii}}{dt^3}\right)^2,} \quad \text{where } Q_{ij} = q_{ij} - \frac{1}{3}q^k{}_k\delta_{ij}.$$

$$\tag{4.5.8}$$

The function $L = L(t)$ is called the *luminosity* of the gravitational wave at time t, with the minus sign inserted in (4.5.8) so that a positive L corresponds to energy emitted by the system.

4.6   Backreaction, the Chirp Mass

Consider a binary system of celestial objects, be it stars, planets, or black holes, in a near-Newtonian configuration. We would like to estimate the influence of the emission of energy in gravitational waves on the evolution of the system. Our presentation parallels [23].

We model the system as two point masses m_1 and m_2 positioned at $\vec{x}_1$ and $\vec{x}_2$. Newton's equations of motion read

$$m_1\ddot{\vec{x}}_1 = \frac{Gm_1m_2(\vec{x}_2 - \vec{x}_1)}{|\vec{x}_2 - \vec{x}_1|^3}, \qquad m_2\ddot{\vec{x}}_2 = \frac{Gm_1m_2(\vec{x}_1 - \vec{x}_2)}{|\vec{x}_2 - \vec{x}_1|^3}. \tag{4.6.1}$$

Passing to new variables $\vec{x}$ and $\vec{X}$ defined as

$$\vec{x} := \vec{x}_2 - \vec{x}_1, \qquad \vec{X} := \frac{m_1\vec{x}_1 + m_2\vec{x}_2}{m_1 + m_2} \tag{4.6.2}$$

(thus $\vec{X}$ is the center of mass), and setting

$$M := m_1 + m_2, \qquad \mu := \frac{m_1m_2}{m_1 + m_2} \equiv \frac{m_1m_2}{M},$$

one finds

$$\ddot{\vec{X}} = 0, \qquad \ddot{\vec{x}} = -\frac{MG\vec{x}}{|\vec{x}|^3}. \tag{4.6.3}$$

We can, and will, work in the center of mass reference frame, in which $\vec{X}(t) \equiv 0$. For simplicity we assume that the orbits are circular:

$$\vec{x}(t) = R(\cos(\omega t), \sin(\omega t), 0), \qquad \omega = \sqrt{\frac{GM}{R^3}} \tag{4.6.4}$$

(compare (3.8.16)). Since $\vec{X} = 0$ we have $m_1\vec{x}_1 = -m_2\vec{x}_2$, hence

$$\vec{x}(t) = \underbrace{\vec{x}_2}_{-m_1\vec{x}_1/m_2} - \underbrace{\vec{x}_1}_{-m_2\vec{x}_2/m_1} = \left(1 + \frac{m_2}{m_1}\right)\vec{x}_2 = -\left(1 + \frac{m_1}{m_2}\right)\vec{x}_1. \tag{4.6.5}$$

The quadrupole moment of the system is obtained by adding the contributions from each star:

$$q_{ij} = m_1 x_1^i x_1^j + m_2 x_2^i x_2^j = m_1 \frac{x^i x^j}{\left(1 + \frac{m_1}{m_2}\right)^2} + m_2 \frac{x^i x^j}{\left(1 + \frac{m_2}{m_1}\right)^2}$$

$$= \frac{m_1 m_2}{m_1 + m_2} x^i x^j \equiv \mu x^i x^j = \mu R^2 \begin{pmatrix} \cos^2(\omega t) & \sin(\omega t)\cos(\omega t) & 0 \\ \sin(\omega t)\cos(\omega t) & \sin^2(\omega t) & 0 \\ 0 & 0 & 0 \end{pmatrix}$$

$$= \mu R^2 \begin{pmatrix} \frac{1}{2}(\cos(2\omega t) + 1) & \frac{1}{2}\sin(2\omega t) & 0 \\ \frac{1}{2}\sin(2\omega t) & \frac{1}{2}(-\cos(2\omega t) + 1) & 0 \\ 0 & 0 & 0 \end{pmatrix}. \tag{4.6.6}$$

We see that the trace of q_{ij} is time-independent, so that

$$\frac{d^3 Q_{ij}}{dt^3} = \frac{d^3 q_{ij}}{dt^3} = 4\mu\omega^3 R^2 \begin{pmatrix} \sin(2\omega t) & -\cos(2\omega t) & 0 \\ -\cos(2\omega t) & -\sin(2\omega t) & 0 \\ 0 & 0 & 0 \end{pmatrix}. \tag{4.6.7}$$

Inserting this into (4.5.8) we obtain the gravitational luminosity, denoted by L, of the system:

$$L := -\frac{dp^0}{dt} = \frac{G}{5c^5} \sum_{ij} \left(\frac{d^3 Q_{ij}}{dt^3}\right)^2 = \frac{32G\mu^2 R^4 \omega^6}{5c^5} = \frac{32G^4 \mu^2 M^3}{5c^5 R^5}, \tag{4.6.8}$$

where in the last step we have used

$$\omega^2 = \frac{GM}{R^3}, \tag{4.6.9}$$

cf. (4.6.4).

The peak luminosity is reached when R is the black hole radius, $R = 2GM/c^2$, leading to

$$L_{\max} = \frac{c^5 \mu^2}{5GM^2}. \tag{4.6.10}$$

For m_1 of the same order as m_2, this is of the order of the enormous number

$$\frac{c^5}{G} = \frac{3.62837 \times 10^{52} \text{ kg m}^2}{s^3} = \frac{203894}{s} M_\odot c^2,$$

two hundred thousand solar masses per second, as first pointed-out by Dyson [29]. This result should be taken with a grain of salt, since the weak-field hypotheses made in all calculations so far are badly violated when setting $R = 2GM/c^2$.

In the center-of-mass frame, the total Newtonian energy E_N of the system reads

$$E_N = \frac{1}{2}m_1(\dot{\vec{x}}_1)^2 + \frac{1}{2}m_2(\dot{\vec{x}}_2)^2 - \frac{Gm_1m_2}{|\vec{x}_1 - \vec{x}_2|} = \frac{1}{2}\mu\dot{\vec{x}}^2 - \frac{Gm_1m_2}{|\vec{x}|}$$

$$= \frac{1}{2}\mu R^2\omega^2 - \frac{Gm_1m_2}{R}. \tag{4.6.11}$$

Using (4.6.9) we obtain

$$E_N = \frac{G\mu M}{2R} - \frac{Gm_1m_2}{R} = -\frac{Gm_1m_2}{2R}. \tag{4.6.12}$$

The energy outflow by gravitational waves will cause the radius to shrink. A change of orbital radius R by δR leads to a change of energy

$$\delta E_n = \frac{\partial E_N}{\partial R}\delta R = \frac{Gm_1m_2}{2R^2}\delta R.$$

Balancing this against the flux of energy $\frac{dp^0}{dt}$ lost by gravitational waves,

$$\frac{dE_N}{dt} = \frac{dp^0}{dt}, \tag{4.6.13}$$

and using (4.6.8) one is led to

$$\frac{Gm_1m_2}{2R^2}\dot{R} = -\frac{32G^4\mu^2M^3}{5c^5R^5} \quad\Longleftrightarrow\quad \dot{R} = -\frac{64G^3\mu M^2}{5c^5R^3}. \tag{4.6.14}$$

It should be kept in mind that this is an approximate equation, justified for small velocities and large distances, hence for small $\dot{R}$, so that the Newtonian approximation holds, with the orbit remaining approximately circular during the evolution. This appears to be the case for the Taylor–Hulse pulsar, for which the formula fits exceedingly well the observational data, see ◻ Fig. 4.1, p. 120.

Using (4.6.9), the ODE (4.6.14) can be rewritten as an equation for ω:

$$\dot{\omega} = -\frac{3}{2}\sqrt{\frac{GM}{R^5}}\dot{R} = \frac{3\omega}{2R} \times \frac{64G^3\mu M^2}{5c^5R^3}$$

$$= \frac{96}{5}\left(\frac{G\mu^{\frac{3}{5}}M^{\frac{2}{5}}}{c^3}\right)^{\frac{5}{3}}\omega^{\frac{11}{3}}. \tag{4.6.15}$$

Note that $\dot{\omega}$ and ω are directly measurable by Fourier-transforming the signal observed in the wave detectors, allowing one to determine the so-called *chirp mass* $\mathcal{M}$ from the data,

$$\mathcal{M} := \mu^{\frac{3}{5}} M^{\frac{2}{5}} \equiv \frac{(m_1 m_2)^{\frac{3}{5}}}{(m_1 + m_2)^{\frac{1}{5}}} . \tag{4.6.16}$$

In terms of $\mathcal{M}$ we thus have

$$\dot{\omega} = \frac{96}{5} \left(\frac{G\mathcal{M}}{c^3} \right)^{\frac{5}{3}} \omega^{\frac{11}{3}} \tag{4.6.17}$$

$$\approx 2.72 \times 10^{-8} \left(\frac{\mathcal{M}}{M_\odot} \right)^{\frac{5}{3}} (\omega \times s)^{\frac{11}{3}} s^{-2} . \tag{4.6.18}$$

Solving the ODE (4.6.17) one finds the time-evolution of the frequency of the wave:

$$\omega^{-8/3}(t) = \omega^{-8/3}(0) - \frac{8}{3} \times \frac{96}{5} \left(\frac{G\mu^{\frac{3}{5}} M^{\frac{2}{5}}}{c^3} \right)^{\frac{5}{3}} t . \tag{4.6.19}$$

Using $\omega = \sqrt{\frac{GM}{R^3}}$, or solving directly (4.6.14), one obtains an equation for the radius $R(t)$ of the orbit

$$R^4(t) = R^4(0) - \frac{256 G^3 \mu M^2}{5c^5} t , \tag{4.6.20}$$

valid for large R and small t.

If, rather outrageously, we extrapolate the considerations so far to all values of R and $\dot{R}$, we conclude from (4.6.20) that a binary system with initial radius R and initial orbital frequency ω will collapse to zero in finite time t_*:

$$t_* = \frac{5c^5 R^4}{256 G^3 \mu M^2} = \frac{5c^5}{256 \omega^{8/3} (\mathcal{M}G)^{5/3}}$$

$$= 1.4 \times 10^7 \times \left(\frac{M_\odot}{\mathcal{M}} \right)^{5/3} \frac{1}{(\omega \times s)^{8/3}} \times s . \tag{4.6.21}$$

Having determined the chirp mass $\mathcal{M}$ from $\dot{\omega}$ and ω, e.g., after having assumed that m_1 and m_2 are roughly equal, we can get an estimate of the time-to-collapse from (4.6.21). In favorable cases it will allow us to direct radio and optical telescopes to the right place at the right time to witness the collapse in all observation channels.

All the arguments so far will certainly stop being valid after the system has reached its Schwarzschild radius $R(t_{BH}) = 2GM_{BH}/c^2$ at time t_{BH}. Inserting this value in the left-hand side of (4.6.20) gives

$$\left(\frac{2GM_{\text{BH}}}{c^2}\right)^4 = \left(\frac{GM}{\omega^2(0)}\right)^{4/3} - \frac{256G^3\mu M^2}{5c^5}t_{\text{BH}}. \tag{4.6.22}$$

Whenever t_{BH} is available from the data, we can use (4.6.22) to obtain an estimate of M_{BH}.

Another, presumably more precise, estimate of M_{BH} is obtained by measuring the frequency of the "black hole ringing" radiation, after the collapse has taken place, using the formula for the quasi-normal-modes ringing frequency of a Schwarzschild black hole,

$$f \approx 1.207 \times \frac{M_\odot}{M_{\text{BH}}}10^4\,\text{Hz}. \tag{4.6.23}$$

4.7   Multipole Expansions

We return to the weak-field retarded solution,

$$\bar{h}_{\mu\nu}(t, \vec{x}) = \kappa \int_{\mathbb{R}^3} \frac{T_{\mu\nu}(t - |\vec{x} - \vec{y}|, \vec{y})}{|\vec{x} - \vec{y}|}d^3y. \tag{4.7.1}$$

We continue to assume that we are located at a very large distance from the sources. So far we made the approximation that

$$|\vec{x} - \vec{y}| \approx |\vec{x}|. \tag{4.7.2}$$

One can instead write down complete expansions of (4.7.1) in terms of inverse powers of $r := |\vec{x}|$; this is called the *multipole expansion*.

Here, in order to describe the Lense-Thirring effect, we will calculate the r^{-2} terms in $\bar{h}_{0\mu}$. For this we write, assuming $|\vec{x}| \gg |\vec{y}|$,

$$|\vec{x} - \vec{y}|^2 = |\vec{x}|^2 - 2\vec{x} \cdot \vec{y} + |\vec{y}|^2$$

$$= r^2\Big(1 - 2\underbrace{\frac{\vec{x} \cdot \vec{y}}{r^2}}_{O(1/r)} + \underbrace{\frac{|\vec{y}|^2}{r^2}}_{O(1/r^2)}\Big)$$

$$= r^2\Big(1 - 2\frac{\vec{x}}{r^2} \cdot \vec{y}\Big) + O(r^{-2}).$$

Hence

$$|\vec{x} - \vec{y}| = r\Big(1 - \frac{\vec{x} \cdot \vec{y}}{r^2}\Big) + O(r^{-2}).$$

This implies

$$\frac{1}{|\vec{x} - \vec{y}|} = \frac{1}{r} \times \frac{1}{1 - \frac{\vec{x} \cdot \vec{y}}{r^2} + O(r^{-2})} = \frac{1}{r} \times \left(1 + \frac{\vec{x} \cdot \vec{y}}{r^2} + O(r^{-2})\right)$$

$$= \frac{1}{r} + \frac{\vec{x} \cdot \vec{y}}{r^3} + O(r^{-3}),$$

$$T_{\mu\nu}(t - |\vec{x} - \vec{y}|, \vec{y}) \approx T_{\mu\nu}(t - |\vec{x}|, \vec{y}) + \partial_t T_{\mu\nu}(t - |\vec{x}|, \vec{y})(|\vec{x} - \vec{y}| - |\vec{x}|)$$

$$\approx T_{\mu\nu}(t - r, \vec{y}) - \partial_t T_{\mu\nu}(t - r, \vec{y}) \frac{\vec{x} \cdot \vec{y}}{r},$$

$$\frac{T_{\mu\nu}(t - |\vec{x} - \vec{y}|, \vec{y})}{|\vec{x} - \vec{y}|} \approx \left(T_{\mu\nu}(t - r, \vec{y}) - \partial_t T_{\mu\nu}(t - r, \vec{y}) \frac{\vec{x} \cdot \vec{y}}{r}\right)\left(\frac{1}{r} + \frac{\vec{x} \cdot \vec{y}}{r^3}\right)$$

$$\approx \frac{T_{\mu\nu}(t - r, \vec{y})}{r} + \frac{(-\partial_t T_{\mu\nu}(t - r, \vec{y})r + T_{\mu\nu}(t - r, \vec{y}))\vec{x} \cdot \vec{y}}{r^3}.$$

At this level of approximation we find now

$$\bar{h}_{00} \approx \frac{\kappa M}{r} + \frac{\kappa}{r^3} \vec{x} \cdot \int_{\mathbb{R}^3} \left(-\partial_t T_{00}(t - r, \vec{y})r + T_{00}(t - r, \vec{y})\right)\vec{y}, \tag{4.7.3}$$

$$\bar{h}_{0i} \approx -\frac{\kappa p_i}{r} + \frac{\kappa}{r^3} \vec{x} \cdot \int_{\mathbb{R}^3} \left(-\partial_t T_{0i}(t - r, \vec{y})r + T_{0i}(t - r, \vec{y})\right)\vec{y}, \tag{4.7.4}$$

with error terms which are $O(r^{-3})$ for time-independent configurations.

To proceed, one could use the divergence equation satisfied by the energy-momentum tensor to get rid of the time-derivative terms. We will instead consider the simplest case where the matter distribution is time-independent, $\partial_t T_{\mu\nu} = 0$.

Let $\vec{d}$ be the *center of mass* of matter fields,

$$\vec{d} := \int_{\mathbb{R}^3} T_{00}(0, \vec{y})\vec{y}. \tag{4.7.5}$$

The formula for $\bar{h}_{00}$ becomes

$$\bar{h}_{00} = \frac{\kappa M}{r} + \frac{\kappa \vec{d} \cdot \vec{x}}{r^3} + O(r^{-3}) \tag{4.7.6}$$

Note that $\vec{d}$ can always be set to zero by a shift of coordinates when $M \neq 0$: indeed, if $\vec{y}$ is replaced by $\vec{y}' = \vec{y} + \vec{a}$, the new center of mass becomes

$$\vec{d}' = \int_{\mathbb{R}^3} T_{00}(0, \vec{y}')\vec{y}' d^3 y' = \int_{\mathbb{R}^3} T_{00}(0, \vec{y})(\vec{y} + \vec{d})d^3 y = \vec{d} + M\vec{a}.$$

Choosing $\vec{a} = -\vec{d}/M$ brings $\vec{d}'$ to zero. Dropping primes, in the new coordinates we will have

$$\bar{h}_{00} = \frac{\kappa M}{r} + O(r^{-3}).$$

(4.7.7)

Consider, next, the formula (4.7.4) for $\bar{h}_{0i}$ in a zero-momentum frame:

$$\bar{h}_{0i} = \frac{\kappa}{r^3} x^j \int_{\mathbb{R}^3} T_{0i}(0, \vec{y}) y^j d^3 y + O(r^{-3})$$

(4.7.8)

(summation over j). Now,

$$\int_{\mathbb{R}^3} T_{0i}(0, \vec{y}) y^j d^3 y = \frac{1}{2} \int_{\mathbb{R}^3} \left(T_{0i}(0, \vec{y}) y^j + T_{0j}(0, \vec{y}) y^i\right) d^3 y$$
$$+ \frac{1}{2} \int_{\mathbb{R}^3} \left(T_{0i}(0, \vec{y}) y^j - T_{0j}(0, \vec{y}) y^i\right) d^3 y.$$

For time-independent matter fields the symmetric part

$$\int_{\mathbb{R}^3} \left(T_{0i}(0, \vec{y}) y^j + T_{0i}(0, \vec{y}) y^j\right) d^3 y$$

above vanishes, which follows from (4.5.6): indeed, as shown there, using conservation of matter and Stokes' theorem it holds that

$$\frac{d}{dt} \int_{\mathbb{R}^3} T_{00}(t, \vec{y}) y^i y^j d^3 y \Big|_{t=0} = - \int_{\mathbb{R}^3} \left(T_{0i}(0, \vec{y}) y^j + T_{0i}(0, \vec{y}) y^j\right) d^3 y.$$

(4.7.9)

So, whenever the left-hand side of the last equation vanishes one can conclude that (4.7.8) becomes

$$\bar{h}_{0i} = \frac{\kappa}{2r^3} x^j \int_{\mathbb{R}^3} \left(T_{0i}(0, \vec{y}) y^j - T_{0j}(0, \vec{y}) y^i\right) d^3 y + O(r^{-3}).$$

(4.7.10)

Let $\vec{J}$ be the total angular momentum of the system,

$$J^k = \epsilon^k{}_{ij} \int_{\mathbb{R}^3} y^i T_0{}^j(0, \vec{y}) d^3 y.$$

(4.7.11)

Then

$$\int_{\mathbb{R}^3} \left(T_{0i}(0, \vec{y}) y^j - T_{0j}(0, \vec{y}) y^i\right) d^3 y = -\epsilon_{ijk} J^k,$$

and we have established

$$\bar{h}_{0i} = -\frac{\kappa}{2r^3}\epsilon_{ijk}x^j J^k + O(r^{-3}).$$ (4.7.12)

The inclusion of this term in the calculation of the gyroscope precession leads to an effect called the *Lense-Thirring effect*, also called *frame-dragging*, compare ▶ Sect. 3.10.5, p. 114. The "frame-dragging" terminology is due to the fact that (4.7.12) gives further corrections to the motion of gyroscopes, which can be interpreted as the fact that *rotation of the sources leads to rotation of the inertial frames throughout space-time*. This general relativistic correction has been confirmed within an experimental error of order of 20% by the Gravity Probe B experiment discussed on p. 110 in ▶ Sect. 3.10.3.

Stars

Piotr T. Chruściel

© Springer Nature Switzerland AG 2019
P. T. Chruściel, *Elements of General Relativity*, Compact Textbooks in Mathematics,
https://doi.org/10.1007/978-3-030-28416-9_5

In this chapter we provide an introduction to general relativistic stellar models.

5.1 Perfect Fluids

An excellent matter model for many stars is gas. In the general relativity literature one usually uses the term "perfect fluid" as a synonym.

Consider a fluid, or a gas, in Newtonian mechanics. In the rest frame of the fluid the pressures are isotropic, hence the stress-energy tensor is

$$(T_{\mu\nu}) = \begin{pmatrix} \rho & 0 & 0 & 0 \\ 0 & p & 0 & 0 \\ 0 & 0 & p & 0 \\ 0 & 0 & 0 & p \end{pmatrix}, \tag{5.1.1}$$

where ρ is the rest-frame energy density, and p is the pressure. In this frame the four-velocity of the fluid is $u = \partial_t$, so that (5.1.1) can be written in a manifestly Lorentz covariant form

$$T_{\mu\nu} = (\rho + p)u_\mu u_\nu + p g_{\mu\nu}, \tag{5.1.2}$$

where $g_{\mu\nu}$ is the Minkowski metric $\eta_{\mu\nu}$. Invoking the correspondence principle, Eq. (5.1.2) provides then the relevant energy-momentum tensor for perfect fluids, or gases, for general metrics.

The reader unconvinced by the analogy can simply accept (5.1.2) as the definition of the energy-momentum tensor of a perfect fluid.

The Newtonian equations of fluid dynamics, in the absence of forces other than gravitational, consist of Euler's equation

$$\partial_t v^i + v^k \partial_k v^i = -\frac{1}{\rho} \partial_i p - \partial_i \phi, \tag{5.1.3}$$

where ϕ is the gravitational potential, together with the law of conservation of mass:

$$\partial_t \rho + \partial_i (\rho v^i) = 0 \, . \tag{5.1.4}$$

For *isentropic processes* we further have the conservation equation for the entropy per particle s:

$$\partial_t s + v^k \partial_k s = 0 \, . \tag{5.1.5}$$

Whether in special or in general relativity, the corresponding equations can be obtained from the conservation equation $\nabla_\mu T^\mu{}_\nu = 0$. Written out in detail, this equation reads

$$\nabla_\mu ((\rho + p)u^\mu)u_\nu + (\rho + p)u^\mu \nabla_\mu u_\nu + \nabla_\nu p = 0 \, . \tag{5.1.6}$$

Contracting with u^ν, and keeping in mind that

$$u^\nu \nabla_\mu u_\nu = \frac{1}{2} \nabla_\mu (\underbrace{u_\nu u^\nu}_{-1}) = 0 \, ,$$

we obtain

$$-\nabla_\mu ((\rho + p)u^\mu) + u^\nu \nabla_\nu p = 0 \, . \tag{5.1.7}$$

Equivalently,

$$u^\mu \nabla_\mu \rho + (\rho + p)\nabla_\mu u^\mu = 0 \, . \tag{5.1.8}$$

Inserting (5.1.7) into (5.1.6) we conclude that

$$\boxed{(\rho + p)u^\mu \nabla_\mu u_\nu + u^\mu \nabla_\mu p \, u_\nu + \nabla_\nu p = 0 \, .} \tag{5.1.9}$$

To obtain a closed evolution system we need to impose a relationship between ρ and p. An equation of the form

$$p = p(\rho) \, , \tag{5.1.10}$$

which will be called *equation of state*, will lead to a closed system of equations for, say, p and v^i, provided the equation of state can be inverted to give $\rho = \rho(p)$. In view of the implicit function theorem, this will be the case if one assumes

$$\partial_\rho p > 0 \, . \tag{5.1.11}$$

The system (5.1.8)–(5.1.9), together with (5.1.10)–(5.1.11), and *assuming that the metric is given*, provides a closed system of evolution equations for ρ and u^μ. One can check that this system has a well-posed Cauchy problem in regions where $\rho > 0$. A similar fact remains true with a dynamical metric when the Einstein equations with fluid matter sources are added to the above; a justification requires considerably more work.

Incidentally Actually, an equation of state in textbook thermodynamics is an equation of the form

$$\rho = \rho(n, s),$$

where ρ is the energy density, n is the density of particles, and s is the entropy per particle. One can then calculate p using (5.1.22) below:

$$n d\rho - (\rho + p)dn = n^2 T ds \quad \Longrightarrow \quad p = n\frac{\partial \rho}{\partial n}\Big|_s - \rho. \tag{5.1.12}$$

We will see shortly that s is constant along the flow of u^μ, so changing (5.1.10) to $p = p(\rho, s)$ will still be compatible with the current set-up, keeping in mind that $u^\mu \partial_\mu s = 0$.

Example 5.1.1

As an example of equation of state, consider a Newtonian ideal gas: we then have the classical relations

$$pV = nRT, \quad U = c_V nRT, \tag{5.1.13}$$

where p is the pressure of the gas, V is the volume, n is the amount of substance in moles, R is the gas constant, T is the temperature, and U is the internal energy. Finally, c_V is the heat capacity at constant V, equal to 3/2 for monoatomic gas, 5/2 for diatomic gas, and 3 for complex molecules.

From (5.1.13) we find

$$p = \frac{U}{c_V V} = \frac{\varepsilon}{c_V} \quad \Longleftrightarrow \quad \varepsilon = c_V p,$$

where we have denoted by $\varepsilon := U/V$ the density of internal energy. The total energy density ρ is the sum of the rest-mass energy ρ_0 and the internal energy density:

$$\rho = \rho_0 + \varepsilon = \rho_0 + c_V p.$$

We conclude that

$$p = \frac{\rho - \rho_0}{c_V}. \tag{5.1.14}$$

For monoatomic gas this gives

$$p = \frac{2}{3}(\rho - \rho_0).$$

■

It turns out that (5.1.8) can be rewritten as a *continuity equation*, as follows: Assuming that an equation of state $p = p(\rho)$ has been prescribed, define

$$n = n_0 \exp\left(\int \frac{d\rho}{\rho + p}\right), \tag{5.1.15}$$

with a function n_0 which is constant along the flow of the vector field u:

$$u^\mu \partial_\mu n_0 = 0. \tag{5.1.16}$$

Then we have

$$u^\mu \partial_\mu n = n \frac{u^\mu \partial_\mu \rho}{\rho + p}.$$

Hence, using the chain rule and (5.1.8),

$$\nabla_\mu(nu^\mu) = u^\mu \nabla_\mu n + n \nabla_\mu u^\mu = nu^\mu \frac{\nabla_\mu \rho}{\rho + p} + n \nabla_\mu u^\mu = 0. \tag{5.1.17}$$

It follows that the vector field nu^μ has vanishing divergence. The physical meaning of the associated conservation law is that of conservation of the number of particles; we will return to this shortly, compare (5.1.25) below.

While we will be interested in the above equations for general curved metrics, it is useful to first have a look at the special relativistic version of the equations. Consider thus (5.1.9) in Minkowski spacetime. To put back the c-factors in, recall that $x^0 = ct$, so that

$$u = \gamma\left(\partial_0 + \frac{v^i}{c}\partial_i\right) = \frac{\gamma}{c}\left(\partial_t + v^i \partial_i\right), \quad \gamma = \frac{1}{\sqrt{1 - \frac{|\vec{v}|^2}{c^2}}},$$

then the space-components of (5.1.9) read

$$(\rho c^2 + p)\frac{\gamma}{c}\left(\partial_t + v^i \partial_i\right)\left(\gamma \frac{v_k}{c}\right) + \frac{v_k}{c}\frac{\gamma}{c}\left(\partial_t + v^i \partial_i\right)p + \partial_k p = 0.$$

Assuming

$$\gamma \approx 1, \quad |p| \ll \rho c^2, \quad \frac{1}{c^2}\left(|\partial_t p| + |v^i \partial_i p|\right)|\vec{v}| \ll |Dp|,$$

where $|Dp|$ denotes the Euclidean norm of the space-gradient of p, the space-components of (5.1.9) become

$$\rho(\partial_t v^i + v^k \partial_k v^i) \approx -\partial_i p\,,$$

which approximates the Euler equation in the absence of external forces if ρ is approximated by the mass density.

To understand the Newtonian meaning of (5.1.8) it is necessary to recall some facts from thermodynamics; this will be done in the next section.

5.1.1 Some Newtonian Thermodynamics

Let us start with a warning: We will use elementary Newtonian thermodynamic to derive an *exact statement* concerning the relativistic Euler equations above. Extrapolating the validity of this Newtonian treatment to the relativistic regime might be questionable. The fact that one gets exact conservation of entropy using the Newtonian thermodynamical relation (5.1.18) below sounds too good to be true, so some healthy skepticism concerning the range of applicability of the treatment below is in order.

Incidentally A field theoretical relativistic treatment of the problem at hand can be found in [50, 51, 53] and references therein.

In any case, imagine a gas or fluid consisting of a very large number of interacting identical particles, each with rest mass m_0. Suppose that, in the region under consideration, and for some interval of time, there exists a (space-and-time independent) number N so that at every spacetime point $(t, \vec{x})$ the global properties of the nearest N particles are well described by Newtonian thermodynamics. In particular a local temperature $T(t, \vec{x})$ is well defined at every spacetime point. We let $V = V(t, \vec{x})$ be the volume centered at $\vec{x}$ at time t and occupied by the N nearest particles.

Note that thermodynamics does not make sense for a small number of particles, while if N—and therefore the associated volume V—is chosen too large there will be no well-defined temperature $T(t, \vec{x})$. Thus whether or not the above applies will very much depend upon the physical situation at hand. When no number N as above exists, the considerations below break down.

Assuming the setting just described, we thus consider N particles occupying a volume V, and undergoing a process during which the number of particles does not change:

$$dN = 0\,.$$

The fundamental relation of Newtonian thermodynamics reads

$$dU = -p\,dV + T\,dS\,, \tag{5.1.18}$$

where U is the internal energy of the system, and S the entropy.

Let $s = S/N$ be the entropy per particle, and let $n = N/V$ be the density of particles, hence

$$V = \frac{N}{n}, \quad S = Ns.$$

We also define $\varepsilon = U/V$ to be the internal energy density, thus

$$U = \varepsilon V = N\frac{\varepsilon}{n}.$$

Using these variables, (5.1.18) reads

$$d\left(\frac{N\varepsilon}{n}\right) = -pd\left(\frac{N}{n}\right) + Td(Ns). \tag{5.1.19}$$

Equivalently, since $dN = 0$,

$$d\left(\frac{\varepsilon}{n}\right) = -pd\left(\frac{1}{n}\right) + Tds. \tag{5.1.20}$$

The total energy density ρ is the sum of the rest-mass energy density $\rho_0 = Nm_0c^2/V$, where the rest mass of the particles m_0 is also assumed to be unchanged during the process, and of the internal energy density:

$$\rho = \rho_0 + \varepsilon = m_0c^2n + \varepsilon.$$

Then

$$d\left(\frac{\rho}{n}\right) = d\left(\frac{\varepsilon}{n} + m_0c^2\right) = d\left(\frac{\varepsilon}{n}\right) = -pd\left(\frac{1}{n}\right) + Tds; \tag{5.1.21}$$

equivalently

$$nd\rho - (\rho + p)dn = n^2Tds. \tag{5.1.22}$$

Let us assume now that the above applies *as is* to special relativity and thus, invoking the correspondence principle, to general relativity. Let $u = u^\mu\partial_\mu$ be the four-velocity field of the gas. The number N of particles will be constant along the flow, as desired, if the current nu^μ is conserved

$$\nabla_\mu(nu^\mu) = 0 \quad \Longleftrightarrow \quad \nabla_\mu u^\mu = -\frac{1}{n}u^\mu\partial_\mu n. \tag{5.1.23}$$

We have:

Proposition 5.1.2
Any two of the following properties imply the third:
1. *The matter current nu^μ is conserved, $\nabla_\mu(nu^\mu) = 0$.*
2. *The entropy per particle is constant along the flow, $u^\mu \partial_\mu s = 0$.*
3. *We have $u^\mu \nabla_\mu \rho + (\rho + p)\nabla_\mu u^\mu = 0$ (compare (5.1.8)).*

Proof
The identity (5.1.22) gives

$$nT \underbrace{u^\mu \partial_\mu s}_{} = u^\mu \partial_\mu \rho - \frac{(p+\rho)}{n} \underbrace{u^\mu \partial_\mu n}_{\nabla_\mu(nu^\mu) - n\nabla_\mu u^\mu}$$

$$= \underbrace{u^\mu \nabla_\mu \rho + (\rho + p)\nabla_\mu u^\mu}_{} - \frac{(p+\rho)}{n} \underbrace{\nabla_\mu(nu^\mu)}_{},$$

and the result readily follows. □

The usual interpretation of Proposition 5.1.2 is that the energy-momentum tensor

$$T_{\mu\nu} = (p + \rho)u_\mu u_\nu + pg_{\mu\nu}$$

is suitable for describing *isentropic thermodynamical processes*. Then (5.1.8) is the law of conservation of the number of particles.

Alternatively, when combined with Newtonian thermodynamics, (5.1.8) and the *supplementary hypothesis* of conservation of the number of particles imply that the flow is *isentropic*.

Incidentally To make contact with (5.1.15), consider a process where the entropy-per-particle s is constant during evolution, $ds \equiv 0$, Eq. (5.1.22) gives

$$nd\rho - (\rho + p)dn = 0 \quad \Longleftrightarrow \quad \frac{dn}{n} = \frac{d\rho}{\rho + p}. \tag{5.1.24}$$

Integrating, one obtains

$$n = n_0 \exp\left(\int \frac{d\rho}{\rho + p}\right). \tag{5.1.25}$$

Here the constant of integration n_0 might depend upon its initial value (e.g., $n_0 = n_0(\vec{x})$ in a coordinate system in which $u^\mu = u^0 \partial_t$). This reproduces (5.1.15).

5.2   Spherically Symmetric Static Stars

There exist various types of stars—e.g., main sequence stars, red giants, white dwarfs, neutron stars—but all of them seem to be well described by fluid models; the difference lying in the equation of state used. In these lectures we will be interested in time-independent stars in general relativity. This leads us to study static perfect fluid solutions of Einstein equations.

A fundamental theorem, first proved by Beig and Simon [15] under more restrictive assumptions, and then in considerable generality by Masood-ul-Alam [60], asserts that

asymptotically flat, static perfect fluid solutions of the Einstein equations of finite extent with $\rho \geq 0$ and $\partial_\rho p \geq 0$ are spherically symmetric, with the metric taking the form

$$g = -e^\nu dt^2 + e^\lambda dr^2 + r^2 d\Omega^2 \,, \tag{5.2.1}$$

for some smooth functions $\nu = \nu(r)$ and $\lambda = \lambda(r)$, with

$$\lambda(0) = 0, \, \lambda(r) \to_{r \to \infty} 0, \, \nu(r) \to_{r \to \infty} 0, \text{ and } u = u^0(r)\partial_t.$$

Incidentally Using elementary facts from the theory of group actions, the above form of the metric is easily derived once stationarity and spherical symmetry are known, provided that the area of the orbits of the rotation groups has no critical points. It can be shown that critical points of the area of $SO(3)$-orbits correspond to event horizons, which is not possible if the Killing vector ∂_t is strictly timelike. So, in stationary spacetimes without event horizons one is necessarily led to (5.2.1), once spherical symmetry has been established.

Alternatively, the reader can view (5.2.1) as an ansatz for the metric, which is assumed to satisfy the Einstein equations with perfect fluid source with energy density ρ, pressure p, and four-velocity $u = u^t \partial_t$.

A calculation leads to the following form of the Einstein equations $G^\mu{}_\nu = 8\pi T^\mu{}_\nu$ (recall that $G_{\mu\nu} = R_{\mu\nu} - \frac{1}{2} R g_{\mu\nu}$):

$$G^0{}_0 = e^{-\lambda} \left(\frac{1}{r^2} - \frac{\lambda'}{r} \right) - \frac{1}{r^2} = -8\pi\rho \,, \tag{5.2.2}$$

$$G^1{}_1 = e^{-\lambda} \left(\frac{1}{r^2} + \frac{\nu'}{r} \right) - \frac{1}{r^2} = 8\pi p \,. \tag{5.2.3}$$

The remaining components of the Einstein tensor will not be needed in what follows, and we simply note that $G^0{}_1 \equiv 0$ holds for metrics under current consideration.

If $\rho = p = 0$, we see that $\nu = -\lambda$ is compatible with the above. This is a reassuring consistency check, as this last relation is satisfied in the Schwarzschild metric.

The aim of what follows is to see how to construct solutions of these equations. We will see that the solutions are governed by an equation first derived by Tolman, Oppenheimer, and Volkov, namely (5.2.8) below.

5.2.1 The Tolman–Oppenheimer–Volkov Equation

Equation (5.2.2) can be integrated as follows: we calculate

$$(re^{-\lambda})' = e^{-\lambda}(1 - r\lambda') = r^2 e^{-\lambda}(\frac{1}{r^2} - \frac{1}{r}\lambda') = 1 - 8\pi\rho r^2,$$

leading to

$$e^{-\lambda} = 1 - \frac{2m(r)}{r}, \quad \text{where } m(r) = m_0 + 4\pi \int_0^r \rho r^2 dr. \tag{5.2.4}$$

In particular, when $\rho \equiv 0$ we recover the Schwarzschild metric.

Assuming that we have a regular center, we have $m(0) = m_0 = 0$. The metric is Schwarzschild at large distances if it is vacuum there, and then has total mass

$$m = 4\pi \int_0^\infty \rho r^2 dr.$$

Incidentally Note that we have proved a very special case of the *positive energy theorem*: under the current hypotheses $m \geq 0$, with $m = 0$ if and only if the spacetime is Minkowski.

The difference of (5.2.3) and (5.2.2) gives

$$(\lambda' + \nu')\frac{e^{-\lambda}}{r} = 8\pi(p + \rho),$$

leading to

$$\nu = -\lambda - 8\pi \int_r^\infty re^\lambda(p + \rho) \, dr, \tag{5.2.5}$$

where we have assumed that the integrals over $\mathbb{R}$ that appear in the definitions of ν and λ converge, and that the coordinate t approaches a Minkowskian time at large distances, $\lim_{r\to\infty} e^\nu = 1$.

One more equation will be provided by the general relativistic Euler equation (5.1.9), p. 168, with the subscript ν there equal to one:

$$(\rho + p)u^\mu \nabla_\mu u_1 + u^\mu \nabla_\mu p \, u_1 + \nabla_1 p = 0. \tag{5.2.6}$$

From $u \sim \partial_t$ and the normalization condition $g(u, u) = -1$ we find $u = e^{-\nu/2}\partial_t$, and also $u^\mu \nabla_\mu u_1 = u^0 \nabla_0 u_1$. Now

$$
\begin{aligned}
\nabla_0 u_1 &= \partial_0 u_1 - \Gamma^\mu{}_{01} u_\mu = -\Gamma^0{}_{01} u_0 \\
&= -\frac{1}{2} g^{00} (\partial_0 g_{01} + \partial_1 g_{00} - \partial_0 g_{01}) u_0 = -\frac{u_0}{2} g^{00} \partial_1 g_{00} \\
&= -\frac{u_0}{2} \partial_1 \nu .
\end{aligned}
$$

Since $u^0 u_0 = -1$, (5.2.6) reads

$$
\frac{(\rho + p)}{2} \nu' + p' = 0 . \tag{5.2.7}
$$

Using (5.2.3) we find

$$
\begin{aligned}
p' &= -\frac{(\rho + p)}{2} \nu' = -\frac{(\rho + p)}{2r} \left(e^\lambda \left(8\pi p r^2 + 1 \right) - 1 \right) \\
&= -\frac{(\rho + p) e^\lambda}{2r} \left(8\pi p r^2 + 1 - e^{-\lambda} \right) \\
&= -\frac{(\rho + p) e^\lambda}{r^2} \left(4\pi p r^3 + m(r) \right) \\
&= -\frac{(\rho + p)}{e^{-\lambda} r^2} \left(4\pi p r^3 + m(r) \right) .
\end{aligned}
$$

We have obtained the *Tolman–Oppenheimer–Volkov (TOV) equation,*

$$
\boxed{ p' = -\frac{(\rho + p)(4\pi p r^3 + m(r))}{r(r - 2m(r))} . } \tag{5.2.8}
$$

One obtains a closed system of equations by adding an equation of state $p = p(\rho)$, together with the defining equation

$$
m' = 4\pi \rho r^2 , \tag{5.2.9}
$$

We note the obvious consequence of (5.2.8) that

there are no static dust solutions. $\tag{5.2.10}$

Indeed, for dust we have $p \equiv 0$, which inserted in (5.2.8) gives $\rho m \equiv 0$. This contradicts $\rho \geq 0$ unless there is no matter at all.

Of course, (5.2.10) should be clear by physical considerations: without pressure there is nothing to counteract the gravitational attraction between the dust particles, and so no time-independent configurations are possible.

Stellar models can now be constructed as follows: Given an equation of state and a central value of density $\rho(0) = \rho_0$, one can integrate the coupled system of equations (5.2.8)–(5.2.9). The metric function e^λ is then obtained from $e^{-\lambda} = 1 - 2m(r)/r$, *as long as* $2m(r) < r$. The metric function ν can then be obtained from (5.2.5). Alternatively we can integrate (5.2.7), taking into account the *asymptotic flatness condition* $\lim_{r \to \infty} e^\nu = 1$:

$$\nu = \int_r^\infty \frac{2p'}{(\rho + p)} dr = -2 \int_r^\infty \frac{4\pi p r^3 + m(r)}{r(r - 2m(r))} dr .\tag{5.2.11}$$

5.2.2 The Lane–Emden Equation

It is of interest to compare the above with the Newtonian case. Recall that the (Newtonian) Euler equation in the presence of a gravitational potential ϕ reads

$$\partial_t v^i + v^k \partial_k v^i = -\frac{1}{\rho} \partial_i p - \partial_i \phi .\tag{5.2.12}$$

For a static fluid we have $v^i \equiv 0$, and assuming spherical symmetry (5.2.12) reduces to

$$p'(r) = -\rho(r)\phi'(r) .\tag{5.2.13}$$

Incidentally In spherical symmetry this can be derived directly as follows: consider a spherically symmetric shell of height dr and surface S at radius r, hence with volume $S dr$ and mass $\rho S dr$. There is a pressure force $p(r)S$ sustaining the lower face of the shell and $p(r + dr)S$ acting on the upper face, resulting in a net outwards pointing force $-p'S dr$. This has to counterbalance the gravitational force $-\rho S dr \phi'$, leading to the equilibrium equation (5.2.13).

In a spherically symmetric configuration it holds that

$$\phi' = Gm(r)/r^2 ,\tag{5.2.14}$$

where $m(r)$ is defined by (5.2.9), leading to

$$p'(r) = \rho(r)\phi'(r) = -\frac{G\rho(r)m(r)}{r^2} .\tag{5.2.15}$$

In order to derive (5.2.14), recall the equation satisfied by the Newtonian potential ϕ:

$$\Delta\phi = 4\pi G\rho .\tag{5.2.16}$$

For spherically symmetric ρ we can integrate the above over a ball $B(r)$ of radius r to obtain

$$\int_{B(r)} \Delta\phi \, d^3x = 4\pi G \int_{B(r)} \rho \, d^3x =: 4\pi Gm(r) .\tag{5.2.17}$$

The left-hand side can be written as a boundary integral over the sphere $S(r)$,

$$\int_{B(r)} \Delta\phi \, d^3x = \int_{S(r)} \phi'(r) \, d^2S = 4\pi r^2 \phi'(r), \tag{5.2.18}$$

where in the last step we have assumed spherical symmetry of ϕ. Equation (5.2.14) readily follows.

We see that (5.2.8) reduces to (5.2.13) with $G = 1$ when p is considered negligible compared to ρ, and when $2m(r) \ll r$.

From (5.2.13) and (5.2.9) we obtain a second order equation for p:

$$\frac{1}{r^2} \frac{d}{dr} \left(-\frac{r^2}{\rho} \frac{dp}{dr} \right) = 4\pi G\rho. \tag{5.2.19}$$

An astrophysics favorite is the *polytropic equation of state*:

$$p = K\rho^{\frac{n+1}{n}} \quad \Longleftrightarrow \quad \rho = (p/K)^{\frac{n}{n+1}}, \tag{5.2.20}$$

for some constant K. The constant n is called the *polytropic index*, or polytropic exponent. In this case (5.2.19) becomes the *Lane–Emden equation*:

$$\boxed{\frac{1}{r^2} \frac{d}{dr} \left(-\frac{r^2}{p^{\frac{n}{n+1}}} \frac{dp}{dr} \right) = \frac{4\pi G}{K^{\frac{2n}{n+1}}} p^{\frac{n}{n+1}}.} \tag{5.2.21}$$

5.2.3 The Buchdahl–Heinzle Inequality

Let us return now to the relativistic equations. A significant consequence of the TOV equation is contained in the following result:

> **Theorem 5.2.1** (Heinzle [22, 43])
> *Consider a spherically symmetric static perfect fluid star with $p \geq 0$, $\rho \geq 0$, and $dp/d\rho \geq 0$. Then*
>
> $$\frac{2M}{R} \leq \frac{6}{7}. \tag{5.2.22}$$

The inequality (5.2.22) with $\frac{6}{7}$ replaced by $\frac{8}{9}$ has been originally proved by Buchdahl [22], compare [52]. Since $\frac{6}{7} \approx 0.857 < .889 \approx \frac{8}{9}$, Heinzle's inequality (5.2.22) is a slight improvement. Both inequalities are somewhat stronger that the requirement that there is no black hole,

$$\frac{2M}{R} < 1. \tag{5.2.23}$$

We will shortly give a proof of Buchdahl's inequality. (Strictly speaking, the proof that we are about to provide will be complete only in the special case where ρ is constant.) Before doing this, in order to understand the relevance of inequalities such as (5.2.22) let us consider a general inequality of the form

$$\frac{2M}{R} \leq \alpha, \tag{5.2.24}$$

for some $\alpha > 0$. Suppose that in the region $0 \leq r \leq R$ the density of the star is bounded from below,

$$\rho \geq \rho_0,$$

for some $\rho_0 > 0$. Then the mass of the star satisfies

$$M \geq \frac{4}{3} \rho_0 \pi R^3,$$

and (5.2.24) gives

$$\frac{8}{3} \rho_n \pi R^2 \leq \alpha \quad \Longleftrightarrow \quad R < R_0 := \sqrt{\frac{3\alpha}{8\pi\rho_0}}.$$

Inserting this again into (5.2.24) one finds

$$M \leq \frac{\alpha R}{2} \leq \alpha^{\frac{3}{2}} \sqrt{\frac{3}{32\pi\rho_0}}. \tag{5.2.25}$$

To get an idea of the masses involved, let ρ_n be the density of an atomic nucleus,

$$\rho_n \approx 2.3 \times 10^{11} \, \text{kg/cm}^3.$$

(This is a considerable density: one millimeter cube of a neutron star will weigh 2.3×10^5 tons.) Using the mass $M_\odot$ of the Sun as a reference, we can rewrite (5.2.25) as

$$M \leq \alpha^{\frac{3}{2}} \sqrt{\frac{\rho_n}{\rho_0}} \sqrt{\frac{3}{32\pi\rho_n}} \approx \alpha^{\frac{3}{2}} \sqrt{\frac{\rho_n}{\rho_0}} 9M_\odot. \tag{5.2.26}$$

For a neutron star one expects $\rho_0 \approx \rho_n$ throughout the star, and this appears to be the physically correct model for many pulsars. Using Heinzle's bound $\alpha = 6/7$ and $\rho_0 = \rho_n$ in (5.2.26) gives

$$M \lesssim 7.1 M_\odot.$$

We conclude that static neutron stars of constant density with a mass larger than that of about seven solar masses cannot sustain themselves against gravitational collapse.

A sharper bound for the mass of neutron stars—the Chandrasekhar mass—will be derived in ▶ Sect. 5.3, based on completely different considerations.

Let us now pass to an argument which leads to the Buchdahl inequality. For this we consider a star of finite radius described by an equation of state $p = p(\rho)$ with

$$\frac{dp}{d\rho} > 0 . \tag{5.2.27}$$

It follows that the relation $p = p(\rho)$ can be inverted to an equation $\rho = \rho(p)$. Now, a solution of the TOV equation (5.2.8) with $\rho \geq 0$ and $r > m(r)$ satisfies $\frac{dp}{dr} \leq 0$, which implies

$$\frac{d\rho}{dr} = \underbrace{\frac{d\rho}{dp}}_{>0} \underbrace{\frac{dp}{dr}}_{\leq 0} \leq 0 . \tag{5.2.28}$$

Let ρ_0 denote the pressure at the center of the star. Letting R be the coordinate radius of the star, (5.2.28) shows that the total mass m satisfies

$$m = 4\pi \int_0^R \rho(r) r^2 dr \leq 4\pi \int_0^R \rho_0 r^2 dr . \tag{5.2.29}$$

Thus, the mass is maximal for a model where the density of the star is constant, $\rho(r) \equiv \rho(0)$ for $0 \leq r \leq R$.

Remark 5.2.2 Strictly speaking, a constant ρ does not fit into our model-building prescription, where p is a function of ρ: Then p would have to be constant, which is not compatible with the TOV equation. This is, however, irrelevant to the question of deriving an upper bound for the total mass.

We thus consider the TOV equation with a constant ρ. From (5.2.29) we obtain

$$m(r) = \frac{4\pi \rho r^3}{3} ,$$

and (5.2.4) gives

$$e^{-\lambda(r)} = 1 - \frac{8\pi \rho r^2}{3} .$$

Exercise 5.2.3 Show that the space-metric so obtained within the star coincides with the metric on a three-dimensional sphere, with a radius that you should determine. [*Hint: write the sphere* $S^3 \subset \mathbb{R}^4$ *as* $w = \sqrt{R^2 - x^2 - y^2 - z^2} = \sqrt{R^2 - r^2}$, *and calculate the metric induced on* S^3 *by the Euclidean metric* $dx^2 + dy^2 + dz^2 + dw^2 = dr^2 + r^2 d\Omega^2 + dw^2$.]

The TOV equation becomes now

$$p' = -\frac{(\rho + p)(4\pi pr^3 + \frac{4\pi \rho r^3}{3})}{r(r - \frac{8\pi \rho r^3}{3})}$$

$$= -\frac{4\pi(\rho + p)(p + \frac{\rho}{3})r}{1 - \frac{8\pi \rho r^2}{3}}.$$

To solve the last equation, it is convenient to first replace r by a new variable

$$x := \underbrace{\sqrt{\frac{8\pi \rho}{3}}}_{=:a} r .$$

Then

$$\frac{dp}{dx} = \frac{dr}{dx}\frac{dp}{dr} = \frac{1}{a}\frac{x}{ra}\frac{dp}{dr}$$

$$= -\frac{(\rho + p)(3p + \rho)x}{2\rho(1 - x^2)} . \tag{5.2.30}$$

Equivalently

$$\frac{2\rho dp}{(\rho + p)(3p + \rho)} = -\frac{xdx}{(1 - x^2)} = d\ln\sqrt{1 - x^2} \tag{5.2.31}$$

for $0 \le x < 1$.

The boundary of the star is defined by the condition $p(R) = 0$, where R is the coordinate radius of the star. Let us denote by $X = x(R)$ the corresponding value of x. We integrate (5.2.31) between x and X, $0 \le x \le X$, keeping in mind that the convergence of the integral in x requires $X \le 1$. Using

$$\frac{2\rho}{(\rho + p)(3p + \rho)} = \frac{3}{(3p + \rho)} - \frac{1}{(\rho + p)}$$

we find

$$\ln\left(\frac{\rho + p}{\rho + 3p}\right)(x) = \ln\left(\frac{\sqrt{1 - X^2}}{\sqrt{1 - x^2}}\right) =: \ln f(x) . \tag{5.2.32}$$

We note that f is increasing, with $0 < f(0) \le 1$ and $f = 1$ at the boundary of the star, hence $0 < f \le 1$. Equation (5.2.32) can be rewritten as

$$\frac{\rho + p}{\rho + 3p} = f \quad \Longleftrightarrow \quad \rho(1 - f) = p(3f - 1) . \tag{5.2.33}$$

◻ Fig. 5.1 p/ρ as a function of $x = r/\sqrt{\frac{8\pi\rho}{3}}$, for various values of the radius of the star. The first zero of p determines the surface of the star

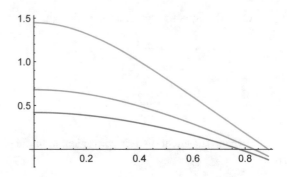

Keeping in mind that $0 < f \le 1$, $\rho > 0$, and $p \ge 0$, this can be solved for p if and only if

$$f > \frac{1}{3}.$$ (5.2.34)

One then obtains

$$\frac{p}{\rho} = \frac{1 - f}{3f - 1} = \frac{\sqrt{1 - x^2} - \sqrt{1 - X^2}}{3\sqrt{1 - X^2} - \sqrt{1 - x^2}}.$$

This determines the function $p(r)$ as a function of the radius of the star, see ◻ Fig. 5.1. The function $r \mapsto p(r)$ is decreasing, with the central pressure $p_0 := p(0)$ given by

$$p_0 = \frac{1 - \sqrt{1 - X^2}}{3\sqrt{1 - X^2} - 1}\rho.$$ (5.2.35)

We have $p(0) \to \infty$ as $X^2 \to 8/9$ from below.

As a byproduct of the analysis above, we have obtained an explicit solution for the pressure profile when ρ is constant. It should, however, be kept in mind that this is not, strictly speaking, a valid stellar model since there is no equation $p = p(\rho)$.

Note that (5.2.35) can be inverted to express X, equivalently the radius of the star, as a function of p_0.

Since the function $[0, 1] \ni x \mapsto f(x)$ is increasing, the condition (5.2.34) will hold for $x \in [0, X]$ if and only if it holds at $x = 0$. This leads to the condition $3\sqrt{1 - X^2} - 1 > 0$, which can be rewritten as

$$\frac{1}{9} < 1 - x(R)^2 = 1 - \frac{8\pi\rho}{3}R^2 = 1 - \frac{2M}{R},$$

where M is the total mass of the star. In other words,

$$\boxed{\frac{2M}{R} \le \frac{8}{9},}$$ (5.2.36)

as desired.

5.3   End State of Stellar Evolution: The Chandrasekhar Mass

The above considerations were rather general. One expects that more restrictive bounds can be obtained when a specific star model is considered. In particular one would like to answer the question, how big can be a star at the end of its evolution. For this, a few words about evolution of stars are in order.

Recall that stars are formed by collapse of interstellar clouds, which are higher density regions of interstellar gas and dust. The nearest such molecular cloud, in which the hydrogen is mostly in the H_2 form, is the Orion Nebula, see ▢ Fig. 5.2.

The star formation mechanism can be initiated by a collision of molecular clouds, or a shock wave from a Supernova explosion. Since gravity is attractive, the cloud starts to contract, with density and temperature rising, forming a protostar. When a protostar has become sufficiently dense, deuterium fusion begins, slowing the collapse. Depending upon the masses of both the star and its core after each of the evolutionary stages, the star will then go through one or several further phases, in which either (a) the end of the fuel for the current nuclear reaction leads to a final steady state, or the contraction continues until (b) a new nuclear reaction starts, possibly producing a catastrophic change in the structure of the star, or until (c) a black hole is left behind. The catastrophic change in step (b) might be a Supernova (see ▢ Fig. 5.3), or a milder shedding of outer layers of the star. The question then arises, what are the possible end states in (a).

If the initial mass of the star is less than one-half of the solar mass, the star will never become hot enough to start burning Helium, and will either slowly evolve to a white dwarf, or convection will turn it into a red giant.

(An example of white dwarf is provided by Sirius B, the binary companion of Sirius A, the latter being the brightest star in the sky. Sirius B is smaller than Earth, with a radius of about 4000 km, and has a mass similar to that of the Sun: a rather dense object. See ▢ Fig. 5.4.)

For heavier masses several intermediate scenarios are possible, leading to white dwarfs, neutron stars, or black holes. It is believed that the last resort mechanism that prevents further collapse is the electron degeneracy pressure for white dwarfs, and the neutron degeneracy pressure for neutron stars.

▢ **Fig. 5.2** Star nurseries: from left to right: the Eagle, Horseshoe, the Orion Nebulae, all from www. spacetelescope.org, credit ESA/Hubble, and the Rosetta Nebula, from spaceimages.esa.int, credit ESA/NASA, reproduced with permission courtesy NASA/JPL-Caltech

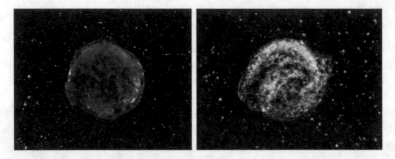

■ Fig. 5.3   Chandra satellite images of remnants of the Tycho Supernova (left) and the Kepler Supernova (right), images © NASA/CXC/SAO from http://chandra.harvard.edu, reproduced with permission

■ Fig. 5.4   Sirius A and B in optical (left, from hubblesite.org, credit NASA/ESA) and in X-ray (right, from http://chandra.harvard.edu, credit NASA/CXC/SAO). Sirius B is the small dot in the lower left quarter of the left image, and the central bright object in the right image. Reproduced with permission

White dwarfs are believed to originate from light stars with masses $M \leq 4M_\cdot$, after a mass ejection which creates a planetary nebula. In 1930 Chandrasekhar calculated the maximum mass of a stable white dwarf star. As already pointed out, it is widely accepted (cf., e.g., [79]) that white dwarfs resist gravitational collapse primarily through electron degeneracy pressure, arising from Pauli's exclusion principle which forbids occurrence of electrons in identical states, rather than thermal pressure. The Chandrasekhar limit is the mass above which the electron degeneracy pressure in the core is insufficient to balance the star's own gravitational self-attraction. Consequently, white dwarfs with masses greater than the limit undergo further gravitational collapse, evolving into a different type of stellar remnant, such as a neutron star or black hole. Those with masses under the limit remain stable as white dwarfs.

The very simple Newtonian argument which will be presented shortly reproduces Chandrasekhar's upper limit of 1.4 solar masses for a neutron star which supports itself against collapse by degeneracy pressure. Alternatively, and more fundamentally, one can proceed as follows: As derived in [54], the equation of state for an ultrarelativistic degenerate electron, or neutron, gas is

$$p = \left(\frac{3}{2\pi}\right)^{\frac{1}{3}} hc \left(\frac{\rho}{4\mu_e m_p}\right)^{\frac{4}{3}},$$ (5.3.1)

where m_p is the mass of the proton, with $\mu_e = 1$ for a hydrogen plasma and $\mu_e = 2$ for a Helium plasma.

In Newtonian's gravity one can solve numerically the corresponding Lane–Emden equation with $\mu_e = 2$, obtaining

$$M = M_C := 1.44 M_\odot.$$ (5.3.2)

In Einstein's theory one can instead numerically solve the TOV equation with the equation of state (5.3.1) obtaining a limit mass between 2 and 3 solar masses.

Hence, up to a multiplicative factor of two, the general relativistic effects do not lead to essentially new phenomena in the problem at hand.

We pass now to the promised simple derivation of Chandrasekhar's mass. As our model we consider a spherically symmetric configuration of self-gravitating gas of N neutrons in a ball of radius R with constant density. We assume that the particles have relativistic velocities, hence rest and kinetic energy equal to

$$E = Nc\sqrt{m^2 c^2 + p^2} \approx Ncp.$$ (5.3.3)

Here p is the length of the space-part of the four-momentum vector. Note that the mass m appearing here is irrelevant for the remaining calculations, in view of the approximation above. For a neutron star one would use $m = m_n$, the mass of a nucleon. For a white dwarf the pressure, equivalently the kinetic energy, would come from electrons, and thus m in (5.3.3) would be taken as the mass of the electron. In any case the value of m in (5.3.3) is clearly irrelevant because of the approximation made.

As an exercise, the reader will check that the gravitational potential energy of the configuration is

$$U = -\frac{3GN^2 m^2}{5R}.$$ (5.3.4)

Here m is the mass of a nucleon, regardless of whether a white dwarf or a neutron star is considered. The total energy E_T of the star will therefore be

$$E_T = E + U = Ncp - \frac{3GN^2 m^2}{5R}.$$ (5.3.5)

Proof of (5.3.4)

The assumption that the mass density n is constant gives

$$n = Nm/(\frac{4}{3}\pi R^3).$$

Next, a spherical shell at radius r of thickness dr and mass density n has potential energy

$$dU = 4\pi n r^2 dr \times \psi(r), \tag{5.3.6}$$

where $\psi(r) = -Gm(r)/r$ is the work needed to bring the shell from infinity to its current location through empty space. Thus

$$dU = 4\pi n r^2 \psi(r) dr = -4\pi n r^2 \times \frac{4G\pi n r^3}{3r} \times dr = -\frac{1}{3}(4\pi n)^2 G r^4 dr,$$

which integrates to

$$U = -\frac{3}{5}\left(\frac{4\pi n}{3}\right)^2 G R^5 = -\frac{3}{5}\frac{GM^2}{R},$$

giving (5.3.4).   □

Incidentally Note that ψ in (5.3.6) is *not* the Newtonian potential ϕ of the configuration. While this is not necessary for the remainder of the calculations here, to make the distinction clear we calculate the last potential: Integrating once the spherically symmetric Laplace equation gives $\phi' = Gm(r)/r^2$ where, as before, $m(r)$ is the mass contained within the radius r. Hence, within the star,

$$\phi' = \frac{Gm(r)}{r^2} = \frac{G}{r^2} \times \frac{4}{3}\pi n r^3 = \frac{4G n \pi}{3} r. \tag{5.3.7}$$

We also know that $\phi = -GM/r$ for $r \geq R$, where R is the radius of the start and M is its total mass. Integrating (5.3.7) from $r \leq R$ to R, for $r \leq R$ one finds

$$\phi = \frac{2G n \pi}{3}(r^2 - R^2) - \frac{GM}{R},$$

which is *not* $-Gm(r)/r$.

Let

$$d = (V/N)^{1/3} \tag{5.3.8}$$

be the average distance between the fermions. At the heart of the argument that follows lie (1) the Heisenberg uncertainty relation which states that

$$\Delta p \, \Delta d \geq \hbar, \tag{5.3.9}$$

and (2) the principle that *stable stationary systems minimize energy*. Since p cannot be less than Δp, and since the total energy E given by (5.3.5) has a contribution Ncp from the fermions, the total energy will be minimized in a configuration where p is replaced by Δp. Hence

$$p\,\Delta d \geq \hbar \quad \Longleftrightarrow \quad \Delta d \geq \frac{\hbar}{p}. \tag{5.3.10}$$

Since $d \geq \Delta d$, the densest configuration will be obtained if $d = \Delta d$ and if equality in (5.3.10) holds:

$$pd = \hbar.$$

The last equation, together with (5.3.8) for a spherical configuration where $V = 4\pi R^3/3$, gives

$$p = \frac{\hbar}{d} = \hbar(\frac{3N}{4\pi R^3})^{1/3}.$$

This leads to

$$E + U = Ncp - \frac{3GN^2m^2}{5R} = \left(c\hbar \frac{3^{1/3}N^{1/3}}{4^{1/3}\pi^{1/3}} - \frac{3GNm^2}{5}\right)\frac{N}{R}. \tag{5.3.11}$$

If the coefficient in front of R^{-1} above was negative, the system would achieve smaller energy by reducing R. Equivalently, the derivative with respect to R of the total energy should be negative or zero. So, stability leads to the requirement that

$$c\hbar \frac{3^{1/3}N^{1/3}}{4^{1/3}\pi^{1/3}} - \frac{3GNm^2}{5} \geq 0 \quad \Longleftrightarrow \quad N^{2/3} \leq \frac{3^{1/3}5c\hbar}{4^{1/3}\pi^{1/3}3Gm^2}.$$

When m is the mass of a nucleon, the corresponding limiting mass

$$M_C = Nm = \left(\frac{5c\hbar}{4^{1/3}\pi^{1/3}3^{2/3}Gm^2}\right)^{\frac{3}{2}}m = \frac{1}{6\sqrt{\pi}m^2}\left(\frac{5c\hbar}{G}\right)^{\frac{3}{2}} \approx 1.4M_\odot$$

is the announced *Chandrasekhar mass*.

One concludes that neutron stars, supported by a pressure arising from the Fermi exclusion principle, and with mass larger than Chandrasekhar's, cannot sustain themselves against collapse.

Note that a positive coefficient in front of (5.3.11) does not exclude an instability causing an expansion of the star. However, it does prevent collapse, which was the main point of the discussion above.

Cosmology

Piotr T. Chruściel

© Springer Nature Switzerland AG 2019
P. T. Chruściel, *Elements of General Relativity*, Compact Textbooks in Mathematics,
https://doi.org/10.1007/978-3-030-28416-9_6

In this chapter we will introduce the reader to the basic general relativistic cosmological models.

The *Cosmic Microwave Background (CMB) radiation* is a steady flux of radiation reaching us from all directions, with almost the same spectrum in all directions. It was first observed by McKellar in the early 1940s. A theoretical prediction was put forward around 1948 by Gamow, Alpher, and Herman, who were unaware of McKellar's observation. The CMB was accidentally rediscovered in 1971 by Penzias and Wilson, and rewarded by a Nobel Prize for Physics in 1978. Subsequent observations by a series of dedicated satellites, the COsmic Background Explorer (COBE), the Wilkinson Microwave Anisotropy Probe (WMAP), and the Planck Satellite mission, determined the anisotropies in the temperature of the CMB spectrum to be of the order of a few parts in 10,000, see ◘ Fig. 6.1 and 6.2. These anisotropies can be explained by the inflation scenario, the model predictions fitting the observations with impressive accuracy, see ◘ Fig. 6.3.

The CMB data provide a firm observational basis for nowadays cosmology, as summarised in ◘ Fig. 6.4. They suggest very strongly a large scale structure of the universe which is, essentially, the same in all directions of observation.

Assuming that we do not occupy a privileged position in the universe, one is led to assume that the universe is filled by a family of observers who, like us, see the same large scale structure of the universe independently of their direction of observation.

In order to determine the geometries which possess the above property, it is convenient to start with some mathematical preliminaries.

6.1 The Lie Derivative

The Lie derivative is a differentiation operation on tensors, in the direction of a vector field, which does not need any exterior structures such as a connection. We start with an axiomatic approach to the definition of Lie derivatives in ▶ Sect. 6.1.1. A more fundamental and elegant geometric definition will be given in ▶ Sect. 6.1.2.

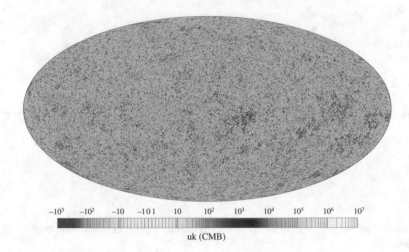

$$-10^3 \quad -10^2 \quad -10 \quad -10\,1 \quad 10 \quad 10^2 \quad 10^3 \quad 10^4 \quad 10^5 \quad 10^6 \quad 10^7$$

uk (CMB)

◻ Fig. 6.1 The sky as seen by the Planck satellite, after the galactic foreground and Doppler shifts due to the motion of earth have been removed (compare ◻ Fig. 6.2), from http://planck.cf.ac.uk/all-sky-images. The temperature of the spectrum of radiation is color-coded, with the temperature of "hottest" red spots and "coldest" blue spots differing from the "green temperature" $T = 2.728$ K by about 2×10^{-4} K. © ESA and the Planck Collaboration. Reproduced with permission

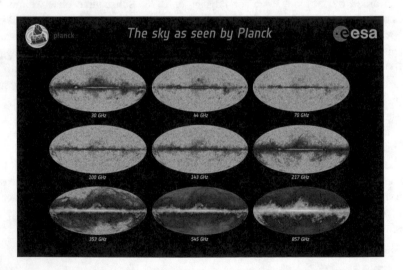

◻ Fig. 6.2 The sky as seen by Planck at various frequencies, from sci.esa.int/planck. © ESA and the Planck Collaboration, reproduced with permission

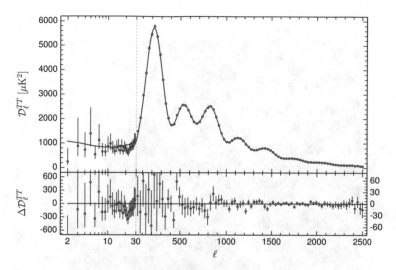

◻ Fig. 6.3 The angular size of the anisotropies of the CMB, from [70]. The upper plot shows the data and a fit to the model, the lower the residuals. © AAS. Reproduced with permission

6.1.1 A Pedestrian Approach

Given a vector field X, the *Lie derivative* $\mathscr{L}_X$ is an operation on tensor fields, defined as follows:

For a function f, one sets

$$\mathscr{L}_X f := X(f). \tag{6.1.1}$$

For a vector field Y, the Lie derivative coincides with the Lie bracket:

$$\mathscr{L}_X Y := [X, Y]. \tag{6.1.2}$$

For a one-form α, $\mathscr{L}_X \alpha$ is defined by imposing the Leibniz rule written the wrong-way round:

$$(\mathscr{L}_X \alpha)(Y) := \mathscr{L}_X(\alpha(Y)) - \alpha(\mathscr{L}_X Y). \tag{6.1.3}$$

(Indeed, the Leibniz rule applied to the contraction $\alpha_i X^i$ would read

$$\mathscr{L}_X(\alpha_a Y^a) = (\mathscr{L}_X \alpha)_a Y^a + \alpha_a (\mathscr{L}_X Y)^a,$$

which can be rewritten as (6.1.3).)

6

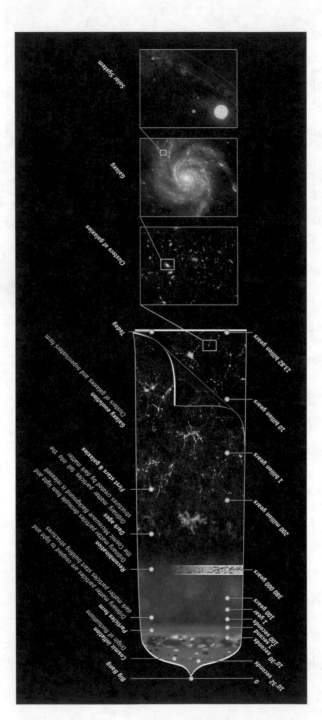

Fig. 6.4 The history of the universe according to the Planck satellite observations. From sci.esa.int/planck. ©ESA—C. Carreau. Reproduced with permission

Let us check that (6.1.3) defines a one-form. Clearly, the right-hand side becomes a sum when Y is replaced by $Y_1 + Y_2$. Next, if we replace Y by fY, where f is a function, then

$$(\mathscr{L}_X\alpha)(fY) = \mathscr{L}_X(\alpha(fY)) - \alpha\big(\underbrace{\mathscr{L}_X(fY)}_{X(f)Y + f\mathscr{L}_X Y}\big)$$

$$= X(f\alpha(Y)) - \alpha\big(X(f)Y + f\mathscr{L}_X Y\big)$$

$$= X(f)\alpha(Y) + fX(\alpha(Y)) - \alpha\big(X(f)Y\big) - \alpha\big(f\mathscr{L}_X Y\big)$$

$$= fX(\alpha(Y)) - f\alpha\big(\mathscr{L}_X Y\big)$$

$$= f\big((\mathscr{L}_X\alpha)(Y)\big) .$$

So $\mathscr{L}_X\alpha$ is a C^∞-linear map on vector fields, hence a covector field.

In coordinate-components notation we have

$$(\mathscr{L}_X\alpha)_a = X^b\partial_b\alpha_a + \alpha_b\partial_a X^b . \tag{6.1.4}$$

Indeed,

$$(\mathscr{L}_X\alpha)_a Y^a := \mathscr{L}_X(\alpha_a Y^a) - \alpha_a(\mathscr{L}_X Y)^a$$

$$= X^b\partial_b(\alpha_a Y^a) - \alpha_a(X^b\partial_b Y^a - Y^b\partial_b X^a)$$

$$= X^b(\partial_b\alpha_a)Y^a + \alpha_a Y^b\partial_b X^a$$

$$= \big(X^b\partial_b\alpha_a + \alpha_b\partial_a X^b\big) Y^a ,$$

as desired

For tensor products, the Lie derivative is defined by imposing linearity under addition together with the Leibniz rule:

$$\mathscr{L}_X(\alpha \otimes \beta) = (\mathscr{L}_X\alpha) \otimes \beta + \alpha \otimes \mathscr{L}_X\beta .$$

Since a general tensor A is a sum of tensor products,

$$A = A^{a_1 \ldots a_p}{}_{b_1 \ldots b_q}\partial_{a_1} \otimes \ldots \partial_{a_p} \otimes dx^{b_1} \otimes \ldots \otimes dx^{a_p} ,$$

the requirement of linearity with respect to addition of tensors and what has been said so far provide a definition of Lie derivative for any tensor.

For example, consider a tensor field $T = T^a{}_b\, dx^a \otimes \partial_b$. Writing $\mathscr{L}_X T^a{}_b$ for $(\mathscr{L}_X T)^a{}_b$, we claim that

$$\mathscr{L}_X T^a{}_b = X^c\partial_c T^a{}_b - T^c{}_b\partial_c X^a + T^a{}_c\partial_b X^c . \tag{6.1.5}$$

To see this, call a tensor $T^a{}_b$ *simple* if it is of the form $Y \otimes \alpha$, where Y is a vector and α is a covector. Using indices, this corresponds to $Y^a \alpha_b$ and so, by the Leibniz rule,

$$\mathcal{L}_X (Y \otimes \alpha)^a{}_b = \mathcal{L}_X (Y^a \alpha_b)$$
$$= (\mathcal{L}_X Y)^a \alpha_b + Y^a (\mathcal{L}_X \alpha)_b$$
$$= (X^c \partial_c Y^a - Y^c \partial_c X^a) \alpha_b + Y^a (X^c \partial_c \alpha_b + \alpha_c \partial_b X^c)$$
$$= X^c \partial_c (Y^a \alpha_b) - Y^c \alpha_b \partial_c X^a + Y^a \alpha_c \partial_b X^c ,$$

which coincides with (6.1.5) if $T^a{}_b = Y^b \alpha_b$. But a general $T^a{}_b$ can be written as a linear combination with constant coefficients of simple tensors,

$$T = \sum_{a,b} \underbrace{T^a{}_b \partial_a \otimes dx^b}_{\text{no summation, so simple}} ,$$

and the result follows.

Similarly, one has, e.g.,

$$\mathcal{L}_X R^{ab} = X^c \partial_c R^{ab} - R^{ac} \partial_c X^b - R^{cb} \partial_c X^a , \tag{6.1.6}$$

$$\mathcal{L}_X S_{ab} = X^c \partial_c S_{ab} + S_{ac} \partial_b X^c + S_{cb} \partial_a X^c , \tag{6.1.7}$$

etc. Those are all special cases of the general formula for the Lie derivative $\mathcal{L}_X A^{a_1 \ldots a_p}{}_{b_1 \ldots b_q}$:

$$\mathcal{L}_X A^{a_1 \ldots a_p}{}_{b_1 \ldots b_q} = X^c \partial_c A^{a_1 \ldots a_p}{}_{b_1 \ldots b_q} - A^{c a_2 \ldots a_p}{}_{b_1 \ldots b_q} \partial_c X^{a_1} - \ldots$$
$$+ A^{a_1 \ldots a_p}{}_{c b_1 \ldots b_q} \partial_{b_1} X^c + \ldots .$$

A useful property of Lie derivatives is

$$\mathcal{L}_{[X,Y]} = [\mathcal{L}_X, \mathcal{L}_Y] , \tag{6.1.8}$$

where, for a tensor T, the commutator $[\mathcal{L}_X, \mathcal{L}_Y]T$ is defined in the usual way:

$$[\mathcal{L}_X, \mathcal{L}_Y]T := \mathcal{L}_X (\mathcal{L}_Y T) - \mathcal{L}_Y (\mathcal{L}_X T) . \tag{6.1.9}$$

To see this, we first note that if $T = f$ is a function, then the right-hand side of (6.1.9) is the definition of $[X, Y](f)$, which in turn coincides with the definition of $\mathcal{L}_{[X,Y]}(f)$.

Next, for a vector field $T = Z$, (6.1.8) reads

$$\mathcal{L}_{[X,Y]}Z = \mathcal{L}_X (\mathcal{L}_Y Z) - \mathcal{L}_Y (\mathcal{L}_X Z) , \tag{6.1.10}$$

which is the same as

$$[[X, Y], Z] = [X, [Y, Z]] - [Y, [X, Z]] , \tag{6.1.11}$$

which is the same as

$$[Z, [Y, X]] + [X, [Z, Y]] + [Y, [X, Z]] = 0, \tag{6.1.12}$$

which is the Jacobi identity. Hence (6.1.8) holds for vector fields.

We continue with a one-form α, exploiting the fact that we have already established the result for functions and vectors: For any vector field Z we have, by definition

$$([\mathscr{L}_X, \mathscr{L}_Y]\alpha)(Z) = [\mathscr{L}_X, \mathscr{L}_Y](\alpha(Z)) - \alpha([\mathscr{L}_X, \mathscr{L}_Y](Z))$$

$$= \mathscr{L}_{[X,Y]}(\alpha(Z)) - \alpha(\mathscr{L}_{[X,Y]}(Z))$$

$$= (\mathscr{L}_{[X,Y]}\alpha)(Z),$$

which is the desired formula.

Incidentally A direct calculation for one-forms, using the definitions, proceeds as follows: Let Z be any vector field,

$$(\mathscr{L}_X\mathscr{L}_Y\alpha)(Z) = X\Big(\underbrace{(\mathscr{L}_Y\alpha)(Z)}_{Y(\alpha(Z)) - \alpha(\mathscr{L}_Y Z)} \Big) - \underbrace{(\mathscr{L}_Y\alpha)(\mathscr{L}_X Z)}_{Y(\alpha(\mathscr{L}_X Z)) - \alpha(\mathscr{L}_Y\mathscr{L}_X Z)}$$

$$= X\big(Y(\alpha(Z))\big) - X\big(\alpha(\mathscr{L}_Y Z)\big) - Y(\alpha(\mathscr{L}_X Z)) + \alpha(\mathscr{L}_Y\mathscr{L}_X Z).$$

Antisymmetrizing over X and Y, the second and third term above cancel out, so that

$$((\mathscr{L}_X\mathscr{L}_Y - \mathscr{L}_Y\mathscr{L}_X)\alpha)(Z) = X\big(Y(\alpha(Z))\big) + \alpha(\mathscr{L}_Y\mathscr{L}_X Z) - (X \longleftrightarrow Y)$$

$$= [X, Y](\alpha(Z)) - \alpha(\mathscr{L}_X\mathscr{L}_Y Z - \mathscr{L}_Y\mathscr{L}_X Z)$$

$$= \mathscr{L}_{[X,Y]}(\alpha(Z)) - \alpha(\mathscr{L}_{[X,Y]}Z)$$

$$= (\mathscr{L}_{[X,Y]}\alpha)(Z).$$

Since Z is arbitrary, (6.1.8) for covectors follows.

To conclude that (6.1.8) holds for arbitrary tensor fields, we note that by construction we have

$$\mathscr{L}_{[X,Y]}(A \otimes B) = \mathscr{L}_{[X,Y]}A \otimes B + A \otimes \mathscr{L}_{[X,Y]}B. \tag{6.1.13}$$

Similarly

$$\mathscr{L}_X\mathscr{L}_Y(A \otimes B) = \mathscr{L}_X(\mathscr{L}_Y A \otimes B + A \otimes \mathscr{L}_Y B)$$

$$= \mathscr{L}_X\mathscr{L}_Y A \otimes B + \mathscr{L}_X A \otimes \mathscr{L}_Y B + \mathscr{L}_Y A \otimes \mathscr{L}_X B$$

$$+ A \otimes \mathscr{L}_X\mathscr{L}_Y B. \tag{6.1.14}$$

Exchanging X with Y and subtracting, the middle terms drop out:

$$[\mathscr{L}_X, \mathscr{L}_Y](A \otimes B) = [\mathscr{L}_X, \mathscr{L}_Y]A \otimes B + A \otimes [\mathscr{L}_X, \mathscr{L}_Y]B . \tag{6.1.15}$$

Basing on what has been said, the reader should have no difficulties finishing the proof of (6.1.8).

Example 6.1.1 As an example of application of the formalism, suppose that there exists a coordinate system in which $(X^a) = (1, 0, 0, 0)$ and $\partial_0 g_{bc} = 0$. Then

$$\mathscr{L}_X g_{ab} = \partial_0 g_{ab} = 0 .$$

But the Lie derivative of a tensor field is a tensor field, and we conclude that $\mathscr{L}_X g_{ab} = 0$ holds in every coordinate system.

Vector fields for which $\mathscr{L}_X g_{ab} = 0$ are called *Killing vectors*: they arise from symmetries of spacetime. We have the useful formula

$$\boxed{\mathscr{L}_X g_{ab} = \nabla_a X_b + \nabla_b X_a .} \tag{6.1.16}$$

An effortless proof of this proceeds as follows: in adapted coordinates in which the derivatives of the metric vanish at a point p, one immediately checks that equality holds at p. But both sides are tensor fields, therefore the result holds at p for all coordinate systems, and hence also everywhere.

The brute-force proof of (6.1.16) is obtained by the following calculation:

$$\mathscr{L}_X g_{ab} = X^c \partial_c g_{ab} + \partial_a X^c g_{cb} + \partial_b X^c g_{ca}$$
$$= X^c \partial_c g_{ab} + \partial_a (X^c g_{cb}) - X^c \partial_a g_{cb} + \partial_b (X^c g_{ca}) - X^c \partial_b g_{ca}$$
$$= \partial_a X_b + \partial_b X_a + X^c \underbrace{(\partial_c g_{ab} - \partial_a g_{cb} - \partial_b g_{ca})}_{-2g_{cd}\Gamma^d_{ab}}$$
$$= \nabla_a X_b + \nabla_b X_a .$$

6.1.2 The Geometric Approach

We pass now to a geometric definition of Lie derivative. This requires, first, an excursion through the land of push-forwards and pull-backs.

6.1.2.1 Transporting Tensor Fields

We start by noting that, given a point p_0 in a manifold M, every vector $X \in T_{p_0}M$ is tangent to some curve. To see this, let $\{x^i\}$ be any local coordinates near p_0, with $x^i(p_0) = x_0^i$, then X can be written as $X^i(p_0)\partial_i$. If we set $\gamma^i(s) = x_0^i + s X^i(p_0)$, then

$$\frac{d\gamma^i}{ds}(0) = X^i(p_0) ,$$

which establishes the claim. This observation shows that studies of vectors can be reduced to studies of curves.

Let, now, M and N be two manifolds, and let $\phi : M \to N$ be a differentiable map between them. Given a vector $X \in T_p M$, the *push-forward* $\phi_* X$ of X is a vector in $T_{\phi(p)} N$ defined as follows: let γ be any curve for which $X = \dot{\gamma}(0)$, then

$$\phi_* X := \left. \frac{d(\phi \circ \gamma)}{ds} \right|_{s=0} . \tag{6.1.17}$$

In local coordinates y^A on N and x^i on M, so that $\phi(x) = (\phi^A(x^i))$, we find

$$(\phi_* X)^A = \left. \frac{d\phi^A(\gamma^i(s))}{ds} \right|_{s=0} = \left. \frac{\partial \phi^A(\gamma^i(s))}{\partial x^i} \dot{\gamma}^i(s) \right|_{s=0}$$

$$= \frac{\partial \phi^A(x^i)}{\partial x^i} X^i . \tag{6.1.18}$$

The formula makes it clear that the definition is independent of the choice of the curve γ satisfying $X = \dot{\gamma}(0)$.

Equivalently, and perhaps more directly, if X is a vector at p and h is a function on h, then $\phi_* X$ is defined by its action on h as

$$\boxed{\phi_* X(h) := X(h \circ \phi) .} \tag{6.1.19}$$

The map ϕ_* is sometimes called the *tangent map of* ϕ.

Applying (6.1.18) to a *vector field* X defined on M we obtain

$$(\phi_* X)^A(\phi(x)) = \frac{\partial \phi^A}{\partial x^i}(x) X^i(x) . \tag{6.1.20}$$

The equation shows that if a point $y \in N$ has more than one pre-image, say $y = \phi(x_1) = \phi(x_2)$ with $x_1 \neq x_2$, then (6.1.20) might define more than one tangent vector at y in general. This leads to an important caveat: we will be certain that the push-forward of a *vector field* on M defines a *vector field* on N only when ϕ is a diffeomorphism. More generally, $\phi_* X$ defines locally a vector field on $\phi(M)$ if and only if ϕ is a local diffeomorphism. In such cases we can invert ϕ (perhaps locally) and write (6.1.20) as

$$(\phi_* X)^j(x) = \left(\frac{\partial \phi^j}{\partial x^i} X^i \right) (\phi^{-1}(x)) . \tag{6.1.21}$$

When ϕ is understood as a coordinate change rather than a diffeomorphism between two manifolds, this is simply the standard transformation law of a vector field under coordinate transformations.

The push-forward operation can be extended to *contravariant* tensors by defining it on tensor products in the obvious way, and extending by linearity. For example, if X, Y, and Z are vectors and T is a three-contravariant tensor, then

$$\phi_*(X \otimes Y \otimes Z + T) := \phi_*X \otimes \phi_*Y \otimes \phi_*Z + \phi_*T \,.$$

Let $\phi : M \to N$ and $\psi : N \to L$ be two maps between manifolds. Directly from the definition we find

$$(\psi \circ \phi)_* = \psi_*\phi_* \,. \tag{6.1.22}$$

Since the map Id_*, tangent to the identity map $\mathrm{Id} : M \to M$, is the identity on every tangent space, from (6.1.22) we see that

$$(\phi^{-1})_* = (\phi_*)^{-1} \,. \tag{6.1.23}$$

Consider, next, a k-multi-linear map α from $T_{\phi(p_0)}M$ to $\mathbb{R}$. The *pull-back* $\phi^*\alpha$ of α is a multi-linear map on $T_{p_0}M$ defined as

$$T_pM \ni (X_1, \ldots X_k) \mapsto (\phi^*\alpha)(X_1, \ldots, X_k) := \alpha(\phi_*X_1, \ldots, \phi_*X_k) \,.$$

As an example, let $\alpha = \alpha_A dy^A$ be a one-form. If $X = X^i\partial_i$, then

$$\boxed{(\phi^*\alpha)(X) = \alpha(\phi_*X)} \tag{6.1.24}$$

$$= \alpha\left(\frac{\partial\phi^A}{\partial x^i}X^i\partial_A\right) = \alpha_A\frac{\partial\phi^A}{\partial x^i}X^i = \alpha_A\frac{\partial\phi^A}{\partial x^i}dx^i(X) \,.$$

Equivalently,

$$(\phi^*\alpha)_i = \alpha_A\frac{\partial\phi^A}{\partial x^i} \,. \tag{6.1.25}$$

For *one-form fields* on N we have thus obtained:

$$(\phi^*\alpha)_i(x) = \alpha_A(\phi(x))\frac{\partial\phi^A(x)}{\partial x^i} \,. \tag{6.1.26}$$

Note that $\phi^*\alpha$ is a field of one-forms on M, *irrespective of injectivity or surjectivity* properties of ϕ.

Similarly, pull-backs of *covariant* tensor fields of higher rank are smooth tensor fields.

For a function f Eq. (6.1.26) reads

$$(\phi^*df)_i(x) = \frac{\partial f}{\partial y^A}(\phi(x))\frac{\partial\phi^A(x)}{\partial x^i} = \frac{\partial(f \circ \phi)}{\partial x^i}(x) \,, \tag{6.1.27}$$

which can be succinctly written as

$$\phi^* df = d(f \circ \phi).$$
(6.1.28)

Using the notation

$$\phi^* f := f \circ \phi,$$
(6.1.29)

we can write (6.1.28) as

$$\phi^* d = d\phi^* \text{ when acting on functions.}$$
(6.1.30)

If $\phi : M \to N$ and $\psi : N \to L$ are differentiable maps between manifolds, one easily checks

$$(\psi \circ \phi)^* = \phi^* \psi^*, \qquad (\phi^{-1})^* = (\phi^*)^{-1}$$
(6.1.31)

(compare (6.1.22)–(6.1.23).)

Summarizing:
1. Pull-backs of covariant tensor fields define covariant tensor fields. In particular the metric can *always* be pulled back.
2. Push-forwards of contravariant tensor fields can be used to define contravariant tensor fields when ϕ is a diffeomorphism.

In this context it is thus clearly of interest to consider diffeomorphisms ϕ, as then tensor products can now be transported in the following way; we will denote by $\hat{\phi}$ the associated map:

$$\hat{\phi} := \begin{cases} \phi^*, & \text{for functions and covariant fields;} \\ (\phi^{-1})_*, & \text{for contravariant tensor fields.} \end{cases}$$
(6.1.32)

We use the rule

$$\hat{\phi}(A \otimes B) = \hat{\phi} A \otimes \hat{\phi} B$$

for tensor products.

So, for example, if X is a vector field and α is a field of one-forms, one has

$$\hat{\phi}(X \otimes \alpha) := (\phi^{-1})_* X \otimes \phi^* \alpha.$$
(6.1.33)

The definition is extended by linearity under addition and multiplication by functions to any tensor fields. Thus, if f is a function and T and S are tensor fields, then

$$\hat{\phi}(f T + S) = \hat{\phi} f \, \hat{\phi} T + \hat{\phi} S \equiv (f \circ \phi) \hat{\phi} T + \hat{\phi} S.$$

Since everything was fairly natural so far, one would expect that contractions transform in a natural way under transport. To make this clear, we start by rewriting (6.1.24) with the base-points made explicit:

$$((\hat{\phi}\alpha)(X))(x) = (\alpha(\phi_* X))(\phi(x)). \tag{6.1.34}$$

Replacing X by $(\phi_*)^{-1}Y = (\phi^{-1})_* Y$ this becomes

$$((\hat{\phi}\alpha)(\hat{\phi}Y))(x) = (\alpha(Y))(\phi(x)). \tag{6.1.35}$$

Equivalently

$$(\hat{\phi}\alpha)(\hat{\phi}Y) = \hat{\phi}(\alpha(Y)). \tag{6.1.36}$$

We note that in way similar to (6.1.31) we have

$$\widehat{\psi \circ \phi} = \hat{\phi}\hat{\psi}, \qquad \widehat{\phi^{-1}} = (\hat{\phi})^{-1}. \tag{6.1.37}$$

6.1.2.2   Flows of Vector Fields

Let X be a differentiable vector field on M. For every $p_0 \in M$ consider the solution to the problem

$$\frac{dx^i}{dt} = X^i(x(t)), \quad x^i(0) = x_0^i. \tag{6.1.38}$$

(Recall that there always exists a *maximal* interval I containing the origin on which (6.1.38) has a solution. Both the interval and the solution are unique. This will always be the solution $I \ni t \mapsto x(t)$ that we will have in mind.) The map

$$(t, x_0) \mapsto \phi_t[X](x_0) := x(t),$$

where $x^i(t)$ is the solution of (6.1.38), is called *the local flow of* X. We say that X generates $\phi_t[X]$.

We will write ϕ_t for $\phi_t[X]$ when X is unambiguous from the context.

The interval of existence of solutions of (6.1.38) depends upon x_0 in general.

Example 6.1.2   As an example, let $M = \mathbb{R}$ and $X = x^2 \partial_x$. We then have to solve

$$\frac{dx}{dt} = x^2, \; x(0) = x_0 \quad \Longrightarrow \quad x(t) = \begin{cases} 0, & x_0 = 0; \\ \frac{x_0}{1-x_0 t}, & x_0 \neq 0, \; 1 - x_0 t > 0. \end{cases}$$

Hence

$$\phi_t(x) = \frac{x}{1 - xt},$$

with $t \in \mathbb{R}$ when $x = 0$, with $t \in (-\infty, 1/x)$ when $x > 0$, and with $t \in (1/x, \infty)$ when $x < 0$.

We say that X is *complete* if $\phi_t[X](p)$ is defined for all $(t, p) \in \mathbb{R} \times M$.

The following standard facts follow immediately from uniqueness of solutions of ODEs and are left as exercises to the reader:

1. ϕ_0 is the identity map.
2. $\phi_t \circ \phi_s = \phi_{t+s}$.
   In particular $\phi_t^{-1} = \phi_{-t}$, and thus:
3. The maps $x \mapsto \phi_t(x)$ are local diffeomorphisms; global if for all $x \in M$ the maps ϕ_t are defined for all $t \in \mathbb{R}$.
4. $\phi_{-t}[X]$ is generated by $-X$:

$$\phi_{-t}[X] = \phi_t[-X].$$

A family of diffeomorphisms satisfying property 2 above is called a *one parameter group of diffeomorphisms*. Thus, *complete* vector fields generate one-parameter families of diffeomorphisms via (6.1.38).

Reciprocally, suppose that a local or global one-parameter group ϕ_t is given, then the formula

$$X = \left.\frac{d\phi_t}{dt}\right|_{t=0}$$

defines a vector field, said to be *generated by* ϕ_t.

6.1.2.3 The Lie Derivative Revisited

The idea of the *Lie transport*, and hence of the *Lie derivative*, is to be able to compare objects along integral curves of a vector field X. This is pretty obvious for scalars: we just compare the values of $f(\phi_t(x))$ with $f(x)$, leading to a derivative

$$\mathscr{L}_X f := \lim_{t \to 0} \frac{f \circ \phi_t - f}{t} \equiv \lim_{t \to 0} \frac{\phi_t^* f - f}{t} \equiv \lim_{t \to 0} \frac{\hat{\phi}_t f - f}{t} \equiv \left.\frac{d(\hat{\phi}_t f)}{dt}\right|_{t=0}.$$

$$(6.1.39)$$

We wish, next, to compare the value of a vector field Y at $\phi_t(x)$ with the value at x. For this, we move from x to $\phi_t(x)$ following the integral curve of X, and produce a new vector at x by applying $(\phi_t^{-1})_*$ to $Y|_{\phi_t(x)}$. This makes it perhaps clearer why we introduced the transport map $\hat{\phi}$, since $(\hat{\phi}Y)(x)$ is precisely the value at x of $(\phi_t^{-1})_*Y$. We can then calculate

$$\mathscr{L}_X Y(x) := \lim_{t \to 0} \frac{\left((\phi_t^{-1}) * Y\right)(\phi_t(x)) - Y(x)}{t}$$

$$\equiv \lim_{t \to 0} \frac{(\hat{\phi}_t Y)(x) - Y(x)}{t} \equiv \left.\frac{d(\hat{\phi}_t Y(x))}{dt}\right|_{t=0}.$$

$$(6.1.40)$$

In general, let X be a vector field and let ϕ_t be the associated local one-parameter family of diffeomorphisms. Let $\hat{\phi}_t$ be the associated family of transport maps for tensor fields. For any tensor field T one sets

$$\mathscr{L}_X T := \lim_{t \to 0} \frac{\hat{\phi}_t T - T}{t} \equiv \frac{d(\hat{\phi}_t T)}{dt}\Big|_{t=0}. \tag{6.1.41}$$

We want to show that this operation coincides with that defined in ▶ Sect. 6.1.1.

The equality of the two operations for functions should be clear, since (6.1.39) easily implies:

$$\mathscr{L}_X f = X(f).$$

Consider, next, a vector field Y. From (6.1.21), setting $\psi_t := \phi_{-t} \equiv (\phi_t)^{-1}$ we have

$$\hat{\phi}_t Y^j(x) := ((\phi_t^{-1})_* Y)^j(x) = \Big(\frac{\partial \psi_t^j}{\partial x^i} Y^i\Big)(\phi_t(x)). \tag{6.1.42}$$

Since ϕ_{-t} is generated by $-X$, we have

$$\psi_0^i(x) = x^i, \quad \frac{\partial \psi_t^j}{\partial x^i}\Big|_{t=0} = \delta_i^j,$$

$$\dot{\psi}_t^j\big|_{t=0} := \frac{d\psi_t^j}{dt}\Big|_{t=0} = -X^j, \quad \frac{\partial \dot{\psi}_t^j}{\partial x^i}\Big|_{t=0} = -\partial_i X^j. \tag{6.1.43}$$

Hence

$$\frac{d(\hat{\phi}_t Y^j)}{dt}(x)|_{t=0} = \frac{\partial \dot{\psi}_0^j}{\partial x^i}(x) Y^i(x) + \partial_k \underbrace{\Big(\frac{\partial \psi_0^j}{\partial x^i} Y^i\Big)}_{Y^j}(x)\dot{\phi}^k(x)$$

$$= -\partial_i X^j(x) Y^i(x) + \partial_j Y^i(x) X^j(x)$$

$$= [X, Y]^j(x),$$

and we have obtained (6.1.2), p. 191.

For a covector field α, it seems simplest to calculate directly from (6.1.26):

$$(\hat{\phi}_t \alpha)_i(x) = (\phi_t^* \alpha)_i(x) = \alpha_k(\phi_t(x)) \frac{\partial \phi_t^k(x)}{\partial x^i}.$$

Hence

$$\mathscr{L}_X \alpha_i = \frac{d(\phi_t^* \alpha)_i(x)}{dt}\Big|_{t=0} = \partial_j \alpha_i(x) X^j(x) + \alpha_k(x) \frac{\partial X^k(x)}{\partial x^i}(x), \tag{6.1.44}$$

as in (6.1.4).

The formulae just derived show that the *Leibniz rule under duality* holds by inspection:

$$\mathcal{L}_X(\alpha(Y)) = \mathcal{L}_X\alpha(Y) + \alpha(\mathcal{L}_X(Y)).\tag{6.1.45}$$

Incidentally Alternatively, one can start by showing that the Leibniz rule under duality holds for (6.1.41), and then use the calculations in ▶ Sect. 6.1.1 to derive (6.1.44): Indeed, by definition we have

$$\phi_t^*\alpha(Y) = \alpha((\phi_t)_*Y),$$

hence

$$\alpha(Y)|_{\phi_t(x)} = \alpha((\phi_t)_*(\phi_t^{-1})_*Y)|_{\phi_t(x)} = \phi_t^*\alpha|_x((\phi_t^{-1})_*Y|_{\phi_t(x)}) = \hat\phi_t\alpha(\hat\phi_t Y)|_x.$$

Equivalently,

$$\hat\phi_t(\alpha(Y)) = (\hat\phi_t\alpha)(\hat\phi_t Y),$$

from which the Leibniz rule under duality immediately follows.

A similar calculation leads to the Leibniz rule under tensor products.

The reader should have no difficulties checking that the remaining requirements set forth in ▶ Sect. 6.1.1 are satisfied.

Incidentally For the sake of the readers familiar with differential forms, we note the following formula of Cartan which provides a convenient tool for calculating the Lie derivative of a differential form α:

$$\mathcal{L}_X\alpha = X\rfloor d\alpha + d(X\rfloor\alpha).\tag{6.1.46}$$

We also note that the commuting of d and $\mathcal{L}_X$ is an immediate consequence of (6.1.46) and of the identity $d^2 = 0$:

$$\mathcal{L}_X d\alpha = d(\mathcal{L}_X\alpha).\tag{6.1.47}$$

6.2 Killing Vectors and Isometries

Let (M, g) be a pseudo-Riemannian manifold. A map ψ is called an *isometry* if

$$\psi^*g = g,\tag{6.2.1}$$

where ψ^* is the pull-back map defined in ▶ Sect. 6.1.2.

A standard fact is that the group of isometries of (M, g), denoted by $\mathrm{Iso}(M, g)$, carries a natural manifold structure. Such groups, when non-discrete, are called *Lie groups*. If (M, g) is Riemannian and compact, then $\mathrm{Iso}(M, g)$ is compact.

It is also a standard fact that any element of the connected component of the identity of a Lie group G belongs to a one-parameter subgroup $\{\phi_t\}_{t \in \mathbb{R}}$ of G. This allows one to study actions of Lie groups by studying the generators of one-parameter subgroups, which are defined as

$$X(f)(x) = \frac{d(f(\phi_t(x)))}{dt}\bigg|_{t=0} \qquad \Longleftrightarrow \qquad X = \frac{d\phi_t}{dt}\bigg|_{t=0}.$$

When the ϕ_t's are isometries, the vector fields X obtained in this way are called *Killing vectors*. The knowledge of Killing vectors provides a considerable amount of information about the isometry group, and we thus continue with an analysis of their properties. We will see shortly that the collection of Killing vectors forms a Lie algebra: by definition, a *Lie algebra* is a vector space equipped with a bilinear bracket operation $[\cdot, \cdot]$ such that

$$[X, Y] = -[Y, X],$$

and

$$[X, [Y, Z]] + [Y, [Z, X]] + [Z, [X, Y] = 0.$$

In the case of Killing vectors, the bracket operation will be the usual bracket (1.1.7), p. 6 of vector fields.

Recall that the integral curve of a vector field X through a point x_0 is defined as a solution of the initial value problem

$$\frac{dx}{dt} = X(x(t)), \quad x(0) = x_0, \tag{6.2.2}$$

A vector field X is said to be *complete* if its integral curves are defined for all values of parameter $t \in \mathbb{R}$ for all initial points x_0.

Of key importance to us will be the fact that *the dimension of the isometry group of $(\mathcal{M}, g)$ equals the dimension of the space of complete Killing vectors.*

Since there exist manifolds on which some Killing vector fields are not complete, we conclude that *the dimension of the isometry group of $(\mathcal{M}, g)$ is smaller than or equal to the dimension of the space of Killing vectors.*

It turns out that all Killing vectors are complete on *complete* Riemannian manifolds, so the distinction does not arise on such manifolds.

6.2.1 Killing Vectors

Let ϕ_t be a one-parameter group of isometries of $(\mathcal{M}, g)$, thus

$$\phi_t^* g = g \quad \Longrightarrow \quad \mathscr{L}_X g = 0. \tag{6.2.3}$$

Recall that (see (6.1.7), p. 194)

$$\mathscr{L}_X g_{\mu\nu} = X^\alpha \partial_\alpha g_{\mu\nu} + \partial_\mu X^\alpha g_{\alpha\nu} + \partial_\nu X^\alpha g_{\mu\alpha}.$$

In a coordinate system where the partial derivatives of the metric vanish at a point p, the right-hand side equals $\nabla_\mu X_\nu + \nabla_\nu X_\mu$. But the left-hand side is a tensor field, and two tensor fields equal in one coordinate system coincide in all coordinate systems. It follows that generators of isometries satisfy the equations

$$\nabla_\alpha X_\beta + \nabla_\beta X_\alpha = 0, \tag{6.2.4}$$

called *Killing equations*.

Conversely, consider a solution of (6.2.4); any such solution is called a *Killing vector field*. From the calculation just carried out, the Lie derivative of the metric with respect to X vanishes. This means that the local flow of X preserves the metric. In other words, X generates local isometries of g.

Actually, some work is needed to establish the implication reverse to (6.2.3):

$$\phi_t^* g = g \quad \Longleftarrow \quad \mathscr{L}_X g = 0. \tag{6.2.5}$$

In order to establish (6.2.5), note that

$$\frac{d(\phi_t^* g)}{dt} = \frac{d(\phi_{t+s}^* g)}{ds}\bigg|_{s=0} = \frac{d(\phi_t^* \phi_s^* g)}{ds}\bigg|_{s=0} = \phi_t^* \frac{d(\phi_s^* g)}{ds}\bigg|_{s=0} = \phi_t^* (\mathscr{L}_X g). \tag{6.2.6}$$

The result follows now by integration:

$$\phi_t^* g - g = \int_0^t \frac{d(\phi_t^* g)}{dt} dt = \int_0^t \phi_t^* (\mathscr{L}_X g) dt = 0. \tag{6.2.7}$$

To make sure that X generates a one-parameter group of isometries one needs moreover to make sure that X is *complete*. This might be difficult to establish, often requiring further global hypotheses.

The map $(t, x_0) \mapsto x(t)$, where $x(t)$ is the solution of (6.2.2), is often denoted by $\phi_t(x_0)$, and is called the *flow of* X. We will sometimes write $\phi_t[X]$ when more than one vector X is involved.

Recall the identity (6.1.8), p. 194:

$$\mathscr{L}_{[X,Y]} = [\mathscr{L}_X, \mathscr{L}_Y]. \tag{6.2.8}$$

This implies that the commutator of two Killing vector fields is a Killing vector field:

$$\mathscr{L}_{[X,Y]}g = \mathscr{L}_X(\underbrace{\mathscr{L}_Y g}_{0}) - \mathscr{L}_Y(\underbrace{\mathscr{L}_X g}_{0}) = 0.$$

Thus, and as already pointed-out, the collection of all Killing vector fields, equipped with the Lie bracket, forms a *Lie algebra*.

Remark 6.2.1 Let p be a fixed point of an isometry ϕ. Then ϕ_* maps T_pM to T_pM; we will refer to this action as the *tangent action*.

For $W \in T_pM$ let $s \mapsto \gamma_W(s)$ be an affinely parameterized geodesic with $\gamma_W(0) = p$ and $\dot{\gamma}_W(0) = W$. Since isometries map geodesics to geodesics, the curve $s \mapsto \phi(\gamma_W(s))$ is a geodesic that passes through p and has tangent vector ϕ_*W there. As the affine parameterization condition is also preserved by isometries, we conclude that

$$\phi(\gamma_W(s)) = \gamma_{\phi_* W}(s). \tag{6.2.9}$$

In particular, in the Riemannian case ϕ maps the metric spheres and balls

$$S_p(r) := \{q \in M : d(p,q) = r\}, \quad B_p(r) := \{q \in M : d(p,q) < r\}$$

to themselves. (Similarly the $S_p(r)$'s and $B_p(r)$'s are invariant in the Lorentzian case as well, or for that matter in any signature, but these sets are not topological spheres, respectively topological balls, anymore.)

Actually, in Riemannian geometry the sets $S_p(r)$ are always spheres for r small enough, but may fail to be so when r is too large. For example, consider the unit sphere S^2 in three-dimensional Euclidean space, and let N be its north pole. For $0 < r < \pi$ the $S_N(r)$'s are circles (one-dimensional spheres), while $S_N(\pi)$ is a point, the south pole. This leads one to the notion of *injectivity radius* $i(p)$ *of a point* p in a Riemannian manifold which, very loosely, can be thought of as the largest number so the $S_p(r)$ are spheres for $r < i(p)$.

The action of a group on M is called *transitive* if for every pair $p, q \in M$ there exists an element ϕ of the group such that $q = \phi(p)$.

Given an action of a group G on a manifold M, the *isotropy group of* $p \in M$ is defined as the set of those elements $\phi \in G$ which leave p fixed: $\phi(p) = p$.

Suppose that the tangent action on T_pM is transitive on the set of vectors X at p satisfying $g(X, X) = 1$. It follows from what we just said that, for complete Riemannian metrics, transitivity on unit vectors at p implies that, for any r smaller than the injectivity radius, the action on the $S_p(r)$'s is transitive as well. □

Remark 6.2.2 Let p be a point in a three-dimensional Riemannian manifold (M, g) such that the tangent action of $\mathrm{Iso}(M, g)$ is transitive on unit vectors of T_pM. The group of isometries of M that leave p fixed is then a closed subgroup of $SO(3)$ which acts transitively on S^2, hence of dimension at least two. Now, it is easily seen (exercise) that connected subgroups of $SO(3)$ are $\{e\}$ (which has dimension zero), $U(1)$ (which has dimension one), or $SO(3)$ itself. We conclude that, on three-dimensional Riemannian manifolds, existence of fixed points of the action implies that the group of isometries of (M, g) contains an $SO(3)$ subgroup. □

Remark 6.2.3 In Riemannian geometry, the *sectional curvature* κ of a plane spanned by two vectors $X, Y \in T_p M$ is defined as

$$\kappa(X, Y) := \frac{g(R(X, Y)X, Y)}{g(X, X)g(Y, Y) - g(X, Y)^2}. \tag{6.2.10}$$

One checks that κ depends only upon the plane, and not the choice of the vectors X and Y spanning the plane. The definition extends to pseudo-Riemannian manifolds as long as the denominator does not vanish; equivalently, the plane spanned by X and Y should *not* be null.

As will be explained in more detail below, for maximally symmetric Riemannian manifolds the action of the isometry group on the collection of two-dimensional subspaces of the tangent bundle is transitive, which implies that κ is independent of the plane and of the point p.

Complete Riemannian manifolds with *constant* κ, not necessarily simply connected, are called *space forms*. □

Equation (6.2.4) leads to a second order system of equations, as follows: Taking cyclic permutations of the equation obtained by differentiating (6.2.4) one has

$$-\nabla_\gamma \nabla_\alpha X_\beta - \nabla_\gamma \nabla_\beta X_\alpha = 0,$$

$$\nabla_\alpha \nabla_\beta X_\gamma + \nabla_\alpha \nabla_\gamma X_\beta = 0,$$

$$\nabla_\beta \nabla_\gamma X_\alpha + \nabla_\beta \nabla_\alpha X_\gamma = 0.$$

Adding, and expressing commutators of derivatives in terms of the Riemann tensor, one obtains

$$2\nabla_\alpha \nabla_\beta X_\gamma = (R_{\sigma\gamma\beta\alpha} + R_{\sigma\alpha\beta\gamma} + \underbrace{R_{\sigma\beta\alpha\gamma}}_{=-R_{\sigma\alpha\gamma\beta}-R_{\sigma\gamma\beta\alpha}})X^\sigma$$

$$= 2R_{\sigma\alpha\beta\gamma} X^\sigma.$$

Thus

$$\nabla_\alpha \nabla_\beta X_\gamma = R_{\sigma\alpha\beta\gamma} X^\sigma. \tag{6.2.11}$$

Example 6.2.4 As an example of application of (6.2.11), let (M, g) be flat. In a coordinate system $\{x^\mu\}$ in which the metric has constant entries (6.2.11) reads

$$\partial_\alpha \partial_\beta X_\gamma = 0\,.$$

The solutions are therefore linear,

$$X^\alpha = A^\alpha + B^\alpha{}_\beta x^\beta\,.$$

Plugging this into (6.2.4), one finds that $B_{\alpha\beta}$ must be antisymmetric. Hence, the dimension of the set of all Killing vectors of $\mathbb{R}^{n,m}$, and thus of $\mathrm{Iso}(\mathbb{R}^{n,m})$, is $(n+m)(n+m+1)/2$, independently of signature.

The local calculations above remain valid on the torus $\mathbb{T}^n := S^1 \times \ldots \times S^1$ equipped with a flat metric. However, no locally defined Killing vectors of the form $B^i{}_j x^j$ survive the periodic identifications, so that the dimension of $\mathrm{Iso}(\mathbb{T}^n, \delta)$ is n: Indeed, using (6.2.11) and integration by parts we have

$$\int X^i \underbrace{D_j D_i X^j}_{=0} = -\int D^j X^i \underbrace{D_i X_j}_{=-D_j X_i} = \int |DX|^2\,, \tag{6.2.12}$$

and so $B_{ij} \equiv D_i X_j = 0$: all Killing vectors on a flat Riemannian $\mathbb{T}^n$ are covariantly constant.

Incidentally A simple modification of the calculation (6.2.12) shows that *the isometry group of a compact Riemannian manifold with <u>strictly negative</u> Ricci tensor is finite*, and that *nontrivial Killing vectors of compact Riemannian manifolds with <u>non-positive</u> Ricci tensor are covariantly constant*. Indeed, for such manifolds the left-hand side of (6.2.12) does not necessarily vanish a priori, instead we have

$$\int |DX|^2 = \int X^i D_j D_i X^j = \int X^i R_{kji}{}^j X^j = \int X^i R_{ki} X^j\,, \tag{6.2.13}$$

The left-hand side is always positive. If the Ricci tensor is non-positive, then the right-hand side is non-positive, which is only possible if both vanish, hence $DX = 0$ and $R_{ij} X^i X^j = 0$. If the Ricci tensor is strictly negative, then $X = 0$, and there are no nontrivial Killing vectors, so that the dimension of the group of isometries is zero. Since the group is compact when (M, g) is Riemannian and compact, it must be finite when no Killing vectors exist.

An important consequence of (6.2.11) is:

Proposition 6.2.5

Let M be connected and let $p \in M$. A Killing vector is uniquely defined by its value $X(p)$ and the value at p of the antisymmetric tensor $\nabla X(p)$.

Proof

Consider two Killing vectors X and Y such that $X(p) = Y(p)$ and $\nabla X(p) = \nabla Y(p)$. Let $q \in M$ and let γ be any curve from p to q. Set

$$Z^\beta := X^\beta - Y^\beta, \qquad A_{\alpha\beta} = \nabla_\alpha (X_\beta - Y_\beta).$$

Along the curve γ we have

$$\frac{DZ_\alpha}{ds} = \dot\gamma^\mu \nabla_\mu Z_\alpha = \dot\gamma^\mu A_{\mu\alpha},$$

$$\frac{DA_{\alpha\beta}}{ds} = \dot\gamma^\mu \nabla_\mu \nabla_\alpha Z_\beta = R_{\gamma\mu\alpha\beta} \dot\gamma^\mu Z^\gamma. \tag{6.2.14}$$

This is a linear first order system of ODEs along γ with vanishing Cauchy data at p. Hence the solution vanishes along γ, and thus $X^\mu(q) = Y^\mu(q)$. Since q is arbitrary, the result follows. $\qquad\square$

Note that there are at most n values of X at p and, in view of antisymmetry, at most $n(n-1)/2$ values of ∇X at p. Since the dimension of the space of complete Killing vectors equals the dimension of the group of isometries, as a corollary we obtain:

Proposition 6.2.6

The dimension of the group of isometries of an n-dimensional pseudo-Riemannian manifold (M, g) is less than or equal to $n(n+1)/2$. $\qquad\square$

6.2.2 Maximally Symmetric Manifolds

A manifold (M, g) is called *maximally symmetric* if the dimension of $\mathrm{Iso}(M, g)$ equals the maximum allowed value, hence $n(n+1)/2$ in dimension n.

A manifold (M, g) is called *homogeneous* if the action of $\mathrm{Iso}(M, g)$ is transitive; i.e., for any two points $p, q \in M$ there exists an isometry which maps p to q. Thus the dimension of the orbit of the isometry group through each point is n. This implies that the dimension of $\mathrm{Iso}(M, g)$ is at least n.

A manifold (M, g) is called *isotropic* at $p \in M$ if for all unit vectors $v, w \in T_p M$ there exists an isometry ϕ such that $\phi(p) = p$ and $\phi_*(v) = w$. In other words, the isometry group acts transitively on the set of *unit* vectors at p.

Remark 6.2.7 A Riemannian manifold (M, g) is said to be *complete* if it is complete as a metric space, with the metric defined by the distance function of g. A complete Riemannian manifold which is isotropic at every point is necessarily homogeneous. This can be seen to be a consequence of Remark 6.2.1, as follows: Given two points $p, q \in M$, let d be the distance between them. Let γ be any shortest geodesic connecting p and q, and let r be the middle point of γ. Then $p, q \in S_r(d/2)$, and transitivity of the action of the isometry group on $S_r(d/2)$ shows that there exists an isometry mapping p to q.

Our observations so far lead to the following:

Proposition 6.2.8

Let (M, g) be a complete three-dimensional Riemannian manifold such that either
1. *(M, g) is homogeneous, and isotropic at some point p; or*
2. *(M, g) is isotropic at every point.*

Then (M, g) is maximally symmetric.

Proof

Suppose we are in case 1, thus (M, g) is isotropic at some point p and homogeneous. We claim that (M, g) is isotropic at every point. Indeed, let q in M, by homogeneity there exists an isometry such that $\phi(p) = q$. Consider two unit vectors $X, Y \in T_q M$, then $\phi_*^{-1} X$ and $\phi_*^{-1} Y$ are unit vectors at $T_p M$. Since M is isotropic at p there exists an isometry ψ such that $\psi(p) = p$ and

$$\psi_* \phi_*^{-1} X = \phi_*^{-1} Y .$$

Hence

$$\phi_* \psi_* \phi_*^{-1} X = Y .$$

Since $\phi_* \psi_* \phi_*^{-1} = (\phi \circ \psi \circ \phi^{-1})_*$, and since $\phi \circ \psi \circ \phi^{-1}$ is an isometry which maps q to q, we infer that M is isotropic at q.

In case 2, by Remark 6.2.7 every manifold which is isotropic at all points is homogeneous.

Thus, in both cases, (M, g) is homogeneous and isotropic at every point.

Let $p \in M$ and let H be the subgroup of the group of isometries of (M, g) that leaves p invariant. Then H is three dimensional by Remark 6.2.2. Since the orbits of $\mathrm{Iso}(M, g)$ are three dimensional, the dimension of $\mathrm{Iso}(M, g)$ is larger than or equal to $\dim H + 3$, hence six. But six is the maximum dimension $3 \times (3+1)/2$ of $\mathrm{Iso}(M, g)$ in dimension three, whence the result. $\square$

Consider a maximally symmetric Riemannian manifold (M, g), let $p \in M$. Introducing coordinates at p so that the metric is diagonal at p shows that the action of the isotropy group of p on $T_p M$ is identical to the usual action of $SO(n)$ on Euclidean $\mathbb{R}^n$. This implies that every ON basis of $T_p M$ can be mapped to every other one by an isometry. It follows that every two-dimensional plane can be mapped to any other one by an isometry. This shows that the sectional curvatures $\kappa(X, Y)$ defined in (6.2.10), p. 207 are the same for all linearly independent pairs (X, Y). Homogeneity implies that the sectional curvatures are further independent of p. We say that maximally symmetric Riemannian manifolds are *spaces of constant sectional curvature*.

A partial converse holds: every space of constant sectional curvature has, *locally*, the maximal number of Killing vectors. As already pointed out, care is needed when interpreting this statement, as locally defined Killing vectors do not necessarily extend to globally defined ones.

Examples of flat maximally symmetric manifolds are provided by

$$\mathbb{R}^{n,m} := (\mathbb{R}^{n+m}, \eta),$$

where η is a quadratic form of signature (n, m). It follows, e.g., from the (pseudo-Riemannian version of the) *Hadamard-Cartan* theorem that these are the only *simply connected geodesically complete* such manifolds.

Curved examples of maximally symmetric manifolds can be constructed as follows: For $a \in \mathbb{R} \setminus \{0\}$ consider the submanifold of $\mathbb{R}^{n,m}$ defined as

$$\mathscr{S}_a := \{\eta_{\alpha\beta} x^\alpha x^\beta = a\}. \tag{6.2.15}$$

Note that the covector field

$$N_\alpha := \frac{\eta_{\alpha\beta} x^\alpha}{\sqrt{|a|}}$$

is conormal to $\mathscr{S}_a$, i.e., $N_\alpha X^\alpha = 0$ for all vectors X^α tangent to $\mathscr{S}_a$. Since $\eta(N, N) = a/|a| \neq 0$, the covector N is timelike if $a < 0$ and spacelike if $a > 0$. In any case N is not null, which implies that the tensor field η induces on $\mathscr{S}_a$ a pseudo-Riemannian metric which will be denoted by h, with signature of h determined by (n, m) and the sign of a. (Compare (4.3.13), p. 145 and the discussion that follows.)

Both η and $\mathscr{S}_a$ are invariant under the defining action of $SO(n, m)$ on $\mathbb{R}^{n+m}$, which implies that the metric h is also invariant under this action.

Further, the action is *effective*; this means that the only element $g \in SO(n, m)$ leaving all points invariant is the identity map. Indeed, suppose that $\phi \in SO(n, m)$ leaves invariant every point $p \in \mathscr{S}_a$. Now, every point $x = (x^\mu)$ for which $\eta_{\alpha\beta} x^\alpha x^\beta$ has the same sign as a can be written in the form $x^\mu = \beta x_0^\mu$, where $x_0 \in \mathscr{S}_a$ and $\beta \in \mathbb{R}$. By linearity of the action, ϕ leaves x invariant. Hence ϕ is a linear map which is the identity on an open set, and thus the identity everywhere.

We conclude that the dimension of the isometry group of $(\mathscr{S}_a, h)$ equals that of $SO(n, m)$, namely $(n + m)(n + m - 1)/2$. Since the dimension of $\mathscr{S}_a$ is $n + m - 1$, this implies that $(\mathscr{S}_a, h)$ is maximally symmetric.

As a first explicit example, let η be either the Minkowski or the flat metric in dimension $n + 1$. To accommodate both cases simultaneously we will write η in the form

$$\eta = \epsilon dw^2 + \delta = \epsilon dw^2 + dr^2 + r^2 d\Omega_{n-1}^2, \qquad \epsilon \in \{\pm 1\},$$

where $w \in \mathbb{R}$, δ is the Euclidean metric on $\mathbb{R}^n$, and $d\Omega^2_{n-1}$ is the canonical metric on the $(n-1)$-dimensional unit round sphere (e.g., $d\Omega^2_2 = d\theta^2 + \sin^2\theta d\varphi^2$ when $n = 3$ and when using spherical coordinates on S^2). Hence $\mathscr{S}_a$ is given by the equation

$$\epsilon w^2 = a - r^2 .$$

Differentiation gives $2\epsilon w\, dw = -2r\, dr$, so that away from the set $\{a = r^2\}$ we find

$$dw^2 = \frac{r^2}{w^2} dr^2 = \epsilon \frac{r^2}{a - r^2} dr^2 .$$

This shows that the induced metric h can be written as

$$h = \frac{r^2}{a - r^2} dr^2 + dr^2 + r^2 d\Omega^2_{n-1} = \frac{a}{a - r^2} dr^2 + r^2 d\Omega^2_{n-1} . \tag{6.2.16}$$

When $\epsilon = 1$ and $a = R^2$, $\mathscr{S}_a$ is a sphere of radius R in Euclidean space, $w^2 = R^2 - r^2 \leq R^2$, and we obtain the following representation of the upper-hemisphere metric, cf. the left ◘ Fig. 6.5:

$$h = \frac{1}{1 - \frac{r^2}{R^2}} dr^2 + r^2 d\Omega^2_{n-1} , \quad 0 < r < R . \tag{6.2.17}$$

Note that the same formula defines a Lorentzian metric for $r > R$:

$$h = -\frac{1}{\frac{r^2}{R^2} - 1} dr^2 + r^2 d\Omega^2_{n-1} , \quad r > R . \tag{6.2.18}$$

This corresponds to the case $a = R^2$ and $\epsilon = -1$ of our construction, in which case $\mathscr{S}_a$ is the timelike hyperboloid $r^2 = R^2 + t^2 \geq R^2$ in Minkowski spacetime, cf., the

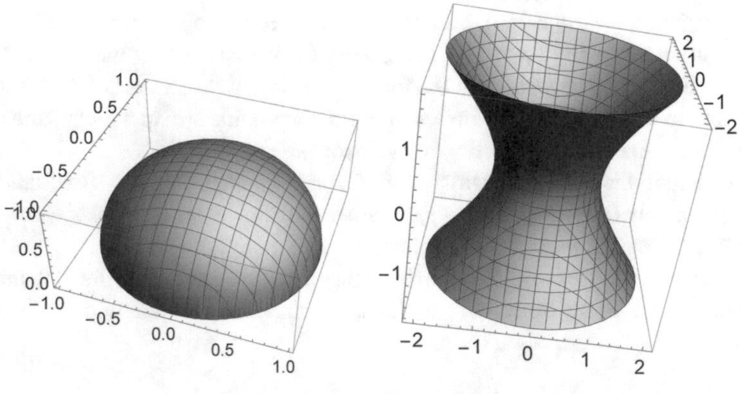

◘ **Fig. 6.5** The north hemisphere in Euclidean space as a graph over the (x, y)-plane (left figure), and a timelike hyperboloid in Minkowski space, with the time coordinate running along the vertical axis (right figure)

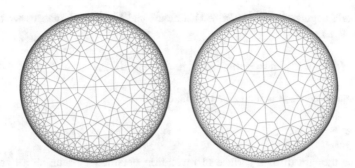

Fig. 6.6 Compact hyperbolic quotient manifolds can be obtained by constructions illustrated in the figures. Figures by Dmitry Brant from https://dmitrybrant.com/, reproduced here with kind permission from the author

right ■ Fig. 6.5. This is the *de Sitter metric* coordinatized in an unusual way (compare (6.4.52) and (6.4.60), p. 232). It solves the vacuum Einstein equations with a positive cosmological constant.

When $\epsilon = -1$ and $a = -R^2$, $\mathscr{S}_a$ consists of two copies of a spacelike hyperboloid in Minkowski space, $t^2 = R^2 + r^2$, with induced metric of constant negative curvature:

$$h = \frac{1}{1 + \frac{r^2}{R^2}} dr^2 + r^2 d\Omega_{n-1}^2 \,. \tag{6.2.19}$$

This is the metric on *hyperbolic space*.

The manifold in this last model is $\mathbb{R}^3$. By taking quotients one can obtain compact manifolds where the metric takes, locally, the form (6.2.19); compare ■ Fig. 6.6. Many non-isometric such quotients are possible, and note that the resulting manifold is only locally maximally symmetric, as no Killing vectors survive the passage to a compact quotient in negative curvature.

Introducing $\hat{r} = r/R$, the above Riemannian metrics can be written in a more standard form

$$h = R^2 \left(\frac{d\hat{r}^2}{1 + k\hat{r}^2} + \hat{r}^2 d\Omega_{n-1}^2 \right), \quad k \in \{0, \pm 1\}, \tag{6.2.20}$$

where k determines whether the metric is flat, or positively curved, or negatively curved.

Exercise 6.2.9 By a direct calculation, or by symmetry considerations, show that the curvature tensor of the metric h of (6.2.20) equals

$$R_{ijk\ell} = \frac{k}{R^2} (h_{ik}h_{j\ell} - h_{i\ell}h_{jk}). \tag{6.2.21}$$

Check that this implies that h has all sectional curvatures equal to k/R^2, Ricci tensor R_{ij} equal to $k(n-1)h_{ij}/R^2$, and Ricci scalar equal to $kn(n-1)/R^2$.

To obtain negatively curved $(n + 1)$-dimensional Lorentzian metrics we take η of signature $(2, n)$:

$$\eta = -dw^2 - dz^2 + \delta = -dw^2 - dz^2 + dr^2 + r^2 d\Omega_{n-1}^2 .$$

The hypersurface $\mathscr{S}_{-R^2}$ is then given by the equation

$$w^2 + z^2 = R^2 + r^2 ,$$

and can be thought of as a surface of revolution obtained by rotating a hyperboloid $w = 0$, $z^2 = R^2 + r^2$, around the $w = z = 0$ axis. Setting

$$w = \rho \cos t , \quad z = \rho \sin t ,$$

one has

$$dw^2 + dz^2 = d\rho^2 + \rho^2 dt^2 ,$$

and the equation for $\mathscr{S}_{-R^2}$ becomes

$$\rho^2 = r^2 + R^2 .$$

Hence

$$d\rho^2 = \frac{r^2}{\rho^2} dr^2 = \frac{r^2 dr^2}{r^2 + R^2} ,$$

and since

$$\eta = -dw^2 - dz^2 + dr^2 + r^2 d\Omega_{n-1}^2 = -(d\rho^2 + \rho^2 dt^2) + dr^2 + r^2 d\Omega_{n-1}^2$$

we obtain

$$h = -\left(\frac{r^2 dr^2}{r^2 + R^2} + (r^2 + R^2) dt^2 \right) + dr^2 + r^2 d\Omega_{n-1}^2$$

$$= -(r^2 + R^2) dt^2 + \frac{dr^2}{1 + \frac{r^2}{R^2}} + r^2 d\Omega_{n-1}^2 . \tag{6.2.22}$$

Note that the slices $\{t = \text{const}\}$ are maximally symmetric hyperbolic. Somewhat perplexingly, in the embedded model the time coordinate t is periodic. We can get rid of this by passing to the *universal cover* of $\mathscr{S}_{-R^2}$ where, by definition, t runs over $\mathbb{R}$ instead of S^1. The resulting spacetime is called the *anti-de Sitter spacetime*.

It is a standard fact that all Riemannian maximally symmetric metrics with the same sectional curvatures are *locally isometric* [32, §27]. So the above formulae give the local form for all maximally symmetric Riemannian metrics.

6.3   The Cosmological Principle, Congruences

The *cosmological principle* is the hypothesis that

There exists a time-function t such that on each of its level sets

$$\mathscr{S}_\tau := \{p : t(p) = \tau\} \tag{6.3.1}$$

the universe looks the same at every point and in every direction.

In many textbooks it is asserted that the cosmological principle implies existence of coordinates so that the metric takes the form (6.4.1), p. 219. This is wrong, as shown in Remark 6.4.1 below: this principle does not suffice to enforce the form (6.4.1) of the metric.

Our starting point to arrive at (6.4.1) therefore must, and will, be different. Namely, following the line of thought exposed at the beginning of this chapter, see p. 189, we will assume the existence of a family of freely falling observers filling the universe, called "congruence," for which *the universe looks the same in all directions*.

Let, thus, $u = u^\alpha \partial_\alpha$ be the field of unit-tangents to the world lines of the congruence filling the spacetime $(\mathscr{M}, g)$. By hypothesis the curves are timelike geodesics, and we parameterize them by proper time:

$$u^\alpha \nabla_\alpha u^\beta = 0, \quad g(u, u) = -1. \tag{6.3.2}$$

The assumption that "the universe looks the same in all directions" will be made precise as the requirement that

there exists a group of isometries of $(\mathscr{M}, g)$ which, at every $p \in \mathscr{M}$, acts transitively on the set of directions normal to u.

In other words, there exist no preferred directions orthogonal to u at any point in spacetime.

To proceed, we start with an analysis of the tensor field $\nabla_{[\alpha} u_{\beta]}$; we wish to show that it vanishes. For this, we first note that

$$u^\alpha \nabla_{[\alpha} u_{\beta]} = 0. \tag{6.3.3}$$

Indeed, $u^\alpha \nabla_\alpha u_\beta = 0$ by the geodesic equation, while

$$u^\alpha \nabla_\beta u_\alpha = \frac{1}{2} \nabla_\beta (u^\alpha u_\alpha) = 0$$

because $u^\alpha u_\alpha = -1$. So, in fact,

$$u^\alpha \nabla_\alpha u_\beta = u^\alpha \nabla_\beta u_\alpha = 0, \tag{6.3.4}$$

which in particular gives (6.3.3).

Next, let e_i, $i = 1, 2, 3$ be any orthonormal basis of the space $u^\perp$ of vectors orthogonal to u. Set

$$Y^i = \frac{1}{2}\epsilon^{ijk}\nabla_{[j}u_{k]},$$

where $\nabla_{[j}u_{k]}$ denotes the space-components of $\nabla_{[\alpha}u_{\beta]}$ in the basis

$$\{e_0 := u, e_1, e_2, e_3\},$$

and ϵ^{ijk} is the usual alternating tensor. The requirement that there are no preferred vectors orthogonal to u implies that Y^i must vanish. From the identity

$$\nabla_{[j}u_{k]} = \epsilon_{jk\ell}Y^\ell$$

we conclude that $\nabla_{[j}u_{k]} \equiv 0$. This, together with (6.3.4), gives the promised equation

$$\nabla_{[\alpha}u_{\beta]} \equiv 0. \tag{6.3.5}$$

(Using differential-forms terminology, this is the statement that *the one-form $u_\alpha dx^\alpha$ is closed*.)

Assuming for simplicity that $\mathscr{M}$ is simply connected, it is a standard fact that (6.3.5) implies existence of a function t such that

$$u_\alpha dx^\alpha = dt \quad\Longleftrightarrow\quad u_\alpha = \partial_\alpha t. \tag{6.3.6}$$

Recall that $\mathscr{S}_\tau$ has been defined in (6.3.1) as the τ-level set of t. Let x^i be any set of local coordinates on $\mathscr{S}_0$. We can extend these coordinates away from $\mathscr{S}_0$ by requiring that the extended coordinates be constant along the integral curves of u:

$$u^\mu\partial_\mu x^i = 0. \tag{6.3.7}$$

Using (t, x^i) as local coordinates on $\mathscr{M}$, (6.3.6) gives

$$u_\alpha = \partial_\alpha t = \delta^0_\alpha, \tag{6.3.8}$$

while (6.3.8) is equivalent to

$$u^\mu\partial_\mu = u^0\partial_t.$$

The normalization condition $g(u, u) = -1$ together with (6.3.8) gives

$$-1 = g_{\mu\nu}u^\mu u^\nu = -u^\mu u_\mu = -u^0,$$

whence $u^0 = 1$ so that

$$u^\mu\partial_\mu = \partial_t. \tag{6.3.9}$$

This implies

$$-1 = g_{\mu\nu}u^\mu u^\nu = g_{00}(u^0)^2 = g_{00} \,. \tag{6.3.10}$$

Furthermore,

$$0 = u^i = g^{i\nu}u_\nu = g^{0i} \,. \tag{6.3.11}$$

Equations (6.3.10)–(6.3.11) and straightforward algebra imply

$$g_{0i} = 0 \,. \tag{6.3.12}$$

In these coordinates the metric therefore takes the form

$$g = -dt^2 + \underbrace{g_{ij}(t, x)\, dx^i dx^j}_{=:h(t)} \,. \tag{6.3.13}$$

It should be clear from this form of the space-time metric that the isometry group of $(\mathcal{M}, g)$ preserves the metric $h(t) \equiv g_{ij}(t, x)\, dx^i dx^j$ induced on the level sets $\mathscr{S}_\tau$ of t by g, and acts transitively on each tangent space of each level set $\mathscr{S}_\tau$. By Proposition 6.2.8, p. 210 these level sets are therefore maximally symmetric.

We now turn our attention to the symmetric tensor field $\nabla_{(\alpha}u_{\beta)}$. From $u^0 = 1$ and (6.3.4) we have

$$\nabla_{(0}u_{0)} = 0 = \nabla_{(0}u_{i)} \,, \tag{6.3.14}$$

so that the only-nonzero components of $\nabla_{(\alpha}u_{\beta)}$ are those which are present in the symmetric space-tensor

$$A_{ij} := \nabla_{(i}u_{j)} \,.$$

We claim that this tensor must be proportional to the metric. Indeed, let $p \in \mathscr{S}_\tau$, and let H_p denote the isotropy group of p. Consider the tensor field A defined in components as $A^i{}_j = g^{ik}A_{kj}$, which can be identified with a map from $T_p\mathscr{S}_\tau$ to $T_p\mathscr{S}_\tau$. Since both A and the metric on $\mathscr{S}_\tau$ are invariant, the eigenspaces of the (symmetric, hence diagonalizable) tensor field A are also invariant under the action of H_p. But H_p acts transitively on unit vectors of $T_p\mathscr{S}_\tau$ so the only subspaces invariant under H_p are $T_p\mathscr{S}_\tau$ and $\{0\}$. Hence all eigenvalues of A coincide, and for $p \in \mathscr{S}_\tau$ we find

$$A^i{}_j = f(\tau, p)\delta^i_j \tag{6.3.15}$$

for some function f.

It should be clear from the proof of Proposition 6.2.8, p. 210 that the tensor field $A_{ij} := \nabla_{(i}u_{j)}$ is invariant under the connected component of the isometry group of $\mathscr{S}_\tau$. This allows us to summarize our arguments so far as follows:

Lemma 6.3.1
Let (M, g) be a maximally symmetric Riemannian manifold, and let A be a symmetric two-covariant tensor field on M which is invariant under the action of the connected component of the isometry group of M. Then A is a constant multiple of the metric.

Proof
Transitivity of the action of the isometry group on M implies that the function f of (6.3.15) is independent of $p \in M$, whence the result.   □

Remark 6.3.2 The result remains true for pseudo-Riemannian manifolds but the proof requires more care, as symmetric two-covariant tensors are not necessarily diagonalizable in non-Riemannian signatures.

In the coordinates of (6.3.13) we have

$$\nabla_i u_j = \partial_i u_j - \Gamma^\mu{}_{ij}u_\mu = -\Gamma^0_{ij} = -\frac{1}{2}g^{00}(\partial_i g_{0j} + \partial_j g_{0i} - \partial_0 g_{ij}) = -\frac{1}{2}\partial_t g_{ij}\,.$$

$$(6.3.16)$$

Lemma 6.3.1 leads us to the conclusion that $\partial_t g_{ij}$ is proportional to the metric on $\mathscr{S}_\tau$. We now have:

Proposition 6.3.3
Suppose that the metric induced on each level set of t is maximally symmetric, and that the time derivative of the metric is invariant under the group of isometries of the metric. Then there exists a coordinate system in spacetime in which

$$g = -dt^2 + R^2(t)\underbrace{g_{ij}(x)dx^i dx^j}_{=:h}\,.$$

$$(6.3.17)$$

Proof
Maximal symmetry of the level sets of t shows that for every t there exists a map ψ_t such that the metric $h(t)$ on $\mathscr{S}_t$ defined in (6.3.13) can be written as

$$h(t) = b^2(t)\psi_t^* h_k\,,$$

with a time-independent metric h_k of sectional curvature $k \in \{0, \pm 1\}$, where $b(t)$ is determined by the sectional curvature of $h(t)$. (If $k = 0$, then we set $b(t) \equiv 1$.) Using Lemma 6.3.1, we find that there exists a function, say $c(t)$, such that

$$\frac{dh(t)}{dt} = c(t)h(t).$$

Hence $h(t) = C(t)h(t_0)$, for some t_0, where $\dot{C} = c$. The result follows by setting $R(t) = b(t)\sqrt{C(t)}$ and $h = h(0)$. $\square$

6.4 Friedman-Lemaître-Robertson-Walker Cosmologies

In view of the results so far, we consider a spacetime $(\mathcal{M}, g)$ such that
1. $\mathcal{M} = I \times N$ where I is an open interval and N is a three-dimensional manifold.
2. For each $t \in I$, $(N, g|_{\{t\} \times N})$ is a maximally symmetric Riemannian manifold.

We denote by t the coordinate running along the first factor in $I \times N$.

The analysis in the previous sections shows that the Friedman-Lemaître-Robertson-Walker (FLRW) metric,

$$g = -dt^2 + g_{ij}dx^i dx^j = -dt^2 + R^2(t) \underbrace{\left(\frac{1}{1 - kr^2}dr^2 + r^2 d\Omega^2 \right)}_{=:h_k}, \qquad (6.4.1)$$

where $k \in \{0, \pm 1\}$, satisfies our requirements. We will sometimes write simply h for the metric h_k defined in the last equation.

Remark 6.4.1 It is often claimed in the literature that (6.4.1) is the only metric which satisfies the cosmological principle which, as already mentioned, is wrong. From what has been said in ▶ Sect. 6.2, after imposing the cosmological principle we can only conclude that for every t there exists a diffeomorphism

$$\psi_t : N \to N \text{ such that } g = -dt^2 + b^2(t)\psi_t^* h_k \qquad (6.4.2)$$

for some function $b(t)$ without zeros. If ψ_t were known to be time-independent, thus $\psi_t \equiv \psi_{t_0}$ for any t_0, then the map

$$(t, p) \mapsto \left(t, \psi_{t_0}^{-1}(p) \right) \qquad (6.4.3)$$

would bring the metric to the form (6.4.1). But time-independence of ψ_t *does not* follow from the cosmological principle alone.

Indeed, a simple example of a metric which satisfies 1. and 2. above is

$$g = -dt^2 + \left(\frac{R(t)^2}{1 - kR(t)^2 r^2}dr^2 + r^2 d\Omega^2 \right), \qquad (6.4.4)$$

which is not of the form (6.4.1) when $\dot{R} \neq 0$. Indeed, a redefinition of the radial coordinate $r \to R(t)r$ will bring the space-part of the metric to the desired form but will introduce off-diagonal $dt\, dr$ terms in the metric.

Another example is provided by the Kasner metrics,

$$g = -d\tau^2 + \tau^{2p_1} dw^2 + \tau^{2p_2} dy^2 + \tau^{2p_3} dz^2 , \tag{6.4.5}$$

with $\tau > 0$, which are vacuum solutions of the Einstein equations if

$$p_1 + p_2 + p_3 = 1 , \quad p_1^2 + p_2^2 + p_3^2 = 1 .$$

(When $p_1 = 1$, $p_2 = 0 = p_3$ this is the Minkowski metric where (t, x) have been replaced by $t = \tau \cosh w$, $x = \tau \sinh w$.) The level sets of τ are flat $\mathbb{R}^3$'s (the maps $(w, y, z) \mapsto \psi_\tau(w, y, z) := (\tau^{p_1} w, \tau^{p_2} y, \tau^{p_3} z)$ bring the metric induced on the levels set of τ to the canonical form), hence homogeneous and isotropic, but the metric is not of the form (6.4.1).

These examples show that a further condition is needed to ensure that the maps ψ_t of (6.4.2) are time-independent. One such possibility is to assume the existence of a congruence as we did in ▶ Sect. 6.3. Another one is to invoke the fact that Einstein equations are second order in time, therefore the relevant data for the evolutionary problem at a given time must include both the metric and its first time derivative. Invariance under the isometries could then be imposed on both the metric and its first time-derivative. One can then conclude as before using the fact that, for well-behaved evolutionary systems, symmetries of initial data extend to symmetries of solutions.

We wish to understand the geometry of a universe described by a metric of the form (6.4.1).

We start our analysis by noting that the vector field ∂_t has unit length. This implies that t coincides with the proper time for observers with *constant space coordinates*. Further, such observers are freely falling: Indeed, if $\frac{dx^\mu}{dt} \partial_\mu = \partial_t$, then

$$\frac{d^2 x^\mu}{dt^2} + \Gamma^\mu{}_{\alpha\beta} \frac{dx^\alpha}{dt} \frac{dx^\beta}{dt} = \Gamma^\mu{}_{00} = \frac{1}{2} g^{\mu\sigma} (2\partial_0 g_{\sigma 0} - \partial_\sigma g_{00}) = 0 . \tag{6.4.6}$$

Equivalently, world lines orthogonal to the level-sets of t are timelike geodesics. (Alternatively, we have $g(\nabla t, \nabla t) = -1$, so that the integral curves of $\partial_t = -\nabla t$ are geodesics by Problem 12, p. 263.)

6.4.1 Hubble Law

To continue, some remarks on Riemannian distance are in order. Consider a Riemannian manifold (N, h), the *h-distance* $d_h(p_1, p_2)$ between a pair of points $p_1, p_2 \in N$ is defined as

$$\inf_\gamma \int_0^1 \sqrt{h(\dot{\gamma}, \dot{\gamma})} d\lambda , \tag{6.4.7}$$

where the infimum is taken over all differentiable curves $[0, 1] \ni \lambda \mapsto \gamma(s) \in N$ such that $\gamma(0) = p_1$ and $\gamma(1) = p_2$. The dot in (6.4.7) denotes a derivative with respect to λ.

We will simply say *distance* when the metric is understood. Incidentally, the integral in (6.4.7) is called the *length of* γ with respect to the metric h.

Remark 6.4.2 For further use we note that given any point $p \in N$ there exists a neighborhood $\mathcal{O}$ of p such that for any $p_1, p_2 \in \mathcal{O}$ the infimum in (6.4.7) is attained on a geodesic segment from p_1 to p_2. It should be kept in mind that there may be more than one such geodesic segment, and that not every geodesic segment joining two points realizes the distance.

For example, on Euclidean $\mathbb{R}^n$ every pair of points is joined by a unique shortest geodesic, the straight line-segment between the points.

As another example, let p be the north pole of a sphere $S^2 \subset \mathbb{R}^3$. It follows from Example 1.2.2, p. 26, that the great circles passing through p are geodesics when S^2 is equipped with the metric induced from the Euclidean metric on $\mathbb{R}^3$. This and isometry invariance show that the intersection of S^2 with any plane in $\mathbb{R}^3$ passing through the origin is a geodesic. In particular there are infinitely many geodesics realizing the distance from the north pole to the south pole. Note that any pairs of points on S^2 which are not antipodal lie on a unique great circle. One can reach the second point by following the great circle from the first point in two different ways, one of the resulting geodesic segments realizing the distance, the other providing an example of a longer geodesic segment. □

Let us denote by $d_t^g(p_1, p_2)$ the space-distance in the metric $g_{ij}dx^i dx^j$ between two points $p_1, p_2 \in N$ at a given moment of time. It follows from (6.4.7) that

$$d_t^g(p_1, p_2) = R(t)d_h(p_1, p_2),$$ (6.4.8)

where d_h is the distance in the metric $h \equiv h_k$.

Let us denote by v the *rate of change of* $d_t^g(p_1, p_2)$ *with p_1 and p_2 fixed*. We have

$$v = \frac{\dot{R}}{R}d_t^g(p_1, p_2).$$ (6.4.9)

This formula has the interpretation of a law, first observed by Lemaître [57] in 1927 but attributed to Hubble based on a paper [48] published in 1929: If we define the "time-dependent *Hubble constant*"

$$H(t) = \frac{\dot{R}(t)}{R(t)},$$ (6.4.10)

and if two galaxies have fixed space-locations $p_1, p_2 \in N$, then (6.4.9) shows that the physical distance d between the galaxies at the current time $t = t_0$ changes according to the *Hubble-Lemaître law*

$$\boxed{v = H_0 d}, \quad \text{where } H_0 := H(t_0) = 67.8 \pm 0.9 \, \text{km s}^{-1} \, \text{Mparsec}^{-1}.$$ (6.4.11)

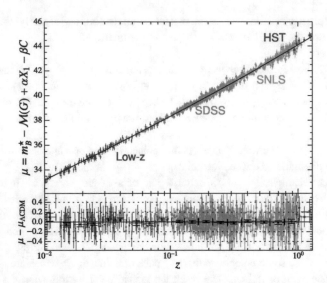

□ Fig. 6.7 The Hubble diagram of galaxies [distance vs. redshift (velocity)] from a large combined Supernova Ia (SNIa) distance-indicator sample [17, 68], together with the residual from the best theoretical fit, which includes accelerated expansion. The graph presents distance (as distance modulus; proportional to log of distance) vs. redshift z (Doppler shift, proportional to velocity for small redshift: $v/c \sim z$). The different SNIa samples are denoted by different colors and are listed by name [low-z sample; Sloan SDSS sample; SN legacy survey, SNLS; and Hubble Space Telescope SNIa, HST; for detail and references, see [17]. The black line represents the $d(z)$ relation expected for the current cosmology (a flat universe with mass density 30% and cosmological constant 70%) and a Hubble Constant of $H_0 = 70$ km/s/Mpc. The slight deviation from a linear graph at large distances is the evidence for acceleration. Hubble's 1929 graph from □ Fig. 6.8 (plotted with reverse axes, v vs. d) will fit in a tiny spot near/below the origin of this diagram. Credit: M. Betoule et al., A&A, 568, A22, 2014 [17], reproduced with permission © ESO

Here we use the value of H_0 from the Planck satellite mission [69], see □ Figs. 6.7, 6.9, and 6.10, compare the original plot of Hubble in □ Fig. 6.8.

Incidentally The value of the Hubble constant obtained from Supernovae Ia observations reads [74]

$$H_0 = (73.48 \pm 1.66) \, \text{km s}^{-1} \, \text{Mpc}^{-1} \, .$$

This differs by 3.7σ from the value, revised in 2018 [74], extracted from the Planck data:

$$H_0 = (67.4 \pm 0.5) \, \text{km s}^{-1} \, \text{Mpc}^{-1} \, .$$

(The value of H_0 determined by Lemaître [57] was 625 km per second per megaparsec. In Hubble's paper [48] one has $H_0 \approx 500$ km per second per megaparsec, cf. the middle Figure 6.8.)

In [21] the Hubble constant is determined by measuring the time delays between multiple images, produced by a foreground lensing object, of distant quasars. The time delays depend on the matter distribution in the lens (galaxy), on the overall matter distribution along the line of sight, and on the cosmological parameters. Assuming the ΛCDM model (namely, Cold Dark Matter with a positive cosmological constant Λ) one obtains

$$H_0 = 69.2^{+1.4}_{-2.2} \text{ km s}^{-1} \text{ Mpc}^{-1} .$$

It therefore seems that we have the first decimal of H_0 right, and that the observations reached a stage where the discrepancies require an explanation, not available so far.

Let us make the simplest possible approximation that R is linear: $R(t) = \alpha t$ for some α. This gives $H(t) = \dot{R}(t)/R(t) = 1/t$, which leads to the age T of the universe equal to $1/H_0$:

$$T \approx 4.5 \times 10^{17} s \approx 14 \underbrace{\text{billion}}_{10^9} \text{ years.}$$

It turns out that this is roughly the time needed for some of the stars to have evolved to the form at which we see them today; any significantly smaller number would lead to serious problems with our understanding of stellar evolution.

The best fit as of 2015 [69] to the more careful model described in ▶ Sect. 6.4.7 below reads

$$T = 13.79 \pm 0.04 \times 10^9 \text{ years,} \tag{6.4.12}$$

compare Table 6.3, p. 247 for the 2018 data.

The creationist's estimate of the age of the universe $T \approx 6000$ years would require a Hubble constant $H_0 \approx 1.63 \times 10^6 \text{ km s}^{-1} \text{ Mpc}^{-1}$, if fitted to a FLRW model. ☺

The "Hubble constant" is actually not a constant, but a function of t. To measure its rate of change, it is customary to introduce the dimensionless *deceleration parameter q* defined as

$$q = -\frac{\ddot{R} R}{\dot{R}^2} . \tag{6.4.13}$$

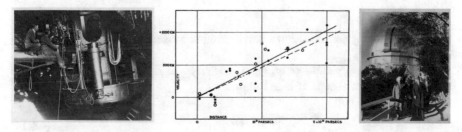

☐ **Fig. 6.8** Edwin Hubble and James Jeans on the Cassegrain platform of the 100-inch telescope, Mount Wilson Observatory; Hubble's original diagram from [48]; and Albert Einstein, Walther Mayer and William W. Campbell at the 100-inch telescope dome, Mount Wilson Observatory. Images courtesy of the Observatories of the Carnegie Institution for Science Collection at the Huntington Library, San Marino, California, reproduced with permission

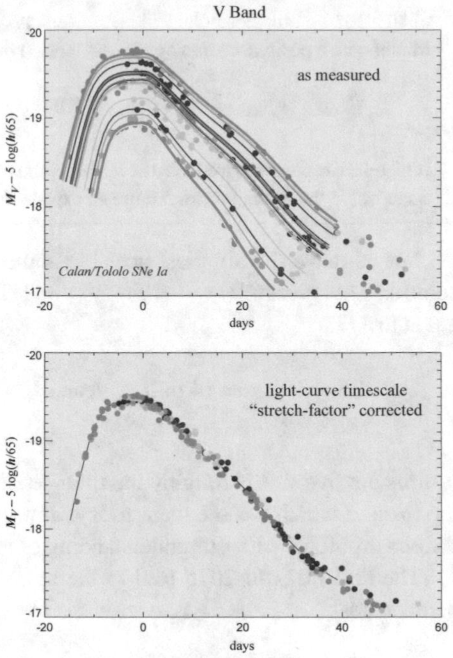

■ **Fig. 6.9** The Phillips law relates the width of the time-evolution of Supernovae Ia to their brightness. The lower plot shows the Supernovae from the upper plot, after normalizing, from [66]. Reproduced with permission

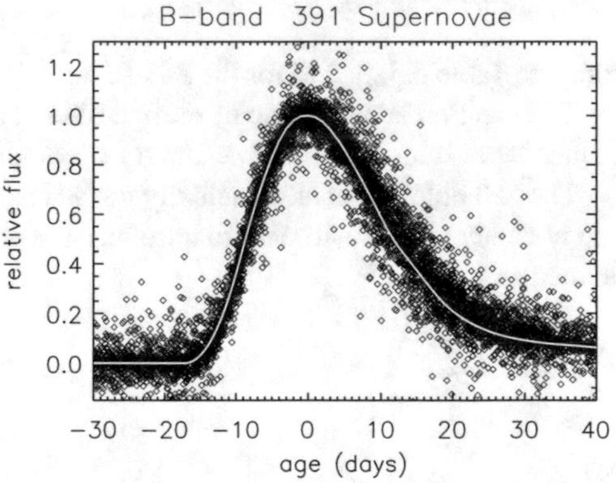

■ **Fig. 6.10** Time evolution of 391 individual SDSS-II Supernovae observations plotted in the rest frame and normalized to a peak flux of unity, from [41]. © AAS. Reproduced with permission

Positive values of q correspond to a decelerating universe; the universe is accelerating for negative ones.

The deceleration parameter or, more precisely, the *deceleration function* can be related to the rate of change of H as follows:

$$\dot{H} = \frac{\ddot{R}}{R} - \frac{\dot{R}^2}{R^2} = \frac{\dot{R}^2}{R^2}(\frac{\ddot{R}R}{\dot{R}^2} - 1) = -H^2(q+1)\,. \tag{6.4.14}$$

The value today q_0 of q, based on CMB and Supernova observations [69] (see ◘ Fig. 6.7), is

$$q_0 = -0.0639 \pm 0.009 .$$

Hence, the expansion of the universe is increasing, and the scale factor "accelerates": we say that we are experiencing *accelerated expansion*.

6.4.2 The Cosmological Redshift

Let us now turn our attention to null geodesics. Because the metric is homogeneous, any null geodesic can be mapped, using an isometry, to a null geodesic γ through the origin. Conservation of angular moment along geodesics in spherically symmetric metrics, or a direct calculation, show that a geodesic through the origin is radial. Since γ is null and both θ and φ are constant we have

$$dt^2 = \frac{R^2(t)dr^2}{1 - kr^2} \quad \Longrightarrow \quad \frac{dt}{R(t)} = \frac{dr}{\sqrt{1 - kr^2}} ,$$

where γ has been chosen to be future-directed. Let

$$\tau = \int \frac{dt}{R(t)} .$$

Supposing that the crest of a wave is emitted at the origin p_1 and observed at p_2, we find

$$\tau_{\text{observation,crest}} - \tau_{\text{emission,crest}} = \int_{0=r(p_1)}^{r(p_2)} \frac{dr}{\sqrt{1 - kr^2}} = d_h(p_1, p_2) . \tag{6.4.15}$$

One has an identical equation for the emission and observation of the next crest. Subtracting, one is led to

$$\Delta\tau_{\text{o}} = \Delta\tau_{\text{e}} , \tag{6.4.16}$$

where we have denoted by $\Delta\tau_{\text{o}}$ the τ-time elapsed between the observations of two successive crests, and $\Delta\tau_{\text{e}}$ the τ-time elapsed between the emissions of two successive crests. Hence, for Δt_{o} small compared to the rate of change of $R(t)$, using obvious notation,

$$\Delta\tau_{\text{o}} = \int_{t_{\text{o}}}^{t_{\text{o}}+\Delta t_{\text{o}}} \frac{dt}{R(t)} \approx \frac{\Delta t_{\text{o}}}{R(t_{\text{o}})} . \tag{6.4.17}$$

Similarly

$$\Delta\tau_{\text{e}} \approx \frac{\Delta t_{\text{e}}}{R(t_{\text{e}})} . \tag{6.4.18}$$

Inserting this into (6.4.16) we obtain

$$\frac{\Delta t_0}{R(t_0)} \approx \frac{\Delta t_e}{R(t_e)}.$$ (6.4.19)

Let ω be the frequency of the waves, as seen by observers moving orthogonally to the level sets of t. Since t is the proper time of such observers, ω is inversely proportional to Δt,

$$\frac{\omega_0}{\omega_e} \approx \frac{R(t_e)}{R(t_0)}.$$ (6.4.20)

The left-hand side of (6.4.20) describes how the spectrum of a radiating object changes between the time of observation and that of the emission. A convenient way of measuring this is given by the *red-shift factor z*, defined in terms of the wave-length λ as

$$z := \frac{\lambda_0 - \lambda_e}{\lambda_e} = \frac{\lambda_0}{\lambda_e} - 1 = \frac{\omega_e}{\omega_0} - 1$$

$$\approx \frac{R(t_0)}{R(t_e)} - 1.$$ (6.4.21)

Hence, the value of z provides direct information about the ratio of the scale factors of the universe between the observation and the emission times.

For times for which a linear approximation applies we can write

$$R(t_0) \approx R(t_e) + \dot{R}(t_e)(t_0 - t_e)$$

$$= R(t_e) + \underbrace{\frac{\dot{R}(t_e)}{R(t_e)}}_{=H(t_e)} R(t_e)(t_0 - t_e)$$

$$= R(t_e)\big(1 + H(t_e)(t_0 - t_e)\big).$$

From (6.4.21),

$$R(t_0) \approx R(t_e)\big(1 + z\big),$$

and we obtain

$$H(t_e)\big(t_0 - t_e\big) \approx z.$$ (6.4.22)

If $H(t)$ is slowly varying, we can also write

$$H(t_0)\big(t_0 - t_e\big) \approx z.$$ (6.4.23)

Let us return to (6.4.15), which we simply rewrite as

$$\tau_o - \tau_e = d_h(p_1, p_2). \tag{6.4.24}$$

For slowly varying R we have, by definition of τ,

$$\tau_o - \tau_e = \int_{t_e}^{t_o} \frac{dt}{R(t)} \approx \frac{1}{R(t_o)}(t_o - t_e).$$

This leads to

$$
\begin{aligned}
t_o - t_e &\approx R(t_o)\big(\tau_o - \tau_e\big) \\
&= R(t_o)d_h(p_1, p_2) \\
&= d,
\end{aligned}
\tag{6.4.25}
$$

where d is the physical distance between p_1 and p_2 at the time of observation. We conclude that

$$\boxed{H_0 d \approx z}, \tag{6.4.26}$$

This gives an approximate relation between the red-shift factor z and the space-distance between us and the sources, which applies for time scales at which both $R(t)$ and $H(t)$ do not change much.

Comparing with (6.4.11), we also see that

$$\boxed{v \approx z}. \tag{6.4.27}$$

Thus, keeping in mind the provisos above, the velocity v of a comoving galaxy (in units $c = 1$) coincides with its redshift factor.

Formula (6.4.27) can be compared with the special-relativistic Doppler shift for head-on motion:

$$\frac{\lambda_o}{\lambda_e} = \sqrt{\frac{1+\beta}{1-\beta}}, \tag{6.4.28}$$

where $\beta = v/c$ is the velocity of the emitter relative to the observer. Equation (6.4.28) gives

$$
\begin{aligned}
z = \frac{\lambda_o}{\lambda_e} - 1 &= \frac{\sqrt{1+\beta} - \sqrt{1-\beta}}{\sqrt{1-\beta}} = \frac{2\beta}{(\sqrt{1+\beta} + \sqrt{1-\beta})\sqrt{1-\beta}} \\
&\approx \beta,
\end{aligned}
\tag{6.4.29}
$$

for small β. Not unexpectedly, this coincides with (6.4.27) for small $v = \beta$, with the origin of the effect being the changing scale factor of the universe.   □

Exercise 6.4.3   Derive the second-order-approximate formula

$$H_0 d \approx z - \frac{1 + q_0}{2} z^2 .$$ (6.4.30)

The deceleration parameter q_0 can be determined by fitting (6.4.30) to ◻ Fig. 6.7.

6.4.3 The Einstein Equations for a FLRW Metric

Consider a metric of the form (6.4.1),

$$g = -dt^2 + g_{ij}dx^i dx^j = -dt^2 + R^2(t) \underbrace{\left(\frac{1}{1 - kr^2}dr^2 + r^2 d\Omega^2 \right)}_{=:h_k} .$$ (6.4.31)

A calculation gives the following nonvanishing components of the Einstein tensor

$$-G^0{}_0 = 3\frac{\dot{R}^2 + k}{R^2} , \quad G^1{}_1 = G^2{}_2 = G^3{}_3 = -\frac{2\ddot{R}}{R} - \frac{\dot{R}^2 + k}{R^2} .$$ (6.4.32)

The above is easily seen to be compatible with a perfect fluid energy-momentum tensor,

$$T_{\mu\nu} = (p + \rho)u_\mu u_\nu + pg_{\mu\nu} ,$$ (6.4.33)

with $u^\mu \partial_\mu = \partial_t$, $\rho = \rho(t)$ and $p = p(t)$, in which case the Einstein equations $G_{\mu\nu} + \Lambda g_{\mu\nu} = \kappa T_{\mu\nu}$, where $\kappa := 8\pi G/c^4$, read

$$\boxed{3\frac{\dot{R}^2 + k}{R^2} = \kappa\rho + \Lambda ,}$$ (6.4.34)

$$-\frac{2\ddot{R}}{R} - \frac{\dot{R}^2 + k}{R^2} = \kappa p - \Lambda .$$ (6.4.35)

We will often refer to (6.4.34) as the *Friedman equation*.

The combination

$$\frac{1}{3} \times (6.4.34) + (6.4.35)$$

gives

$$-\frac{2\ddot{R}}{R} = \frac{\kappa}{3}(\rho + 3p) - \frac{2\Lambda}{3} .$$ (6.4.36)

Recall that the contracted Bianchi-identity implies that the divergence of the Einstein tensor vanishes identically; this can be thought of as an integrability condition for the Einstein equations. An example of this has already been seen in ▶ Sect. 5.1, p. 167, where we used this divergence identity to derive the field equations for the fluid variables. In the current case the divergence identity reads

$$\frac{d}{dt}\left\{R^3 \times \text{(left-hand side of (6.4.34))}\right\} + \text{(left-hand side of (6.4.35))} \times 3R^2\dot{R} = 0,$$

(6.4.37)

which can also be checked by a direct calculation. But then we must also have

$$\frac{d}{dt}\left\{R^3 \times \text{(right-hand side of (6.4.34))}\right\} + \text{(right-hand side of (6.4.35))} \times 3R^2\dot{R} = 0,$$

(6.4.38)

which is equivalent to

$$\boxed{\frac{d}{dt}\left(\rho R^3\right) + p\frac{d}{dt}\left(R^3\right) = 0.}$$

(6.4.39)

In particular we have shown

$$\boxed{\text{If (6.4.34) holds then (6.4.35) implies (6.4.39).}}$$

(6.4.40)

Given an equation of state, say $p = p(\rho)$ and constants Λ and k, Eqs. (6.4.34) and (6.4.39) provide a closed set of first-order evolution equations for ρ, p, and R.

For further purposes it is useful to enquire about the validity of the implication reverse to (6.4.40). If (6.4.34) and (6.4.39) hold we can rewrite (6.4.38) as

$$\frac{d}{dt}\left(R^3 \times \text{(left-hand side of (6.4.34))}\right) + \text{(right-hand side of (6.4.35))} \times 3R^2\dot{R} = 0.$$

(6.4.41)

Equation (6.4.38) gives

$$-\text{(left-hand side of (6.4.35))} \times 3R^2\dot{R} + \text{(right-hand side of (6.4.35))} \times 3R^2\dot{R} = 0.$$

(6.4.42)

Keeping in mind that we are assuming that $R \neq 0$ except perhaps at isolated points, we see that

$$\boxed{\text{If (6.4.34) holds then (6.4.39) implies (6.4.35) on regions where } R\dot{R} \neq 0.}$$

(6.4.43)

Summarizing, (6.4.39) is equivalent to (6.4.35) on regions where $R\dot{R} \neq 0$ whenever (6.4.34) holds.

6.4.4 Vacuum Solutions

The simplest solutions are the vacuum ones, with $\rho = p = 0$. They are mostly irrelevant for cosmology except for providing the asymptotics for large R, but have some interest in themselves.

In vacuum we have

$$\dot{R}^2 = \frac{\Lambda}{3}R^2 - k \,. \tag{6.4.44}$$

We start by noting that $\boxed{\Lambda < 0}$ is only possible with $k = -1$, in which case we obtain

$$R = \frac{1}{\omega}\sin(\omega t) \,, \quad \omega = \sqrt{\frac{|\Lambda|}{3}} \,.$$

The resulting metric,

$$g = -dt^2 + \frac{\sin^2(\omega t)}{\omega^2}h_{-1} \,, \tag{6.4.45}$$

is the *anti de Sitter metric* (6.2.22) in a different coordinate system.

The case $\boxed{\Lambda = 0}$ leads to the Minkowski metric or subsets thereof. This is clear when $k = 0$, since then R is a constant. When $k = -1$ we find $R = \alpha t$, for some constant $\alpha \in \mathbb{R} \setminus \{0\}$. The resulting metric describes the interior of the light-cone in Minkowski spacetime in nonstandard coordinates.

When $\boxed{\Lambda > 0}$ we obtain three different foliations of (subsets of) the same spacetime, namely the *de Sitter* spacetime, with a metric already seen in (6.2.18) in a different coordinate system. More precisely:

1. When $k = 0$ one obtains a foliation of (part of) de Sitter spacetime by flat submanifolds:

$$g = -dt^2 + e^{2\sqrt{\frac{\Lambda}{3}}t}(dx^2 + dy^2 + dz^2) \,. \tag{6.4.46}$$

We note the exponential expansion of this universe.

2. The case $k = -1$ leads to

$$R(t) = \frac{1}{\lambda}\sinh(\lambda t) \,, \quad \lambda = \sqrt{\frac{\Lambda}{3}} \,, \tag{6.4.47}$$

giving a foliation of part of de Sitter spacetime by exponentially expanding hyperboloidal surfaces.

3. Finally, the case $k = 1$ gives a global foliation of de Sitter spacetime by spheres:

$$R(t) = \frac{1}{\lambda}\cosh(\lambda t), \quad \lambda = \sqrt{\frac{\Lambda}{3}}. \tag{6.4.48}$$

Incidentally The case $\Lambda > 0$ appears to be the most relevant for cosmology, so let us show how the various coordinate systems above relate to each other in this case (compare [75, 80]).

In ▶ Sect. 6.2.2 (see p. 212) we derived the $(n + 1)$-dimensional de Sitter metric,

$$h = -\frac{1}{\frac{r^2}{\alpha^2} - 1} dr^2 + r^2 d\Omega_n^2, \quad r > \alpha > 0. \tag{6.4.49}$$

as the metric induced on the timelike hyperboloid

$$\mathscr{S}_{\alpha^2} := \{w^2 + (x^1)^2 + \ldots + (x^n)^2 = \alpha^2 + \tau^2\}$$

$$\equiv \{r^2 = \alpha^2 + \tau^2\} \subset \mathbb{R}^{1,n+1}. \tag{6.4.50}$$

Here we have replaced the symbol R used in ▶ Sect. 6.2.2 by α, the dimension n used there by $n + 1$, and the coordinate t used there by τ, to avoid a clash of notation with the current time coordinate t and the scale factor $R(t)$.

The apparent singularity of the metric h at $r = \alpha$ in (6.4.49) is due to a poor choice of coordinates, as can be seen by setting, for $t > 0$,

$$r = \alpha \cosh(\alpha^{-1}t). \tag{6.4.51}$$

This leads to

$$\boxed{h = -dt^2 + \alpha^2 \cosh^2(\alpha^{-1}t) d\Omega_n^2, \quad \alpha > 0,} \tag{6.4.52}$$

where we can now extend the metric from $t > 0$ to all values of $t \in \mathbb{R}$. After this extension, we obtain a globally defined, manifestly regular, and in fact geodesically complete metric on the cylinder $\mathbb{R} \times S^n$. We recover the scale factor (6.4.48) by setting $\alpha = \lambda^{-1}$.

To obtain (6.4.46), on

$$\mathscr{S}_{\alpha^2} \cap \{\tau + w > 0\}$$

we introduce new coordinates $(\hat{t}, \hat{x}^i)$ by setting

$$\tau = \alpha \sinh(\alpha^{-1}t) - \frac{1}{2}\exp(-\alpha^{-1}t)\sum(\hat{x}^i)^2, \tag{6.4.53}$$

$$w = \alpha \cosh(\alpha^{-1}t) - \frac{1}{2}\exp(-\alpha^{-1}t)\sum(\hat{x}^i)^2, \tag{6.4.54}$$

$$x^i = \exp(-\alpha^{-1}t)\hat{x}^i. \tag{6.4.55}$$

The flat metric

$$-d\tau^2 + dw^2 + (dx^1)^2 + \ldots + (dx^n)^2$$

expressed in the above coordinates takes the form

$$-dt^2 + e^{-2t/\alpha}\left((d\hat{x}^1)^2 + \ldots + (d\hat{x}^n)^2\right),\tag{6.4.56}$$

which coincides with (6.4.46) after removing the hats on the $\hat{x}^i$'s, changing t to its negative, and setting

$$\alpha^{-1} = \sqrt{\frac{\Lambda}{3}}.\tag{6.4.57}$$

Exercise 6.4.4 Let $\rho = \sqrt{x^2 + y^2 + z^2}$, where (x, y, z) are as in (6.4.46), and let us denote by τ the coordinate t appearing in (6.4.46), thus we rewrite (6.4.46) as

$$g = -d\tau^2 + e^{2\sqrt{\frac{\Lambda}{3}}\tau}\left(d\rho^2 + \rho^2(d\theta^2 + \sin^2(\theta)d\varphi^2)\right).\tag{6.4.58}$$

Check that the coordinate transformation

$$r = \rho e^{\tau/\ell}, \quad t = \tau - \frac{\ell}{2}\ln(-\ell^2 + \rho^2 e^{2\tau/\ell}),\tag{6.4.59}$$

where $\ell^{-1} = \sqrt{\Lambda/3}$, brings (6.4.58) to the form

$$g = -V dt^2 + \frac{dr^2}{V} + r^2(d\theta^2 + \sin^2(\theta)d\varphi^2), \quad V = 1 - \frac{r^2}{\ell^2}.\tag{6.4.60}$$

6.4.5 Dust ("Matter-Dominated") Universe

From now on we will assume that $\rho \not\equiv 0$.
Consider a universe where matter satisfies $p = 0$. Equation (6.4.39) gives

$$\rho = \frac{\mathscr{E}}{R^3}.\tag{6.4.61}$$

Positivity of energy density requires $\mathscr{E} \geq 0$ and, since the vacuum models have already been taken care of, we will in fact assume $\mathscr{E} > 0$.

In the simplest case $k = \Lambda = 0$ the Friedman equation (6.4.34) gives

$$\frac{3\dot{R}^2}{R^2} = \kappa\rho = \kappa\frac{\mathscr{E}}{R^3}.\tag{6.4.62}$$

Equivalently, for constants C, C_1, C_2, and C_3 which are irrelevant for our further purposes,

$$R\dot{R}^2 = C \implies \sqrt{R}\dot{R} = C_1 \implies R^{\frac{3}{2}} = C_2 t \implies R = C_3 t^{2/3}.$$

Hence

$$H = \frac{\dot{R}}{R} = \frac{2}{3t}.$$

If this was a correct description of our world, it would give a current age of the universe

$$t_0 = \frac{2}{3}\frac{1}{H_0},$$

which is too small by a factor of about $1/3$: Indeed, in such a universe, the stars would not have had enough time to evolve to the form we see today; for this, they need about 14 Giga years.

Still assuming $\Lambda = 0$ but $k \neq 0$, from (6.4.34) and (6.4.61) one obtains instead

$$\dot{R}^2 + k = \frac{\kappa \mathscr{E}}{3R}. \tag{6.4.63}$$

Incidentally Equation (6.4.63) is reminiscent of the law of conservation of Newtonian energy, compare (3.8.4), p. 73, for a point mass m_0 moving in a central Newtonian gravitational potential with mass m:

$$E_N = \frac{m_0 \dot{r}^2}{2} + \frac{J_N^2}{2m_0 r^2} - \frac{Gmm_0}{r}, \quad \text{with } J_N = m_0 r^2 \dot{\varphi}. \tag{6.4.64}$$

We see that the equations coincide after setting $r = R$, $m_0 = 2$, $E_N = -k$, and $J_N = 0$ and choosing m appropriately. We will return to this analogy in the general case shortly.

Solutions of (6.4.63) can be found in parametric form, with $C = k\kappa \mathscr{E}/6$:

$$k = 1 : R = C(1 - \cos\eta), \quad t = C(\eta - \sin\eta), \tag{6.4.65}$$

$$k = -1 : R = C(\cosh\eta - 1), \quad t = C(-\eta + \sinh\eta). \tag{6.4.66}$$

The plots of the resulting curves can be found in ◻ Fig. 6.11. We see that the so-called "closed" models, with $k = 1$, lead to a *big-bang to big-crunch* scenario, while the "open models" $k = 0$ and -1 lead to a *big-bang to big-freeze* spacetimes.

Incidentally The "closed" terminology is motivated by the fact that when $k = 1$ the space-slices are "closed manifolds", namely a sphere or quotients thereof. It is somewhat misleading for the $k = 0$ and -1 cases, since these values of k admit spatially non-compact models (e.g. a flat $\mathbb{R}^3$ or hyperbolic space as space-slices), but also their compact quotients (e.g. flat three-dimensional tori's, or compact hyperbolic manifolds).

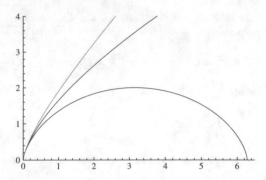

□ Fig. 6.11 The scale factor R/C in a dust-filled universe with $\Lambda = 0$ as a function of t/C, with $k = 1$ (lowest curve), $k = 0$ (middle curve), and $k = -1$

By inspection of the graph, or directly from the parametric solution, we see that the leading behavior of R for small t is independent of k, hence the solution behaves as a constant times $t^{2/3}$ near the big bang.

Exercise 6.4.5 Check, or derive, (6.4.65)–(6.4.66).

For general parameters such that $\mathscr{E}\Lambda \neq 0$ no explicit solutions in terms of elementary functions are known, whether explicit or parametric, but it is easy to obtain a qualitative description. Equation (6.4.34) with $\rho = \mathscr{E}/R^3$ takes the form

$$3\frac{\dot{R}^2 + k}{R^2} = \kappa\rho + \Lambda = \frac{\kappa\mathscr{E}}{R^3} + \Lambda\,. \tag{6.4.67}$$

Equivalently,

$$\dot{R}^2 = \frac{\kappa\mathscr{E}}{3R} + \frac{\Lambda}{3}R^2 - k\,. \tag{6.4.68}$$

For small R the dominant term at the right-hand side is $\frac{\kappa\mathscr{E}}{R^3}$:

$$\left(\frac{dR}{dt}\right)^2 \approx \frac{\kappa\mathscr{E}}{3R}\,. \tag{6.4.69}$$

This coincides with (6.4.62), except for the equality sign there replaced by $\approx$ here. This implies that, for small R, solutions of (6.4.68) behave as solutions of (6.4.62), namely $R \approx Ct^{2/3}$, regardless of the values of k and Λ.

For negative Λ the right-hand side of (6.4.68) is negative when R is sufficiently large, while the left-hand side is positive, which is not possible. We conclude that R is bounded from above for all solutions with negative Λ.

For positive Λ and large R the dominant term at the right-hand side of (6.4.68) is $\Lambda R^2/3$:

$$\left(\frac{dR}{dt}\right)^2 \approx \frac{\Lambda}{3}R^2\,.$$

□ Fig. 6.12 Representative plots of the potential $V(R)$ with $\Lambda < 0, \Lambda = 0, 0 < \Lambda < \Lambda_c$, $\Lambda = \Lambda_c$, and $\Lambda > \Lambda_c$, with Λ_c given by (6.4.73). The order corresponds to a decreasing ordering of the graphs

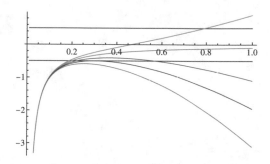

This implies that the corresponding solutions behave as the expanding solutions of

$$\left(\frac{dR}{dt}\right)^2 = \frac{\Lambda}{3}R^2 \implies R = Ae^{\sqrt{\frac{\Lambda}{3}}t}, \tag{6.4.70}$$

for some constant A. We conclude that, for positive Λ, those solutions that become sufficiently large keep expanding forever, exponentially fast.

It turns out to be useful to rewrite (6.4.67) as

$$\frac{1}{2}\left(\frac{dR}{dt}\right)^2 \underbrace{-\frac{\kappa\mathscr{E}}{6R} - \frac{\Lambda}{6}R^2}_{=:V(R)} = -\frac{k}{2}. \tag{6.4.71}$$

Equation (6.4.71) has the form of the law of conservation of energy for a particle of mass equal to one, total energy equal to $-k/2$, and moving in the potential $V(R)$. The trajectories of the system can thus be understood by plotting the potential, see □ Fig. 6.12.

Those trajectories for which the energy level $-k/2$ never intersects the graph of V describe universes which start from a big bang and expand forever.

On the other hand, for parameters such that the energy-level $-k/2$ intersects the graph, the trajectory starting from $R = 0$ will attain a maximum at a point where $V(R) = -k/2$:

$$-\frac{\kappa\mathscr{E}}{6R} - \frac{\Lambda}{6}R^2 = -\frac{k}{2} \in \left\{0, \pm\frac{1}{2}\right\}. \tag{6.4.72}$$

When $\Lambda < 0$ the function V is monotonous, tends to minus infinity as $R \searrow 0$, and tends to infinity as R approaches infinity, and so the equation always has a solution. This can also be clearly seen from the graph.

When $k \in \{-1, 0\}$ and $\Lambda \geq 0$ there is no solution of (6.4.72), since then the left-hand side is negative while the right-hand side is nonnegative. So all solutions are forever expanding in this case.

For $k = 1$ and for positive Λ there exists a critical value, say Λ_c, such that for $\Lambda > \Lambda_c$ no solutions of (6.4.72) exist, and for $\Lambda < \Lambda_c$ two solutions exist. Note that

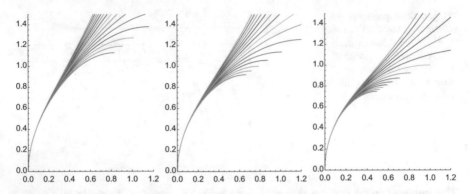

Fig. 6.13 The scale factor $x = 3R/(\kappa\mathscr{E})$ in a dust-filled universe as a function of $3t/(\kappa\mathscr{E})$, with $\lambda = \frac{\kappa^2\mathscr{E}^2}{3^3}\Lambda \in \{-1.4, -1.2, \ldots, 1.2, 1.4\}$ and $k = -1$ (left plot), $k = 0$ (middle plot), and $k = 1$ (right plot). The curves that stop at their maximum should be continued by reflection across the vertical line passing through the maximum

in the range $\Lambda < \Lambda_c$ there exist solutions of the Friedman equations which come-in from infinity, achieve a minimum radius, and return to infinity: this is a "big-freeze to big-freeze scenario."

The value of Λ_c can be found by observing that when $\Lambda = \Lambda_c$, the graph of V is tangent to the horizontal line $k = -1/2$ at the intersection point $R = R_c$. We thus have both $V(R_c) = -1/2$ and $V'(R_c) = 0$. Simple algebra gives then

$$\Lambda_c = \frac{4}{\kappa^2\mathscr{E}^2}, \qquad R_c = \left(\frac{\kappa}{4}\right)^{\frac{1}{3}}\mathscr{E}. \tag{6.4.73}$$

Incidentally It is easy to solve the equation numerically regardless of the value of Λ. Introducing

$$x = \frac{3}{\kappa\mathscr{E}}R, \qquad \tau = \frac{3}{\kappa\mathscr{E}}t, \tag{6.4.74}$$

(6.4.71) implies

$$\frac{dx}{\sqrt{\frac{1}{x} + \lambda x^2 - k}} = \pm d\tau, \qquad \lambda = \frac{\kappa^2\mathscr{E}^2}{3^3}\Lambda, \tag{6.4.75}$$

and we choose the plus sign for expanding solutions. The result of integration of (6.4.75) is plotted in ◘ Fig. 6.13 for several values of the parameter λ. The plots exhibit clearly the fact, already discussed on general grounds, that a negative Λ will always lead to a big-bang—big-crunch scenario, independently of k.

6.4.6 Radiation-Fluid Universe

There exist two ways of approximating radiation by a fluid model. The first one is to assume that $p = 0$ and that

$$T_{\mu\nu} = \rho k_\mu k_\nu \,,\tag{6.4.76}$$

where k is a null vector. This is an exact formula for plane waves in Minkowski spacetime. But the cosmic microwave background is clearly not a single plane wave, so the form (6.4.76) of $T_{\mu\nu}$ is certainly irrelevant for describing the universe in the large.

Incidentally To obtain (6.4.76) we take the Maxwell four-potential to be

$$A_\mu = \Re(a_\mu e^{ik_\alpha x^\alpha}) \,,$$

where a_μ and k_μ are real-valued constant vectors satisfying $a_\mu k^\mu = k_\mu k^\mu = 0$, then

$$F_{\mu\nu} = \partial_\mu A_\nu - \partial_\nu A_\mu = \Re\left(i(k_\mu a_\nu - k_\nu a_\mu)e^{ik_\alpha x^\alpha}\right)$$
$$= -(k_\mu a_\nu - k_\nu a_\mu)\sin(k_\alpha x^\alpha) \,,$$

which gives

$$T_{\mu\nu} = \frac{1}{4\pi}\left(F_{\mu\alpha}F_\nu{}^\alpha - \frac{1}{4}F_{\alpha\beta}F^{\alpha\beta}g_{\mu\nu}\right)$$
$$= \frac{1}{4\pi}a_\alpha a^\alpha \sin^2(k_\beta x^\beta)k_\mu k_\nu \,.\tag{6.4.77}$$

The question arises, what is the energy-momentum tensor describing a background of photons coming isotropically from all directions. Some kind of averaging procedure applied to the above should give us the relevant $T_{\mu\nu}$. Here the key observation is that the Maxwell energy-momentum tensor has vanishing trace, and averaging over many photons should preserve this property. This is the logic behind the expectation that a bath of photons can be approximately described by a perfect fluid energy-momentum tensor as in (6.4.33), p. 228, but with vanishing trace:

$$T^\mu{}_\mu = (\rho + p)\underbrace{u^\mu u_\mu}_{-1} + p\delta^\mu_\mu = -\rho + 3p = 0 \,.\tag{6.4.78}$$

One therefore defines a *radiation fluid* as a fluid for which

$$p = \frac{1}{3}\rho \,.\tag{6.4.79}$$

Using this equation of state, (6.4.39) becomes

$$\frac{d}{dt}\left(\rho R^3\right) + p\frac{d}{dt}\left(R^3\right) = \frac{d}{dt}\left(\rho R^3\right) + \frac{\rho}{3}\frac{d}{dt}\left(R^3\right) = 0 \,.\tag{6.4.80}$$

Setting $f = R^3$, this can be rewritten as

$$0 = \frac{d\rho}{dt} f + \rho \frac{df}{dt} + \frac{\rho}{3} \frac{df}{dt} = \frac{d\rho}{dt} f + \frac{4\rho}{3} \frac{df}{dt} = f^{-1/3} \frac{d}{dt} \left(\rho f^{4/3} \right) .$$

Hence

$$\frac{d}{dt} \left(\rho R^4 \right) = 0 ,$$

thus there exists a constant K such that

$$\rho = \frac{K}{R^4} . \tag{6.4.81}$$

This equation can be understood as the *law of conservation of the number of photons*: Indeed, recall that the energy of each photon is red-shifted by a factor of $1/R$. If there are $n(\omega, t)$ photons with energy ω at time t, then the energy contained in this frequency channel evolves in time as $n(\omega, t)\hbar\omega c / R(t)$. This will coincide with (6.4.81) if $n(\omega, t) = n(\omega)/R^3(t)$. Integration over a volume with fixed space-coordinates introduces a factor $R^3(t)$, giving a constant number of photons at each frequency level.

In the simplest case $k = \Lambda = 0$ the Friedman equation (6.4.34) gives

$$\frac{3\dot{R}^2}{R^2} = \kappa\rho = \kappa \frac{K}{R^4} .$$

Equivalently,

$$R^2 \dot{R}^2 = \frac{\kappa K}{3} \implies R\dot{R} = \sqrt{\frac{\kappa K}{3}} \implies R^2 = 2\sqrt{\frac{\kappa K}{3}} t$$

$$\implies R = \left(\frac{4\kappa K}{3} \right)^{\frac{1}{4}} t^{1/2} .$$

Hence

$$H = \frac{\dot{R}}{R} = \frac{1}{2t} ,$$

which would lead to an existence time t_0 for the universe even shorter than in a universe filled with dust. This is clearly not the right model either.

More generally, assuming $\Lambda = 0$ but $k \neq 0$ the Friedman equation becomes

$$\dot{R}^2 + k = \frac{\kappa K}{3R^2} . \tag{6.4.82}$$

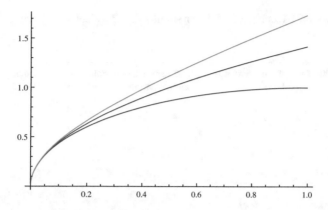

Fig. 6.14 The scale factor R/C in a radiation-filled universe as a function of t/C, with $k = 1$ (lowest curve), $k = 0$ (middle curve), and $k = -1$

The solutions are

$$R = C\sqrt{k\left(1 - \left(1 - \frac{kt}{C}\right)^2\right)},$$

(6.4.83)

where $C^2 = \kappa K/3$. The resulting functions $R(t)$ can be seen in ▢ Fig. 6.14. In a manner similar to that of the dust case, for small t we find a behavior independent of k albeit with a different exponent, namely

$$R \sim C\sqrt{2t}.$$

(6.4.84)

Exercise 6.4.6 Check, or derive, (6.4.83).

The qualitative discussion of the Friedman equation for dust-filled universes coincides, with minor modifications, to that of the radiation model. Equation (6.4.71) is replaced by

$$\frac{1}{2}\left(\frac{dR}{dt}\right)^2 \underbrace{- \frac{\kappa K}{6R^2} - \frac{\Lambda}{6}R^2}_{=:V(R)} = -\frac{k}{2}.$$

(6.4.85)

The general properties of the graph of V are very similar to the dust case, with an identical discussion of the trajectories. In particular the large R behavior of solutions is exponential, as in (6.4.70). The main difference is the small R behavior of solutions, since then (6.4.69) is replaced by

$$\left(\frac{dR}{dt}\right)^2 \approx \frac{\kappa K}{3R^2}.$$

(6.4.86)

This leads to a $C\sqrt{t}$ behavior of R for small t, independently of k and Λ, as already seen in (6.4.84).

Incidentally In contradistinction to the dust case, explicit solutions can be found even when $\Lambda \neq 0$. Equation (6.4.34) with $\rho = K/R^4$ takes the form

$$\left(\frac{dR}{dt}\right)^2 = \frac{\kappa K}{3R^2} + \frac{\Lambda}{3}R^2 - k . \tag{6.4.87}$$

Introducing

$$x = \sqrt{\frac{3}{\kappa K}}R , \quad \tau = \sqrt{\frac{3}{\kappa K}}t , \tag{6.4.88}$$

(6.4.87) implies

$$\frac{dx}{\sqrt{\frac{1}{x^2} + \lambda x^2 - k}} = d\tau , \quad \lambda = \frac{\kappa K}{9}\Lambda . \tag{6.4.89}$$

Imposing the condition that $x = 0$ at $\tau = 0$, this is easily integrated to give, for $\lambda > 0$,

$$2\sqrt{\lambda}\sqrt{-kx^2 + \lambda x^4 + 1} - k + 2\lambda x^2 = \underbrace{(2\sqrt{\lambda} - k)}_{=:\alpha} e^{2\sqrt{\lambda}\tau} . \tag{6.4.90}$$

Solving for x one obtains

$$x = \sqrt{\frac{\alpha^2 e^{2\sqrt{\lambda}\tau} + 2k\alpha + (k^2 - 4\lambda)e^{-2\sqrt{\lambda}\tau}}{4\lambda\alpha}}$$

$$= \frac{1}{2}\sqrt{\frac{e^{-2\sqrt{\lambda}\tau}\left(e^{2\sqrt{\lambda}\tau} - 1\right)\left(-ke^{2\sqrt{\lambda}\tau} + k + 2\sqrt{\lambda}\left(e^{2\sqrt{\lambda}\tau} + 1\right)\right)}{\lambda}}$$

$$= \sqrt{\frac{\sinh(\sqrt{\lambda}\tau)\left(2\sqrt{\lambda}\cosh(\sqrt{\lambda}\tau) - k\sinh(\sqrt{\lambda}\tau)\right)}{\lambda}} . \tag{6.4.91}$$

When $\lambda < 0$ one similarly finds

$$x = \sqrt{\frac{\sin(\sqrt{|\lambda|}\tau)\left(2\sqrt{|\lambda|}\cos(\sqrt{|\lambda|}\tau) - k\sin(\sqrt{|\lambda|}\tau)\right)}{|\lambda|}} .$$

The function x is plotted in ☐ Fig. 6.15 for several values of the parameter λ.

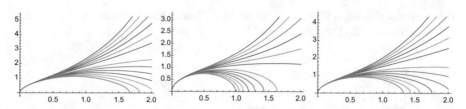

□ Fig. 6.15 The scale factor $x = \sqrt{\frac{3}{\kappa K}} R$ in a radiation-filled universe as a function of $\sqrt{\frac{3}{\kappa K}} t$, with $\lambda = \frac{\kappa K}{9} \Lambda \in \{-1.4, -1.2, \ldots, 1.2, 1.4\}$ and $k = -1$ (left plot), $k = 0$ (middle plot), and $k = 1$ (right plot)

6.4.7 Radiation-and-Dust Universe

It turns out that there is an ansatz which accounts for both dust and radiation, by setting

$$\rho = \rho_{\text{dust}} + \rho_{\text{radiation}} = \frac{\mathscr{E}}{R^3} + \frac{K}{R^4} .$$ (6.4.92)

If we plug this into the continuity equation

$$\frac{d}{dt}(\rho R^3) + p \frac{d}{dt}(R^3) = 0 ,$$

we obtain

$$\frac{d}{dt}\left(\mathscr{E} + K \frac{1}{R}\right) + 3pR^2 \dot{R} = 0 \quad \Longleftrightarrow \quad p = \frac{K}{3R^4} = \frac{1}{3}\rho_{\text{radiation}} = p_{\text{radiation}} .$$

The ansatz (6.4.92) describes thus a universe containing both matter and radiation, where only radiation gives a contribution to the pressure.

Inserting such a ρ into the Friedman equation one obtains

$$\frac{3\dot{R}^2}{R^2} = \kappa\rho + \Lambda - \frac{3k}{R^2} = \frac{\kappa\mathscr{E}}{R^3} + \frac{\kappa K}{R^4} + \Lambda - \frac{3k}{R^2} .$$ (6.4.93)

We have seen that explicit (possibly parametric) solutions of this equation can be obtained when $K\mathscr{E} = 0$. In the remaining cases we can solve the equation numerically, obtaining a family of solutions depending upon Λ, k, $\mathscr{E}$, and K, and fit the results to observations. For this, it is useful to define the *critical density* as the *energy density that one would need today for a universe with* $\Lambda = k = 0$:

$$3H_0^2 = \left.\frac{3\dot{R}^2}{R^2}\right|_{t=t_0} = \kappa \rho_{\text{critical}}.$$

Equation (6.4.93) at $t = t_0$ can be rewritten as

$$\kappa \rho_{\text{critical}} = \left.\frac{3\dot{R}^2}{R^2}\right|_{t=t_0} = \frac{\kappa \mathscr{E}}{R_0^3} + \frac{\kappa K}{R_0^4} + \Lambda - \frac{3k}{R_0^2}. \tag{6.4.94}$$

Equivalently

$$\rho_{\text{critical}} = \underbrace{\frac{\mathscr{E}}{R_0^3}}_{\rho_{\text{matter}}} + \underbrace{\frac{K}{R_0^4}}_{\rho_{\text{radiation}}} + \frac{\Lambda}{\kappa} - \frac{3k}{\kappa R_0^2}. \tag{6.4.95}$$

Introducing

$$\Omega_{\text{matter}} = \frac{\rho_{\text{matter}}}{\rho_{\text{critical}}}, \quad \Omega_{\text{radiation}} = \frac{\rho_{\text{radiation}}}{\rho_{\text{critical}}}, \quad \Omega_\Lambda = \frac{\Lambda}{\kappa \rho_{\text{critical}}} = \frac{\Lambda}{3H_0^2},$$

and

$$\Omega_{\text{curvature}} = -\frac{3k}{\kappa R_0^2 \rho_{\text{critical}}} = -\frac{k}{H_0^2 R_0^2},$$

Eq. (6.4.95) can be rewritten as a balance equation:

$$1 = \Omega_{\text{matter}} + \Omega_{\text{radiation}} + \Omega_\Lambda + \Omega_{\text{curvature}}. \tag{6.4.96}$$

In 2015 the data, based on cosmic microwave background (CMB) observations by the Wilkinson Microwave Anisotropy Probe (WMAP) [46] and Planck [67] satellites, on Baryon Acoustic Oscillations (BAO) [64], and on Supernova Ia data [25, 81] (see ◘ Fig. 6.16) were best fitted with (compare Table 6.1 for the 2013 WMAP data [46], and Tables 6.2 and 6.3 for the 2018 Planck data)

$$\Omega_{\text{matter}} = 0.273 \pm 0.015, \ \Omega_{\text{radiation}} \approx 5 \times 10^{-5}, \ \Omega_\Lambda = 0.726 \pm 0.015, \ \Omega_{\text{curvature}} \approx 0,$$

(the estimate for the last term is $-0.0179 < \Omega_{\text{curvature}} < 0.0081$). The matter contribution Ω_{matter} can be further decomposed as

$$\Omega_{\text{matter}} = \underbrace{\Omega_{\text{baryons}}}_{\approx 0.0456 \pm 0.0015} + \underbrace{\Omega_{\text{dark matter}}}_{\approx 0.228 \pm 0.013}.$$

Supernova Cosmology Project
Suzuki, et al., *Ap.J.* (2011)

◻ Fig. 6.16 A plot in the $\Omega_{\text{matter}} - \Omega_\Lambda$ plane summarizing the cosmic microwave background, baryon acoustic oscillations, and supernova observations [25, 81], from [82]. © AAS. Reproduced with permission

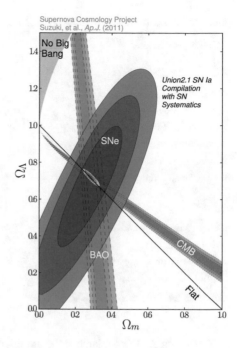

So, it seems that the dynamics of our universe can be described by a model with $k = 0$, $\mathscr{E} > 0$, $K > 0$, and $\Lambda > 0$, with a scale factor evolving according to the equation

$$3\dot{R}^2 = \frac{\kappa\mathscr{E}}{R} + \frac{\kappa K}{R^2} + \Lambda R^2 \,. \tag{6.4.97}$$

6.4.7.1 Asymptotic Behavior

Assuming $K > 0$, for small R the dominant term at the right-hand side of (6.4.97) is $\frac{\kappa K}{R^2}$:

$$3\dot{R}^2 \approx \frac{\kappa K}{R^2} \,. \tag{6.4.98}$$

Solving (6.4.98) with $\approx$ replaced by an equality we find

$$R \approx ct^{\frac{1}{2}} \,,$$

for some constant c. When $K = 0$ but $\mathscr{E} \neq 0$, a similar analysis gives

$$R \approx ct^{\frac{2}{3}} \,.$$

Table 6.1 WMAP cosmological data as of June 2013, from [46]. © AAS. Reproduced with permission

Parameter	Symbol	*WMAP* data	Combined data
Fit ΛCDM parameters			
Physical baryon density	$\Omega_b h^2$	0.02256	0.02240
Physical cold dark matter density	$\Omega_c h^2$	0.1142	0.1146
Dark energy density ($w = -1$)	Ω_Λ	0.7185	0.7181
Curvature perturbations, $k_0 = 0.002\,\mathrm{Mpc}^{-1}$	$10^9 \Delta_\mathcal{R}^2$	2.40	2.43
Scalar spectral index	n_s	0.9710	0.9646
Reionization optical depth	τ	0.0851	0.0800
Derived parameters			
Age of the universe (Gyr)	t_0	13.76	13.75
Hubble parameter, $H_0 = 100\,h\,\mathrm{km\,s}^{-1}\,\mathrm{Mpc}^{-1}$	H_0	69.7	69.7
Density fluctuations @ $8\,h^{-1}\,\mathrm{Mpc}$	σ_8	0.820	0.817
Baryon density/critical density	Ω_b	0.0464	0.0461
Cold dark matter density/critical density	Ω_c	0.235	0.236
Redshift of matter-radiation equality	z_{eq}	3273	3280
Redshift of reionization	z_{reion}	10.36	9.97

Assuming $\Lambda \neq 0$, for large R the dominant term at the right-hand side of (6.4.97) is $\frac{\kappa K}{R^2}$, leading to

$$3\dot{R}^2 \approx \Lambda R^2 \,. \tag{6.4.99}$$

This is clearly not possible if $\Lambda < 0$. We conclude that there exists an upper bound for the scale factor when Λ is negative, namely the largest root, call it $R_{\max}$, of the polynomial obtained by setting $\dot{R} = 0$ in the Friedman equation in its form (6.4.93):

$$0 = R^4 \times \frac{3\dot{R}^2}{R^2} = \Lambda R^4 - 3kR^2 + \kappa(\mathcal{E}R + K) \,. \tag{6.4.100}$$

For $\Lambda > 0$ one can solve (6.4.99) with $\approx$ replaced by an equality, leading to

$$R \approx C \exp(\sqrt{\Lambda/3}\,t)$$

for large times. In this case, if R becomes larger than the largest positive root $R_{\max}$ of (6.4.100), if any, so that the Λ-term in the equation overcomes the sum of the remaining ones, the universe will keep expanding forever, at an exponential rate.

6.4.7.2 Time-Independent Solutions

It turns out that there exist time-independent solutions in this model, $R(t) = R_0$ for all t. Such solutions have to solve the equation $V'(R_0) = 0$, where

$$V(R) = -\frac{\kappa \mathcal{E}}{6R} - \frac{\kappa K}{6R^2} - \frac{\Lambda}{6}R^2 \,. \tag{6.4.101}$$

Table 6.2 Definitions of the cosmological parameters used in Table 6.3, from [67]. ©AAS. Reprinted with permission

Parameter	Prior range	Baseline	Definition
$\omega_b \equiv \Omega_b h^2$	[0.005, 0.11]	...	Baryon density today
$\omega_c \equiv \Omega_c h^2$	[0.001, 0.99]	...	Cold dark matter density today
$100\theta_{MC}$	[0.5, 10.0]	...	100× approximation to r_*/D_A (CosmoMC)
τ	[0.01, 0.8]	...	Thomson scattering optical depth due to reionization
Ω_K	[−0.3, 0.3]	0	Curvature parameter today with $\Omega_{tot} = 1 - \Omega_K$
Σm_v	[0, 5]	0.06	The sum of neutrino masses in eV
$m_{v,sterile}^{eff}$	[0, 3]	0	Effective mass of sterile neutrino in eV
w_0	[−3.0, −0.3]	−1	Dark energy equation of states, $w(a) = w_0 + (1 - a)w_a$
w_a	[−2, 2]	0	As above (perturbations modeled using PPF)
N_{eff}	[0.05, 10.0]	3.046	Effective number of neutrino-like relativistic degrees of freedom (see text)
Y_P	[0.1, 0.5]	BBN	Fraction of baryonic mass in helium
A_L	[0, 10]	1	Amplitude of the lensing power relative to the physical value
n_s	[0.9, 1.1]	...	Scalar spectrum power-law index ($k_0 = 0.05\,\mathrm{Mpc}^{-1}$)
n_t	$n_t = -r_{0.05}/8$	Inflation	Tensor spectrum power-law index ($k_0 = 0.05\,\mathrm{Mpc}^{-1}$)
$dn_s/d\ln k$	[−1, 1]	0	Running of the spectral index
$\ln(10^{10} A_s)$	[2.7, 4.0]	...	Log power of the primordial curvature perturbations ($k_0 = 0.05\,\mathrm{Mpc}^{-1}$)
$r_{0.05}$	[0,2]	0	Ratio of tensor primordial power to curvature power at $k_0 = 0.05\,\mathrm{Mpc}^{-1}$
Ω_Λ		...	Dark energy density divided by the critical density today
t_0		...	Age of the Universe today (in Gyr)
Ω_m		...	Matter density (inc. massive neutrinos) today divided by the critical density
σ_8		...	RMS matter fluctuations today in linear theory
z_{re}		...	Redshift at which Universe is half reionized
H_0	[20,100]	...	Current expansion rate in km s^{-1} Mpc^{-1}
$r_{0.002}$		0	Ratio of tensor primordial power to curvature power at $k_0 = 0.002\,\mathrm{Mpc}^{-1}$
$10^9 A_s$		...	$10^9 \times$ dimensionless curvature power spectrum at $k_0 = 0.05\,\mathrm{Mpc}^{-1}$
$\omega_m \equiv \Omega_m h^2$		...	Total matter density today (inc. massive neutrinos)
z_*		...	Redshift for which the optical depth equals unity (see text)
$r_* = r_s(z_*)$		...	Comoving size of the sound horizon at $z = z_*$
$100\theta_*$		...	100× angular size of sound horizon at $z = z_*(r_*/D_A)$
z_{drag}		...	Redshift at which baryon-drag optical depth equals unity

(continued)

Table 6.2 (continued)

$r_{\text{drag}} = r_s(z_{\text{drag}})$	...	Comoving size of the sound horizon at $z = z_{\text{drag}}$
k_D	...	Characteristic damping comoving wavenumber (Mpc^{-1})
$100\theta_D$	...	$100\times$ angular extent of photon diffusion at last scattering
z_{eq}	...	Redshift of matter-radiation equality (massless neutrinos)
$100\theta_{\text{eq}}$	...	$100\times$ angular size of the comoving horizon at matter-radiation equality
$r_{\text{drag}}/D_V(0.57)$	...	BAO distance ratio at $z = 0.57$

By inspection of the potential V which, when $\mathcal{E}^2 + K^2 \neq 0$, has overall features identical to those in ◘ Fig. 6.12, one readily sees that such solutions exist only if $k = 1$ and $\Lambda > 0$, and are all unstable. In what follows we analyze this in more detail.

The condition $R(t) \equiv R_0$ requires (keeping in mind (6.4.36), p. 228)

$$0 \equiv 3\frac{\dot{R}^2}{R^2}\Big|_{R=R_0} = \frac{\kappa\mathcal{E}}{R_0^3} + \frac{\kappa K}{R_0^4} + \Lambda - \frac{3k}{R_0^2}, \tag{6.4.102}$$

$$0 = -\frac{2\ddot{R}}{R}\Big|_{R=R_0} = \frac{\kappa}{3}(\rho + 3p) - \frac{2\Lambda}{3}$$

$$= \frac{\kappa}{3}\left(\frac{\mathcal{E}}{R^3} + \frac{2K}{R^4}\right) - \frac{2\Lambda}{3}. \tag{6.4.103}$$

Since we assumed $\mathcal{E} \geq 0$ and $K \geq 0$, we conclude from (6.4.103) that we must have $\Lambda \geq 0$ for an equilibrium, with $\Lambda = 0$ being only possible if both $\mathcal{E}$ and K vanish. But then (6.4.102) implies $k \geq 0$, with $k = 0$ if and only if $\mathcal{E} = K = \Lambda = 0$, which gives an empty, flat spacetime. Assuming that we are not in vacuum, we conclude that

> static solutions exist if and only if $k = 1$ and $\Lambda > 0$.

It turns out that all nontrivial static solutions are *linearization unstable*; this means that there exist small perturbations of the static solutions which grow exponentially in time. To see this, we write $R = R_0 + \delta R$, where R_0 is time-independent solution of the equations and δR is assumed to be small. We insert this form of R into the second-order evolution equation, keeping only terms linear in δR. For this, we start by rewriting (6.4.36) as

$$-2\ddot{R} = R\left(\frac{\kappa}{3}(\rho + 3p) - \frac{2\Lambda}{3}\right) = R\left(\frac{\kappa\mathcal{E}}{3R^3} + \frac{2K}{3R^4} - \frac{2\Lambda}{3}\right)$$

$$= \frac{\kappa\mathcal{E}}{3R^2} + \frac{2K}{3R^3} - \frac{2\Lambda R}{3}. \tag{6.4.104}$$

Table 6.3 Cosmological parameters extracted from the Planck Satellite data, from [70]. © AAS. Reproduced with permission

Parameter	TT + lowE 68% limits	TE + lowE 68% limits	EE + lowE 68% limits	TT, TE, EE + lowE 68% limits	TT, TE, EE + lowE + lensing 68% limits	TT, TE, EE + lowE + lensing + BAO 68% limits
$\Omega_b h^2$	0.02212 ± 0.00022	0.02249 ± 0.00025	0.0240 ± 0.0012	0.02236 ± 0.00015	0.02237 ± 0.00015	0.02242 ± 0.00014
$\Omega_c h^2$	0.1206 ± 0.0021	0.1177 ± 0.0020	0.1158 ± 0.0046	0.1202 ± 0.0014	0.1200 ± 0.0012	0.11933 ± 0.00091
$100\theta_{MC}$	1.04077 ± 0.00047	1.04139 ± 0.00049	1.03999 ± 0.00089	1.04090 ± 0.00031	1.04092 ± 0.00031	1.04101 ± 0.00029
τ	0.0522 ± 0.0080	0.0496 ± 0.0085	0.0527 ± 0.0090	$0.0544^{+0.0070}_{-0.0081}$	0.0544 ± 0.0073	0.0561 ± 0.0071
$\ln(10^{10} A_s)$	3.040 ± 0.016	$3.018^{+0.020}_{-0.018}$	3.052 ± 0.022	3.045 ± 0.016	3.044 ± 0.014	3.047 ± 0.014
n_s	0.9626 ± 0.0057	0.967 ± 0.011	0.980 ± 0.015	0.9649 ± 0.0044	0.9649 ± 0.0042	0.9665 ± 0.0038
$H_0 [\mathrm{km\,s^{-1}\,MPc^{-1}}]$	66.88 ± 0.92	68.44 ± 0.91	69.9 ± 2.7	67.27 ± 0.60	67.36 ± 0.54	67.66 ± 0.42
Ω_Λ	0.679 ± 0.013	0.699 ± 0.012	$0.711^{+0.033}_{-0.026}$	0.6834 ± 0.0084	0.6847 ± 0.0073	0.6889 ± 0.0056
Ω_m	0.321 ± 0.013	0.301 ± 0.012	$0.289^{+0.026}_{-0.033}$	0.3166 ± 0.0084	0.3153 ± 0.0073	0.311 ± 0.0056
$\Omega_m h^2$	0.1434 ± 0.0020	0.1408 ± 0.0019	$0.1404^{+0.0034}_{-0.0039}$	0.1432 ± 0.0013	0.1430 ± 0.0011	0.14240 ± 0.00087
$\Omega_m h^3$	0.09589 ± 0.00046	0.09635 ± 0.00051	$0.0981^{+0.0016}_{-0.0018}$	0.09633 ± 0.00029	0.09633 ± 0.00030	0.09635 ± 0.00030
σ_8	0.8118 ± 0.0089	0.793 ± 0.011	0.796 ± 0.018	0.8120 ± 0.0073	0.8111 ± 0.0060	0.8102 ± 0.0060
$s_8 \equiv \sigma_8(\Omega_m/0.3)^{0.5}$	0.840 ± 0.024	0.794 ± 0.024	$0.781^{+0.052}_{-0.060}$	0.834 ± 0.016	0.832 ± 0.013	0.825 ± 0.011
$\sigma_8 \Omega_m^{0.25}$	0.611 ± 0.012	0.587 ± 0.012	0.583 ± 0.027	0.6090 ± 0.0081	0.6078 ± 0.0064	0.6051 ± 0.0058
z_{re}	7.50 ± 0.82	$7.11^{+0.91}_{-0.75}$	$7.10^{+0.87}_{-0.73}$	7.68 ± 0.79	7.67 ± 0.73	7.82 ± 0.71
$10^9 A_s$	2.092 ± 0.034	2.045 ± 0.041	2.116 ± 0.047	$2.101^{+0.031}_{-0.034}$	2.100 ± 0.030	2.105 ± 0.030
$10^9 A_s e^{-2\tau}$	1.884 ± 0.014	1.851 ± 0.018	1.904 ± 0.024	1.884 ± 0.012	1.883 ± 0.011	1.881 ± 0.010
Age [Gyr]	13.830 ± 0.037	13.761 ± 0.038	$13.64^{+0.16}_{-0.14}$	13.800 ± 0.024	13.797 ± 0.023	13.787 ± 0.020
z_*	1090.30 ± 0.41	1089.57 ± 0.42	$1087.8^{+1.6}_{-1.7}$	1089.95 ± 0.27	1089.92 ± 0.25	1089.80 ± 0.21
r_* [Mpc]	144.46 ± 0.48	144.95 ± 0.48	144.29 ± 0.64	144.39 ± 0.30	144.43 ± 0.26	144.57 ± 0.22
$100\theta_*$	1.04097 ± 0.00046	1.04156 ± 0.00049	1.04001 ± 0.00086	1.04109 ± 0.00030	1.04110 ± 0.00031	1.04119 ± 0.00029

z_{drag}	1059.39 ± 0.46	1060.03 ± 0.54	1063.2 ± 2.4	1059.93 ± 0.30	1059.94 ± 0.30	1060.01 ± 0.29
$r_{drag}[\mathrm{Mpc}]$	147.21 ± 0.48	147.59 ± 0.49	146.46 ± 0.70	147.05 ± 0.30	147.09 ± 0.26	147.21 ± 0.23
$k_D[\mathrm{Mpc}^{-1}]$	0.14054 ± 0.00052	0.14043 ± 0.00057	0.1426 ± 0.0012	0.14090 ± 0.00032	0.14087 ± 0.00030	0.14078 ± 0.00028
z_{eq}	3411 ± 48	3349 ± 46	3340^{+81}_{-92}	3407 ± 31	3402 ± 26	3387 ± 21
$k_{eq}[\mathrm{Mpc}^{-1}]$	0.01041 ± 0.00014	0.01022 ± 0.00014	$0.01019^{+0.00025}_{-0.00028}$	0.010398 ± 0.000094	0.010384 ± 0.000081	0.010339 ± 0.000063
$100\theta_{s,eq}$	0.4483 ± 0.0046	0.4547 ± 0.0045	0.4562 ± 0.0092	0.4490 ± 0.0030	0.4494 ± 0.0026	0.4509 ± 0.0020
f_{2000}^{143}	31.2 ± 3.0			29.5 ± 2.7	29.6 ± 2.8	29.4 ± 2.7
$f_{2000}^{143\times217}$	33.6 ± 2.0			32.2 ± 1.9	32.3 ± 1.9	32.1 ± 1.9
f_{2000}^{217}	108.2 ± 1.9			107.0 ± 1.8	107.1 ± 1.8	106.9 ± 1.8

Linearizing around the time-independent solution R_0 one obtains

$$\frac{d^2 \delta R}{dt^2} = \underbrace{\left(\frac{\kappa \mathscr{E}}{3R_0^2} + \frac{K}{R_0^4} + \frac{\Lambda}{3} \right)}_{=:\lambda^2} \delta R \,.$$

The solutions of this equation are linear combinations of $e^{\pm \lambda t}$, and since $\lambda \in \mathbb{R}$ one of the solutions grows without bounds at time increases. This establishes instability.

Appendix A
Some Reminders: Minkowski Spacetime

Piotr T. Chruściel

© Springer Nature Switzerland AG 2019
P. T. Chruściel, *Elements of General Relativity*, Compact Textbooks in Mathematics,
https://doi.org/10.1007/978-3-030-28416-9

A.1 Minkowski Metric, Lorentz Transformations

A *quadratic form* is a homogeneous quadratic polynomial on a vector space, say V. So, in finite dimension n, if $X \in V$ is decomposed as $X^i e_i$ in a basis e_i, $i = 1, \ldots, n$, then a quadratic form Q can be uniquely represented as

$$Q(X) = \sum_{i=1}^{n} Q_{ij} X^i X^j \,,$$

for a collection of numbers Q_{ij}. Note that the antisymmetric part of Q_{ij}, if any, would give zero contribution to the sum, and therefore without loss of generality we can assume that Q_{ij} is symmetric:

$$Q_{ij} = Q_{ji} \,.$$

It is known that every quadratic form can be written as a sum of squares with multiplicative coefficients ± 1. We say that the signature is (p, q) if the resulting formula has p minuses and q pluses. The signature is sometimes also written as

$$(\underbrace{-, \ldots, -}_{p \text{ times}}, \underbrace{+, \ldots, +}_{q \text{ times}}) \,.$$

In the special relativity literature one often thinks of the $(n+1)$-dimensional Minkowski spacetime as the vector space $\mathbb{R}^{n+1}$ equipped with a quadratic form, which will be denoted by η, with signature $(1, n)$, equivalently $(-, +, \ldots, +)$. Another approach is to define Minkowski spacetime as an affine space. We will soon give up those points of view (compare Remark A.1.3 below), but we momentarily accept either of these definitions until the end of this section.

Points $x \in \mathbb{R}^{n+1}$ will often be written as $(t, \vec{x})$.

By the *Sylvester inertia theorem* the signature is coordinate-independent, and (as already pointed out), there exists a choice of coordinates x^μ, called *inertial coordinates*, or *inertial frame*, so that

$$\eta(x, y) = -x^0 y^0 + x^1 y^1 + \ldots + x^n y^n , \tag{A.1.1}$$

or

$$\eta(x, x) = -x^0 x^0 + x^1 x^1 + \ldots + x^n x^n . \tag{A.1.2}$$

(This is the familiar special relativistic "line element," using units in which the speed of light c equals 1. We have implicitly used the obvious one-to-one correspondence between quadratic forms, such as the right-hand side of (A.1.2), with *bilinear forms*, such as the right-hand side of (A.1.1).) One also writes $\eta = (\eta_{\mu\nu})$, representing η by a $(n + 1) \times (n + 1)$ matrix

$$(\eta_{\mu\nu}) = \begin{pmatrix} -1 & 0 & \cdots & 0 \\ 0 & 1 & \cdots & 0 \\ \vdots & 0 & \ddots & 0 \\ 0 & \cdots & 0 & 1 \end{pmatrix},$$

so that

$$\eta(x, y) = \eta_{\mu\nu} x^\mu y^\nu \tag{A.1.3}$$

Here we are using the summation convention, which means that repeated indices, one in subscript and one in superscript position, have to be summed over. For example, if Greek indices are assumed to run from 0 to n, then the right-hand side of (A.1.3) is the same as that of (A.1.1).

A vector X is called *timelike* if $\eta(X, X) < 0$, null if $X \neq 0$ and $\eta(X, X) = 0$, spacelike if $\eta(X, X) > 0$. An example of a null vector is given by the vector with components $(1, 1, 0, 0)^t$ in a coordinate system in which η takes the form (A.1.2).

We write $\mathbb{R}^{1,n}$ for the pair

$$\mathbb{R}^{1,n} := (\mathbb{R}^{n+1}, \eta) .$$

The coordinates in which $\eta(x, y)$ takes the form (A.1.1) are called *inertial*, or *non-accelerating*. Two such coordinate systems x^μ and $\tilde{x}^\mu$ differ by a Lorentz transformation:

$$x^\mu = L^\mu{}_\nu \tilde{x}^\nu , \tag{A.1.4}$$

where $L = (L^\mu{}_\nu)$ is a Lorentz matrix, and a translation.

A useful result about inertial coordinates is the following: given any timelike vector $T = (T^\mu)$, there exists an inertial frame in which the vector T has components $(\tau, 0, \ldots, 0)^t$, for some $\tau \in \mathbb{R} \setminus \{0\}$. (There are in fact many such frames, differing from each other by a rotation, possibly composed with a parity transformation and/or a time reversal.) This is proved by a Gram–Schmidt orthogonalization, using a basis in which T is the first basis vector.

By definition, Lorentz matrices are those matrices which preserve the quadratic form η:

$$\eta_{\mu\nu} = \eta_{\alpha\beta} L^\alpha{}_\mu L^\beta{}_\nu = \underbrace{L^\alpha{}_\mu \eta_{\alpha\beta} L^\beta{}_\nu}_{L^t\eta L \text{ in matrix notation}} .$$

This implies $\det L = \pm 1$.

Examples of Lorentz metrics are:

1. a boost along the x-axis

$$L = (L^\mu{}_\nu) = \begin{pmatrix} \gamma & -\gamma v & 0 & 0 \\ -\gamma v & \gamma & 0 & 0 \\ 0 & 0 & 1 & 0 \\ 0 & 0 & 0 & 1 \end{pmatrix}, \qquad |v| < 1, \quad \gamma = \frac{1}{1 - v^2},$$

2. rotations in space: $t = \tilde{t}, \vec{x} = R\vec{\tilde{x}}$, R – rotation matrix,
3. time reversal: $t = -\tilde{t}, \vec{x} = \vec{\tilde{x}}$,
4. space inversion: $t = \tilde{t}, \vec{x} = -\vec{\tilde{x}}$.

We have:

Theorem A.1.1

Every Lorentz transformation is a composition of a finite number of the transformations above.

Definition A.1.2

Poincaré transformations are compositions of Lorentz transformations and translations.

Remark A.1.3 We shall soon be using the symbol η for the following object:

$$\eta = -dt^2 + (dx^1)^2 + \ldots + (dx^n)^2, \tag{A.1.5}$$

which means something closely related to (A.1.2), but not quite the same thing. In (A.1.2) η is a quadratic form on $\mathbb{R}^{n+1}$, while η in (A.1.5) is a *field of quadratic forms* on $\mathbb{R}^{n+1}$: this means that at each point $x \in \mathbb{R}^{n+1}$, η is a quadratic form on the "tangent space $T_x\mathbb{R}^{n+1}$." The last set is defined as the collection of all vectors tangent to $\mathbb{R}^{n+1}$ at the point x. It is linearly isomorphic to $\mathbb{R}^{n+1}$, with the "tangent bundle" $T\mathbb{R}^{n+1} := \cup_{x \in \mathbb{R}^{n+1}} T_x\mathbb{R}^{n+1}$ being diffeomorphic to $\mathbb{R}^{2(n+1)}$; this is discussed in ▶ Sect. 1.1.3.

A.2   Electromagnetic Fields

In everyday experience, the electromagnetic field is thought of as two vector fields, $\vec{E}$ and $\vec{B}$. If there are no sources, $\vec{E}$ and $\vec{B}$ satisfy the following sourceless Maxwell equations:

$$\nabla \cdot \vec{E} = 0, \qquad \nabla \cdot \vec{B} = 0, \tag{A.2.1}$$

$$\underbrace{\mu_0 \epsilon_0}_{=1} \frac{\partial \vec{E}}{\partial t} = \nabla \wedge \vec{B}, \qquad \frac{\partial \vec{B}}{\partial t} = -\nabla \wedge \vec{E}. \tag{A.2.2}$$

Lorentz discovered his group by studying the invariance properties of those equations under linear coordinate transformations.

The bottom-line of his calculation is as follows: Let us define an antisymmetric matrix F, called *the Maxwell tensor*, by the formula:

$$F = (F^{\mu\nu}) = \begin{pmatrix} 0 & E^1 & E^2 & E^3 \\ -E^1 & 0 & B^3 & -B^2 \\ -E^2 & -B^3 & 0 & B^1 \\ -E^3 & B^2 & -B^1 & 0 \end{pmatrix}.$$

Then, under (A.1.4), the matrix F describing the Maxwell fields in the coordinates system x is related to the matrix $\tilde{F}$ describing the Maxwell fields in the coordinates system $\tilde{x}$ by the formula

$$F = L\tilde{F}L^t \iff F^{\mu\nu} = L^\mu{}_\alpha \tilde{F}^{\alpha\beta} L^\nu{}_\beta.$$

In fact, Lorentz proved:

Theorem A.2.1

The Maxwell equations are preserved by linear transformations of coordinates and of $\vec{E}$ and $\vec{B}$ if and only if the above transformation law holds, where L is the matrix of a Lorentz transformation.

Here a comment on conventions is in order. We will use Greek letters for *spacetime indices*, typically running from 0 to 3 (or from 0 to n in $(n + 1)$-dimensions). Latin indices will be *space indices*, running from 1 to 3 (or from 1 to n).

Incidentally Let ϵ_{ijk} be the three-dimensional alternating symbol, this means that: $\epsilon_{123} = 1$, and $\epsilon_{ijk} = -\epsilon_{jik} = -\epsilon_{jki}$; you should check that these equations define all components of ϵ_{ijk} uniquely. Then we have

$$F_{ij} = \epsilon_{ijk} B^k \,,$$

keeping in mind that $F^{\mu\nu}$ is related to $F_{\mu\nu}$ by "raising indices":

$$F^{\mu\nu} := \eta^{\mu\alpha} \eta^{\nu\beta} F_{\alpha\beta} \,,$$

and that, in the current coordinates system, $B_i = B^i$, $E_i = E^i$.

Exercise A.2.2

Relate the set of equations

$$\partial_\mu F^{\mu\nu} = 0$$

to the Maxwell equations (A.2.1)–(A.2.2). ∎

Appendix B
Exercises

Piotr T. Chruściel

© Springer Nature Switzerland AG 2019
P. T. Chruściel, *Elements of General Relativity*, Compact Textbooks in Mathematics,
https://doi.org/10.1007/978-3-030-28416-9

This appendix contains a collection of exercises. Further exercises can be found sprinkled throughout the main text. Some reminders have been included below for the convenience of the students.

Reminders Let $x^i \mapsto \overline{x}^i$ be a coordinate transformation of $\mathbb{R}^n$, which we write for short as $x \mapsto \overline{x}$. A *scalar field* f on $\mathbb{R}^n$ satisfies the transformation law

$$\overline{f}(\overline{x}) = f(x(\overline{x})) \,.$$

A *vector field* (X^i) on $\mathbb{R}^n$ satisfies the transformation law

$$\overline{X}^i(\overline{x}) = \frac{\partial \overline{x}^i}{\partial x^j} X^j(x(\overline{x})) \,,$$

where the summation convention is used. A *covector field* (X_i) on $\mathbb{R}^n$ satisfies the transformation law

$$\overline{X}_i(\overline{x}) = \frac{\partial x^j}{\partial \overline{x}^i} X_j(x(\overline{x})) \,.$$

We write ∂_i for $\partial/\partial x^i$, and $g(X, Y) = g_{ij} X^i Y^j$.

The Christoffel symbols of the Levi-Civita connection are defined by the formula

$$\Gamma^a_{bc} = \frac{1}{2} g^{ad} (\partial_b g_{cd} + \partial_c g_{bd} - \partial_d g_{bc}) \,.$$

For functions we have $\nabla_i f = \partial_i f$, while for vector fields we have

$$\nabla_i X^j = \partial_i X^j + \Gamma^j_{ki} X^k \,,$$

and it holds that

$$\partial_i(g(X, Y)) = \nabla_i(g(X, Y)) = g(\nabla_i X, Y) + g(X, \nabla_i Y).$$

A curve $s \mapsto \gamma(s)$ is called a geodesic if and only if

$$\frac{d^2\gamma^i}{ds^2} + \Gamma^i_{jk}\frac{d\gamma^j}{ds}\frac{d\gamma^k}{ds} = 0.$$

1   *Changes of coordinates*

Let $x^0 = t$, $x^1 = x$, $x^2 = y$, and $x^3 = z$ be inertial coordinates on flat spacetime, so the Minkowski metric has components

$$(g_{ab}) = \begin{pmatrix} -1 & 0 & 0 & 0 \\ 0 & 1 & 0 & 0 \\ 0 & 0 & 1 & 0 \\ 0 & 0 & 0 & 1 \end{pmatrix}.$$

Let X be the vector field which in the above coordinate system equals $(1, 1, 0, 0)$, and let α be a one-form which in the above coordinate system equals $(1, 1, 0, 0)$. Find the metric coefficients $\tilde{g}_{ab}$, and the components of X and α, in the coordinate system

$$\tilde{x}^0 = \tau, \quad \tilde{x}^1 = \phi, \quad \tilde{x}^2 = y, \quad \tilde{x}^3 = z,$$

where τ, ϕ are "Rindler coordinates," defined by $t = \tau \cosh\phi$, $x = \tau \sinh\phi$. Determine which region of Minkowski space the coordinate system covers. *[Hint: A quick method for the metric is to write it as $ds^2 = -dt^2 + dx^2 + dy^2 + dz^2$ and substitute, for example, $dx = \sinh\phi\, d\tau + \tau \cosh\phi\, d\phi$, and so on.]*

2   *Stereographic projection* Let $(x, y) \in \mathbb{R}^2$ be local coordinates on a two-dimensional sphere obtained by a stereographic projection from the north pole onto the plane tangent to the south pole of the sphere. Find the form of the standard metric on the sphere, namely $g = d\theta^2 + \sin^2(\theta)d\varphi^2$, in these coordinates.

3   *Lie bracket*

1. Show that (X^i) is a vector field in the sense of the definition given above if and only if for every scalar field f the expression

$$X^i \partial_i f$$

is a scalar field.

[Conclusion: vector fields can be identified with homogeneous linear first order partial differential operators $X = X^a \partial_a$ acting on scalar fields as

$$X(f) = X^a \partial_a f.]$$

2. The Lie-bracket $[X, Y]$ of two vector fields X and Y is defined as

$$[X, Y](f) = X(Y(f)) - Y(X(f)).$$

Show that, in local coordinates,

$$[X, Y]^i = X^j \partial_j Y^i - Y^j \partial_j X^i.$$

3. Show that $[X, Y]$ is a vector field if X and Y are.
4. Prove the *Jacobi identity*:

$$[X, [Y, Z]] + [Y, [Z, X]] + [Z, [X, Y]] = 0.$$

4 1. Let X be a vector field which satisfies the following:

   for every affinely parameterized geodesic γ, $g(X, \dot\gamma)$ is constant along γ.

   Show that this implies that X satisfies the following equation, known as *the Killing equation*:

   $$\nabla_\mu X_\nu + \nabla_\nu X_\mu = 0. \tag{B.1.1}$$

   Solutions of the Killing equation are called *Killing vectors*.
   2. Show that a linear combination of Killing vectors with constant coefficients is a Killing vector. Thus, the set of Killing vectors forms a vector space.
   3. Let X be a Killing vector field, and let $s \mapsto x^\mu(s)$ be an integral curve of X: by definition, this means that

   $$\frac{dx^\mu}{ds}(s) = X^\mu(x(s)). \tag{B.1.2}$$

   In other words, the vector field X is tangent to the curves $s \mapsto x^\mu(s)$. Show that the integral curves of X are geodesics if and only if we have

   $$\nabla_\alpha(g(X, X)) = 0 \text{ along the curve } x(s). \tag{B.1.3}$$

5 *Christoffel symbols*

   1. Show that the Euler-Lagrange equations

   $$\frac{d}{ds}\left(\frac{\partial L}{\partial \dot{x}^a}\right) = \frac{\partial L}{\partial x^a} \tag{B.1.4}$$

associated with the Lagrange function

$$L(x^c, \dot{x}^c) = \frac{1}{2} g_{ab} \dot{x}^a \dot{x}^b \tag{B.1.5}$$

(here, a dot denotes d/ds) can be written as

$$\ddot{x}^a + \Gamma^a_{bc} \dot{x}^b \dot{x}^c = 0. \tag{B.1.6}$$

2. Show that if the metric does not explicitly depend upon a coordinate, say x^1, then $g(\dot{x}, \partial_1)$ is constant along every geodesic.
3. Using the above variational principle for geodesics, write down the geodesic equations, and give the obvious constants of motion, for a metric of the form

$$-e^{2f(r)} dt^2 + e^{-2f(r)} dr^2 + r^2 d\varphi^2. \tag{B.1.7}$$

4. Use (B.1.4) to calculate the Christoffel symbols for the metric (B.1.7).
5. Establish the transformation law

$$\Gamma^a_{bc} = \tilde{\Gamma}^d_{ef} \frac{\partial x^a}{\partial \tilde{x}^d} \frac{\partial \tilde{x}^e}{\partial x^b} \frac{\partial \tilde{x}^f}{\partial x^c} + \frac{\partial x^a}{\partial \tilde{x}^d} \frac{\partial^2 \tilde{x}^d}{\partial x^b \partial x^c}$$

by direct calculation. Explain why this implies that the Christoffel symbols do *not* define a tensor.

Reminder Let ∇ be a connection (which we do not assume to be the Levi-Civita one), given two vector fields X and Y the *torsion tensor T* is defined as

$$T(X, Y) = \nabla_X Y - \nabla_Y X - [X, Y].$$

6 *Some tensor manipulations*
1. Show that for all functions f

$$\nabla_a \nabla_b f = \nabla_b \nabla_a f$$

if and only if ∇ is torsion-free.
2. Recall that ∇f is defined as the vector field $g^{ab} \partial_a f \partial_b$. Explain what $df(\nabla f)$ and $\nabla f(f)$ mean, and prove the equalities

$$g(\nabla f, \nabla f) = g^{ab} \partial_a f \partial_b f = df(\nabla f) = \nabla f(f).$$

Conclude that if $f = x^\alpha$ is a coordinate, you can determine whether ∇f is timelike, spacelike, or null, by looking at a component of the inverse metric tensor (which?).

7 *Schwarzschild: Orders of magnitude* Recall that the Schwarzschild metric g takes the form

$$g = -\left(1 - \frac{2MG}{rc^2}\right)c^2 dt^2 + \frac{dr^2}{1 - \frac{2MG}{rc^2}} + r^2(d\theta^2 + \sin\theta^2 d\varphi^2), \qquad (B.1.8)$$

where M is the total mass of the gravitating object, G is Newton's constant, and c is the speed of light. Calculate the deviation of g_{00}/c^2 from minus one, where $x^0 = t$, as well as the deviation of $\partial_i g_{00}/c^2$ from zero, (a) at the surface of the Sun when M is the mass of the Sun, (b) at the orbit of Earth when M is the mass of the Sun, (c) at the surface of Earth when M is the mass of Earth, (d) at the orbit of the moon when M is the mass of Earth, and (e) at the surface of the moon when M is the mass of the moon.

Clearly, something wrong is happening with (B.1.8) at $r = 2MG/c^2$; this is called the *Schwarzschild radius*. Calculate the Schwarzschild radius of (a) the Sun, (b) Earth, and (c) the moon. Calculate the corresponding mass densities, compare the result to the density of a neutron.

8 *Symmetries of the curvature tensor*
   1. What does it mean for a connection ∇ on a spacetime with metric g_{ab} to be (a) a *metric connection*, (b) *torsion-free*?
       Assume henceforth that ∇ is torsion-free.
       In what follows, it is often useful to use the preceding results to do the next ones.
       (c) Given an arbitrary smooth covector field A_a, and a smooth antisymmetric tensor field F_{ab}, show that

$$H_{ab} := \nabla_a A_b - \nabla_b A_a \quad \text{and} \quad \nabla_a F_{bc} + \nabla_b F_{ca} + \nabla_c F_{ab}$$

are both independent of the choice of the connection.
(d) Hence or otherwise show that

$$\nabla_a H_{bc} + \nabla_b H_{ca} + \nabla_c H_{ab} = 0.$$

Recall that the curvature tensor is defined as

$$\nabla_a \nabla_b V^c - \nabla_b \nabla_a V^c = R^c{}_{dab} V^d,$$

(e) Show that

$$\nabla_a \nabla_b A_c - \nabla_b \nabla_a A_c = -R^d{}_{cab} A_d.$$

(f) Hence show that $R^d{}_{abc} + R^d{}_{bca} + R^d{}_{cab} = 0$ for a torsion-free connection.

(g) Show further that, for a tensor T_{ab},

$$\nabla_a \nabla_b T_{cd} - \nabla_b \nabla_a T_{cd} = -R^e{}_{cab} T_{ed} - R^e{}_{dab} T_{ce} \, .$$

(h) Hence show that $R_{abcd} = -R_{bacd}$ if ∇ is metric and torsion-free.

2. Show that the symmetry $R_{abcd} = R_{cdab}$ follows from $R_{abcd} = R_{[ab]cd} = R_{ab[cd]}$ and $R_{a[bcd]} = 0$.

9 Counting components

(1) In four dimensions, a tensor satisfies $T_{abcde} = T_{[abcde]}$. Show that $T_{abcde} = 0$.

(2) A tensor T_{ab} is *symmetric* if $T_{ab} = T_{(ab)}$. In n-dimensional space, it has n^2 components, but only $\frac{1}{2} n(n+1)$ of these can be specified independently—for example the components T_{ab} for $a \leq b$. How many independent components do the following tensors have (in n dimensions)?

  (a) F_{ab} with $F_{ab} = F_{[ab]}$.
  (b) A tensor of type $(0, k)$ such that $T_{ab...c} = T_{[ab...c]}$ (distinguish the cases $k \leq n$ and $k > n$, bearing in mind the result of question (1)).
  (c) R_{abcd} with $R_{abcd} = R_{[ab]cd} = R_{ab[cd]}$.
  (d) R_{abcd} with $R_{abcd} = R_{[ab]cd} = R_{ab[cd]} = R_{cdab}$.

Show that, in four dimensions, a tensor with the symmetries of the Riemann tensor has 20 independent components.

10 (a) Show that for any connection we have

$$\nabla_\alpha \delta^\beta_\gamma = 0.$$

Hence, or otherwise, show that for the Levi-Civita connection it holds that

$$\nabla_\alpha g^{\beta\gamma} = 0.$$

(b) Let $\Delta^{\alpha\beta}$ denote the adjoint matrix of g (recall that $\Delta^{\alpha\beta} = \det g \, g^{\alpha\beta}$). Show that

$$\frac{\partial \det g}{\partial g_{\alpha\beta}} = \Delta^{\alpha\beta} \, .$$

Conclude that

$$g^{\alpha\beta} \partial_\mu g_{\alpha\beta} = \partial_\mu \ln(|\det g_{\alpha\beta}|) \, ,$$

as well as

$$\Gamma^\alpha_{\mu\alpha} = \frac{1}{\sqrt{|\det g_{\alpha\beta}|}} \partial_\mu \left(\sqrt{|\det g_{\alpha\beta}|} \right) .$$

Find likewise a simple expression for $g^{\mu\nu}\Gamma^\alpha_{\mu\nu}$, and deduce that

$$\Box_g f := g^{\mu\nu}\nabla_\mu\nabla_\nu f = \frac{1}{\sqrt{|\det g_{\alpha\beta}|}}\partial_\mu\left(\sqrt{|\det g_{\alpha\beta}|}g^{\mu\nu}\partial_\nu f\right).$$

(c) Show that for a vector field U^α, we have

$$\nabla_\mu U^\mu = \frac{1}{\sqrt{|\det g_{\alpha\beta}|}}\partial_\mu\left(\sqrt{|\det g_{\alpha\beta}|}U^\mu\right).$$

Similarly, for an antisymmetric tensor $F^{\alpha\beta}$, show that

$$\nabla_\mu F^{\mu\nu} = \frac{1}{\sqrt{|\det g_{\alpha\beta}|}}\partial_\mu\left(\sqrt{|\det g_{\alpha\beta}|}F^{\mu\nu}\right).$$

Does this remain true for totally antisymmetric tensors with more indices?

11 *Geodesics* Let $\lambda \mapsto \gamma(\lambda)$ be a curve such that

$$\frac{D\dot\gamma}{d\lambda} = \alpha\dot\gamma, \tag{B.1.9}$$

for some function $\alpha = \alpha(\lambda)$, where $\dot\gamma = d\gamma/d\lambda$. Show that there exists a change of parameter $\tau \mapsto \lambda(\tau)$ so that

$$\frac{D}{d\tau}\frac{d\gamma}{d\tau} = 0. \tag{B.1.10}$$

[It follows that solutions of (B.1.9) are geodesics, but *not* affinely parameterized.]

12 *Integral curves*

(a) Given a vector field X^μ, recall that its integral curves are defined as the solutions of the equations

$$\frac{dx^\mu}{d\lambda} = X^\mu(x(\lambda)).$$

Find the integral curves of the following vector fields on $\mathbb{R}^2$: ∂_x, $x\partial_y + y\partial_x$, $x\partial_y - y\partial_x$, $x\partial_x + y\partial_y$.

(b) Let f be a function satisfying

$$g(\nabla f, \nabla f) = \psi(f),$$

for some function ψ. Show that

$$\nabla^\mu f\,\nabla_\mu\nabla^\alpha f = \frac{1}{2}\psi'\nabla^\alpha f. \tag{B.1.11}$$

Let $\lambda \mapsto \gamma(\lambda)$ be any integral curve of ∇f; thus $d\gamma^\mu/d\lambda = \nabla^\mu f$. Rewrite (B.1.11) as an equation for $D\dot\gamma/d\lambda$. Find the differential equation for a reparameterization $\tau \mapsto \gamma(\lambda(\tau))$ of γ so that

$$\frac{D}{d\tau}\frac{d\gamma^\mu}{d\tau} = 0.$$

[*Hint: use Problem 11.*]

(c) Recall that, for the Schwarzschild metric, we may define the Eddington–Finkelstein coordinate v by

$$dv = dt + \frac{r}{r - 2m}dr.$$

Show that, in the coordinates (v, r, θ, φ), the integral curves of the vector field ∇r meeting $\{r = 2m\}$ are null geodesics.

(d) Let f be one of the coordinates, say $f = x^1$, in a coordinate system $\{x^i\}$. Verify that

$$g(\nabla f, \nabla f) = g^{11}.$$

Using this observation find a family of spacelike geodesics in the (t, r, θ, φ) coordinate system, as well as two distinct families of geodesics in the (v, r, θ, φ) coordinate system. Do any members of the second family coincide with members of the first?

13 Repeat part i of Problem 8 with a connection ∇ that is metric (i.e., $\nabla_\alpha g_{\beta\gamma} = 0$) but has torsion (i.e., $T(X, Y) := \nabla_X Y - \nabla_Y X - [X, Y] \neq 0$). Show that, for such a connection, the first Bianchi identity takes the form

$$R^\delta{}_{[\alpha\beta\gamma]} = \nabla_{[\alpha} T^\delta{}_{\beta\gamma]} - T^\epsilon{}_{[\alpha\beta} T^\delta{}_{\gamma]\epsilon}.$$

[Hint: One way is to let ϕ be an arbitrary function and start from the identity $-R^\delta{}_{\gamma\alpha\beta}\nabla_\delta\phi = \nabla_\alpha\nabla_\beta\nabla_\gamma\phi - \nabla_\beta\nabla_\alpha\nabla_\gamma\phi + T^\delta{}_{\alpha\beta}\nabla_\delta\nabla_\gamma\phi$. Rewrite the second term using $\nabla_\alpha\nabla_\gamma\phi = \nabla_\gamma\nabla_\alpha\phi - T^\delta{}_{\alpha\gamma}\nabla_\delta\phi$. Then skew-symmetrize over α, β, γ. Manipulating what you get and stripping off the $\nabla\phi$ terms (using the fact that ϕ is arbitrary) should give you the desired result.]

14 *Some timelike geodesics in Schwarzschild.* Let $V = 1 - 2m/r$, and consider the Schwarzschild metric:

$$g = -(1 - \frac{2m}{r})dt^2 + \frac{dr^2}{1 - \frac{2m}{r}} + r^2 d\Omega^2, \tag{B.1.12}$$

$$t \in \mathbb{R}, \ r \neq 2m, 0. \tag{B.1.13}$$

Recall that for geodesics in the Schwarzschild metric lying in the equatorial plane $\theta = \pi/2$ we have the following constants of motion (you are strongly encouraged to check that you can reproduce the result):

$$\frac{d}{ds}\left(V\frac{dt}{ds}\right) = 0 \quad \Longrightarrow \quad \frac{dt}{ds} = \frac{E}{1 - \frac{2m}{r}}. \tag{B.1.14}$$

$$\frac{d}{ds}\left(r^2\frac{d\varphi}{ds}\right) = 0 \quad \Longrightarrow \quad \frac{d\varphi}{ds} = \frac{J}{r^2}. \tag{B.1.15}$$

$$\underbrace{V\left(\frac{dt}{ds}\right)^2}_{E^2 V^{-1}} - V^{-1}\left(\frac{dr}{ds}\right)^2 - \underbrace{r^2\left(\frac{d\varphi}{ds}\right)^2}_{J^2/r^2} = \lambda \in \{0, \pm 1\}. \tag{B.1.16}$$

Consider a timelike geodesic in the Schwarzschild metric parameterized by proper time, thus $\lambda = 1$, and lying in the equatorial plane $\theta = \pi/2$.
Q1. Verify that

$$\frac{E^2 - \dot{r}^2}{1 - \frac{2m}{r}} - \frac{J^2}{r^2} = 1.$$

Q2. Deduce that if $E = 1$ and $J = 4m$, then

$$\frac{\sqrt{r} - 2\sqrt{m}}{\sqrt{r} + 2\sqrt{m}} = Ae^{\epsilon\varphi/\sqrt{2}},$$

where $\epsilon = \pm 1$ and A is a constant. Describe the orbit that starts at $\varphi = 0$ in each of the cases (i) $A = 0$, (ii) $A = 1$, $\epsilon = -1$, and (iii) $r(0) = 3m$, $\epsilon = -1$.
Q3: Consider the case $E = 1$ and $J = 0$, and deduce a relation for $r = r(s)$.

15 *Null Geodesics in Schwarzschild* Let γ be an affinely parameterized null geodesic, thus $\lambda = 0$, in the Schwarzschild metric lying in the equatorial plane $\theta = \pi/2$. Assume that $J \neq 0$, hence we can make a change of parameter $\varphi \mapsto s(\varphi)$ using the implicit equation

$$\frac{d\varphi}{ds} = \frac{J}{r^2} \quad \Longrightarrow \quad \frac{ds}{d\varphi} = \frac{r^2}{J}.$$

Set

$$u(\varphi) = \frac{m}{r(s(\varphi))}, \quad p(\varphi) = \frac{du(\varphi)}{d\varphi}.$$

Q1. Show that along γ, for $u \neq \frac{1}{2}$, we have

$$p^2 = 2u^3 - u^2 + \alpha^2 . \tag{B.1.17}$$

Q2. Show that the equations for null geodesics imply the two-dimensional first order autonomous system

$$\frac{du}{d\varphi} = p , \quad \frac{dp}{d\varphi} = 3u^2 - u . \tag{B.1.18}$$

Recall that critical points of a dynamical system $d\vec{x}/d\varphi = \vec{Y}$ (here $\vec{x} = (u, p)$) are defined as points where $\vec{Y}$ vanishes. Find the critical points of (B.1.18). Can you sketch the trajectories in the (u, p) phase-plane?

Q3. For the case $\alpha = 0$ and $u > 1/2$ show that the geodesic has an equation of the form

$$r = 2m \cos^2 \left(\frac{\varphi - \varphi_0}{2} \right) , \tag{B.1.19}$$

with $t = t(\varphi)$ that you should determine.

16 *Penrose diagrams*

(a) Consider the Schwarzschild metric g in Kruskal–Szekeres coordinates $(\hat{u}, \hat{v}, \theta, \varphi)$,

$$g = -\frac{32m^3 \exp(-\frac{r}{2m})}{r} d\hat{u} \, d\hat{v} + r^2 d\Omega^2 . \tag{B.1.20}$$

Rewrite g in the coordinates $(\bar{t}, \bar{r}, \theta, \varphi)$ of (3.5.17). Find the set, say Ω_S, on which the map

$$(\hat{u}, \hat{v}) \mapsto (\bar{t}, \bar{r})$$

is a diffeomorphism.

[*Hint: do not forget that $r > 0$, where r is a function of the product $\hat{u}\hat{v}$ as explained in the lecture.*]

(b) Show that if a vector X is causal for the original Minkowski or Schwarzschild metric, then its two-dimensional projection $X^{\bar{t}}\partial_{\bar{t}} + X^{\bar{r}}\partial_{\bar{r}}$ is causal for the two-dimensional Minkowski metric $-d\bar{t}^2 + d\bar{r}^2$.

17 *A nonphysical black hole metric.* Let a be a strictly positive constant, and for $r > a$ let g be a Lorentzian metric of the form

$$g = -\left(1 - \frac{a}{r}\right)^3 dt^2 + \frac{dr^2}{\left(1 - \frac{a}{r}\right)^3} + r^2 d\theta^2 + r^2 \sin^2 \theta \, d\varphi^2 . \tag{B.1.21}$$

Q1. Proceeding in a way analogous to the analysis of the Schwarzschild metric, replace t by a new coordinate v so that the metric, in the new coordinates, can be smoothly extended from the original manifold $\{t \in \mathbb{R}\} \times \{r > a\} \times S^2$ to a Lorentzian metric on a new manifold $\{v \in \mathbb{R}\} \times \{r > 0\} \times S^2$.

Q2. Show that, with an appropriate choice of the coordinate v, the set $\{r < a\}$ in the extended manifold is a black hole region.

Q3. Find the four-acceleration of stationary observers for this metric.

Q4. Find the gravitational red-shift formula for this metric.

Q5. One can calculate (for the ambitious: use an algebraic manipulation program to carry this out):

$$R = 2a^3 r^{-5},$$

$$R_{\alpha\beta\gamma\delta} R^{\alpha\beta\gamma\delta} = 4a^2 r^{-10}(46a^4 - 150a^3 r + 186a^2 r^2 - 108ar^3 + 27r^4).$$

How does this imply that $\{r = 0\}$ is a singular set for the metric (B.1.21)?

18 *Radial geodesics in Schwarzschild*: Consider the geodesic equation for a massive particle initially at rest in the Schwarzschild metric. The tangent vector $\mathbf{U} = \frac{dy}{ds}$ therefore satisfies $U^a \nabla_a U^b = 0$ and $g_{ab} U^a U^b = -1$ with the initial conditions that $U^r(s = 0) = U^\theta(s = 0) = U^\phi(s = 0) = 0$ and $U^t(s = 0) > 0$.

(a) Show that we have

$$U^t(0) \equiv \frac{dt}{ds}(s = 0) = \frac{1}{\sqrt{1 - \frac{2m}{r(0)}}}.$$

(b) Derive the part of the geodesic equations for $\frac{d}{ds}U^\theta$, $\frac{d}{ds}U^\phi$. [Hint: Repeat the argument presented in the lecture.] Show that the initial conditions $U^\theta(0) = U^\phi(0) = 0$ imply that $U^\theta(s) = U^\phi(s) = 0$ for all s. Therefore, the geodesics are radial (i.e., the only spatial motion is in the radial direction).

(c) Using a formula derived in the lecture (the derivation of which you should reproduce), show that

$$\frac{dt}{ds} = \frac{\sqrt{1 - \frac{2m}{r(0)}}}{1 - \frac{2m}{r(s)}}.$$

(d) From the result of part (c), and the fact that $g_{ab} U^a U^b = -1$, deduce that $r(s)$ satisfies the differential equation

$$\frac{dr}{ds} = -\sqrt{\frac{2m}{r(s)} - \frac{2m}{r(0)}}.$$

[Note: At one point, you will need to take a square root, which involves a choice of sign. In order that $\frac{2m}{r(s)} - \frac{2m}{r(0)}$ is positive, you need $r(s) \le r(0)$, which tells you that you should choose the negative root.] Integrating this equation, show that $r(s)$ is implicitly given by the equation

$$s = -\frac{r(0)^{3/2}}{\sqrt{2m}} \left[\cos^{-1} \sqrt{\frac{r}{r_0}} - \sqrt{\frac{r}{r_0}} \left(1 - \left(\frac{r}{r^0} \right) \right)^{1/2} \right].$$

[*Hint: When doing the integration, you may find it useful to use the substitution* $r = r_0 \cos^2 x$.]

19 Consider a gyroscope with spin vector s circling on a circular timelike geodesic of radius r in the Schwarzschild metric. Let τ denote the proper time along the geodesic and let u be the unit-normalized tangent to the geodesic. Thus

$$\frac{Ds}{d\tau} = 0, \quad g(u, s) = 0.$$

Show that

$$\frac{ds^\theta}{dt} = 0, \quad s^t = \frac{r^2 u^\varphi}{\left(1 - \frac{2m}{r} \right) u^t} s^\varphi, \quad \frac{ds^r}{dt} = (r - 3m)\Omega s^\varphi, \quad \frac{ds^\varphi}{dt} = -\frac{\Omega}{r} s^r,$$

where $\Omega = \sqrt{\frac{m}{r^3}}$, and find $s(t)$, where t is the Schwarzschild coordinate time.

[*Hint: check first which Christoffel symbols you need for this calculation, and calculate them directly from the formula for the Christoffels in terms of the metric and its derivatives.*]

20 Assuming a circular orbit, and using units so that the result is dimensionless, calculate m_s/r, $J_N^2/(m_e r)^2$, and $m_s^2 m_e^2/J_N^2$ for (a) the orbit of Earth around the Sun, and for (b) the orbit of Mercury around the Sun (the orbit of Mercury is actually far from circular, use a radius calculated from the mean of the aphelion and perihelion). Here m_s is the mass of the Sun, and the remaining parameters concern the moving object.

Using the formula from the lecture (you might wish, though, to verify that you can reproduce its derivation), calculate the shift of the perihelion for Mercury and for Earth in orbit around the Sun. Repeat the calculation for an exoplanet with the mass of Jupiter on a circular orbit around a star of ten solar masses with a period of 1 day.

21 Isotropic coordinates for the Schwarzschild metric. We wish to find coordinates $(t, \tilde{r}, \theta, \phi)$, where $\tilde{r} = \tilde{r}(r)$, for the Schwarzschild metric in terms of which the spatial metric is conformally flat. In particular, we wish to find functions $\tilde{r}(r)$, $A(\tilde{r})$, and $B(\tilde{r})$ such that

$$ds^2 = -\left(1 - \frac{2m}{r} \right) dt^2 + \left(1 - \frac{2m}{r} \right)^{-1} dr^2 + r^2 d\Omega^2$$

$$= -A(\tilde{r})^2 dt^2 + B(\tilde{r})^2 (d\tilde{r}^2 + \tilde{r}^2 d\Omega^2). \tag{B.1.22}$$

Q1. Deduce that we require

$$A(\tilde{r})^2 = 1 - \frac{2m}{r}, \tag{B.1.23}$$

$$B(\tilde{r})^2 \left(\frac{d\tilde{r}}{dr}\right)^2 = \frac{r}{r - 2m}, \tag{B.1.24}$$

$$B(\tilde{r})^2 \tilde{r}^2 = r^2. \tag{B.1.25}$$

Q2. Assuming that $d\tilde{r}/dr > 0$ and $\tilde{r} > 0$, show that $\tilde{r}$ satisfies the relation

$$\frac{1}{\tilde{r}} \frac{d\tilde{r}}{dr} = \frac{1}{\sqrt{r(r - 2m)}}. \tag{B.1.26}$$

Integrating with respect to r from $r = 2m$ and letting $\tilde{r}(2m) = m/2$, show that

$$\tilde{r}(r) = \frac{m}{2} \left(\frac{r - m}{m} + \sqrt{\left(\frac{r - m}{m}\right)^2 - 1} \right) \tag{B.1.27}$$

and

$$r = \frac{1}{\tilde{r}} \left(\tilde{r} + \frac{m}{2} \right)^2. \tag{B.1.28}$$

Q3. Show that

$$A(\tilde{r})^2 = \left(\frac{\tilde{r} - m/2}{\tilde{r} + m/2} \right)^2, \quad B(\tilde{r}) = \left(1 + \frac{m}{2\tilde{r}} \right)^2. \tag{B.1.29}$$

giving the metric

$$ds^2 = -\left(\frac{\tilde{r} - m/2}{\tilde{r} + m/2} \right)^2 dt^2 + \left(1 + \frac{m}{2\tilde{r}} \right)^4 (d\tilde{r}^2 + \tilde{r}^2 d\Omega^2). \tag{B.1.30}$$

22 Recall that in one of the previous problems we have derived the identity

$$\nabla^\mu \nabla_\mu f - \frac{1}{\sqrt{|\det g_{\alpha\beta}|}} \partial_\mu \left(\sqrt{|\det g_{\alpha\beta}|} g^{\mu\nu} \partial_\nu f \right).$$

1. Show that, for weak gravitational fields $g_{\mu\nu} = \eta_{\mu\nu} + h_{\mu\nu}$, with $h_{\mu\nu}$ appropriately small, the *wave coordinates* condition

$$\nabla^\mu \nabla_\mu x^\alpha = 0$$

approximately reads

$$\partial_\mu h^\mu{}_\nu = \frac{1}{2}\partial_\nu(h^\alpha{}_\alpha),$$

where $h^\alpha{}_\beta = \eta^{\alpha\gamma}h_{\gamma\beta}$.

23   Let $h_{\alpha\beta} = \Re(A_{\alpha\beta}\exp(ik_\mu x^\mu))$, where $\Re$ denotes the real part, be a linearized gravitational wave in TT gauge (i.e., $A^\alpha{}_\alpha = 0 = A_{\alpha\beta}k^\beta$). Show that, in the linear approximation,

$$R_{\alpha\beta\gamma\delta}k^\delta = 0.$$

(One says that k^α is a *principal null direction* of the Riemann tensor.)

24   Given a timelike unit vector field u, show that a linear plane wave solution $g_{\mu\nu} = \eta_{\mu\nu} + h_{\mu\nu}$ can be put in a gauge where $h^\alpha{}_\alpha = 0 = h_{\mu\nu}u^\nu$.

25   Let φ satisfy the wave equation in a general spacetime with Lorentzian metric g, $\Box_g\varphi := \nabla^\mu\nabla_\mu\varphi = 0$. Set

$$T_{\mu\nu} = \partial_\mu\varphi\partial_\nu\varphi - \frac{1}{2}\nabla^\alpha\varphi\partial_\alpha\varphi g_{\mu\nu}.$$

Show that at the origin of any locally inertial coordinate system we have $T_{00} \geq \sqrt{\sum_i(T_{0i})^2}$. Show that this is equivalent to the statement that for any future-pointing timelike vector u^μ, the vector $T^\mu{}_\nu u^\nu$ is causal past-pointing.

Show that T satisfies the *dominant energy condition*: $T_{\mu\nu}X^\mu Y^\nu \geq 0$ for all timelike future directed X and Y. Is this condition satisfied for the energy-momentum tensor of dust? For that of the Maxwell field?

26   Let $g_{\mu\nu} = \eta_{\mu\nu} + A_{\mu\nu}\cos(k_\alpha x^\alpha)$ be the metric of a linearized plane wave in the gauge $A_{\mu\nu}k^\mu = 0 = A_{\mu\nu}u^\mu = A_\mu{}^\mu$, where $k^\mu\partial_\mu = \omega(\partial_t + \partial_z)$, $u^\mu\partial_\mu = \partial_t$. Introducing the variables $u = t - z$ and $v = t + z$, and using the variational principle for geodesics, write down the geodesic equations.

Show existence of the following two classes of geodesics:

The first class consists of geodesics with $u = $ constant, in which case you should be able to integrate the geodesic equations completely.

For the geodesics for which u is not constant, use u as a parameter along the geodesics, and solve the equations neglecting terms which are quadratic in ϵ, assuming that $|A_{\mu\nu}| \leq \epsilon$ and that the space-velocities are smaller than ϵ.

For the ambitious: Find all geodesics, without assuming the smallness conditions above.

27   *Laser Interferometric Space Observatory (LISA), or: if LIGO mirrors were freely falling.* Consider two freely falling mirrors separated by a coordinate distance L lying in a plane perpendicular to the direction of propagation of a linearized plane wave with $h_\times = 0$. Assuming that the time of flight of light between the mirrors is very small compared to the period of the wave, and neglecting all terms quadratic

in $g_{\mu\nu} - \eta_{\mu\nu}$, calculate the time needed for a photon to travel back and forth from one mirror to another. Compare the result with the calculation in the lecture (which you might wish to repeat), of the space-distance between both mirrors.

[*Hint : Light follows null geodesics. For this, you can use the results of Problem 28. Alternatively, use directly the Killing vectors, there are at least three of them. It is useful to view the solution as a small perturbation of the associated Minkowskian solution.*]

28 Calculate the field of unit normals and the induced metric for
1. S^2 included in a flat $\mathbb{R}^3$ (use polar coordinates on S^2)
2. S^2 viewed as a sphere of constant radius in Schwarzschild (use polar coordinates on S^2)
3. $S = \{x^0 = \sqrt{1+r^2}\}$ in four-dimensional Minkowski; use the Cartesian space coordinates x^i as coordinates on S
4. $S' = \{r = \sqrt{1+t^2}\}$ in four-dimensional Minkowski; use t and polar coordinates as coordinates on S'

29 Find a function f so that the metric induced on the hypersurface $\{t = f(r)\}$ in Schwarzschild spacetime is flat.

[*Hint: the metric induced on $t = f(r)$ is obtained by replacing every occurrence of dt in the metric by df.*]

30 Let φ satisfy the wave equation in a general spacetime with Lorentzian metric g, $\Box_g\varphi := \nabla^\mu\nabla_\mu\varphi = 0$. Set

$$T_{\mu\nu} = \partial_\mu\varphi\partial_\nu\varphi - \frac{1}{2}\nabla^\alpha\varphi\partial_\alpha\varphi g_{\mu\nu} .$$

Check that $\nabla_\mu T^\mu{}_\nu = 0$.
Recall that a vector field X^μ is called *Killing* if

$$\nabla_\mu X_\nu + \nabla_\nu X_\mu = 0 .$$

1. Check that ∂_t and ∂_φ are Killing vectors both in the Minkowski metric and in the Schwarzschild metric.
2. Check that you remember the proof of the following: if γ is a geodesic and X is Killing, then $g(\gamma, X)$ is constant along γ.
3. Show that if $\nabla_\mu T^\mu{}_\nu = 0$ and X is Killing, then the vector field $J^\mu := T^\mu{}_\nu X^\nu$ has vanishing divergence, $\nabla_\mu J^\mu = 0$.

Let $X = \partial_t$ be a Killing vector. Given a spacelike hypersurface $\mathscr{S}$ with unit future-directed timelike normal N^μ, set

$$E := \int_{\mathscr{S}} T_{\mu\nu}X^\nu N^\mu d^3\mu_h ,$$

where μ_h is the measure associate with the metric h induced on $\mathscr{S}$ by g. E is called the total energy E of the field contained in $\mathscr{S}$. Give an explicit expression for E for the surface $\{t = 0\}$ in the Minkowski spacetime, and in the Schwarzschild spacetime.

What is the conservation law associated with

$$J := \int_{\mathscr{S}} T_{\mu\nu} X^\nu n^\mu d^3 x \,,$$

where $X = \partial_\varphi$? Likewise write an explicit formula for J in Minkowski and in Schwarzschild.

31 Consider timelike geodesics in the $\theta = \pi/2$ plane in the Schwarzschild geometry:
   (a) Show that for every $r > 3m$ there exist timelike geodesics for which $\dot{r} \equiv 0$.
   (b) Show that on such geodesics we have

$$\varphi = \varphi_0 + \Omega t \,, \quad \Omega = \frac{m^{1/2}}{r^{3/2}} \,, \quad t = t_0 \pm \frac{Js}{\sqrt{mr}} \,,$$

where $J = r^2 d\varphi/ds$, and s is the proper time along the geodesic.

32 Consider a pointlike body of mass m_0 and angular momentum J_N moving in an elliptic, Kepler orbit of eccentricity e in the field of a central mass of mass m. The density distribution of the central mass is spherically symmetric and time-independent. Assume that, in Cartesian coordinates, this motion takes place in the $z = 0$ plane. Show that the *time-dependent part* of the quadrupole moment of the system takes the form

$$q_{ij} = m_0 r(\varphi(t))^2 \begin{pmatrix} \cos^2 \varphi(t) & \sin \varphi(t) \cos \varphi(t) & 0 \\ \sin \varphi(t) \cos \varphi(t) & \sin^2 \varphi(t) & 0 \\ 0 & 0 & 0 \end{pmatrix} \,.$$

Using the explicit form of $r(\varphi(t))$ for Newtonian orbits seen in the lectures, derive an expression for the linearized metric perturbation $\overline{h}_{xx}$, to first order in the eccentricity e.

33 *Lie derivative*

Given a vector field X, the *Lie derivative* $\mathscr{L}_X$ is an operation on tensors, defined as follows:
   1. For a function f, one sets $\mathscr{L}_X f := X(f)$.
   2. For a vector field Y, one sets $\mathscr{L}_X Y := [X, Y]$, the Lie bracket.
   3. For a one-form α, $\mathscr{L}_X \alpha$ is defined by imposing the Leibniz rule written backwards:

$$(\mathscr{L}_X \alpha)(Y) := \mathscr{L}_X(\alpha(Y)) - \alpha(\mathscr{L}_X Y) \,.$$

Q1: (a) Why is this the same as the Leibniz rule? [*Hint: write this equation using indices.*] (b) Check that $\mathscr{L}_X\alpha$ is a tensor. [*Hint: check that the right-hand side is linear under multiplication of Y by a function.*]

Q2: (a) Show that

$$(\mathscr{L}_X\alpha)_a = X^b\partial_b\alpha_a + \alpha_b\partial_a X^b.$$

For tensor products the Lie derivative is defined again by imposing linearity under addition together with the Leibniz rule:

$$\mathscr{L}_X(\alpha \otimes \beta) = (\mathscr{L}_X\alpha) \otimes \beta + \alpha \otimes \mathscr{L}_X\beta.$$

Since a general tensor A is sum of tensor products,

$$A = A^{a_1\ldots a_p}{}_{b_1\ldots b_q}\partial_{a_1} \otimes \ldots \partial_{a_p} \otimes dx^{b_1} \otimes \ldots \otimes dx^{b_q},$$

requiring linearity with respect to addition of tensors gives thus a definition of Lie derivative for any tensor.

(b) Does the Lie derivative, so defined, commute with contractions?

Q3: Show that

$$\mathscr{L}_X T^a{}_b = X^c\partial_c T^a{}_b - T^c{}_b\partial_c X^a + T^a{}_c\partial_b X^c.$$

Q4: Can you see the general formula for the Lie derivative $\mathscr{L}_X A^{a_1\ldots a_p}{}_{b_1\ldots b_q}$?

Q5: Prove

$$\mathscr{L}_{[X,Y]} = [\mathscr{L}_X, \mathscr{L}_Y],$$

where, for a tensor α, $[\mathscr{L}_X, \mathscr{L}_Y]\alpha$ is defined as

$$[\mathscr{L}_X, \mathscr{L}_Y]\alpha := \mathscr{L}_X(\mathscr{L}_Y\alpha) - \mathscr{L}_Y(\mathscr{L}_X\alpha).$$

[*Hint: note that after checking i, ii, and iv of the definition, iii follows; for assistance with Q5, you may wish to prove first the Jacobi identity:*

$$[[X, Y], Z] + [[Y, Z], X] + [[Z, X], Y] = 0.]$$

Q6: Show that if there exists a coordinate system in which $(X^a) = (1, 0, 0, 0)$ (everywhere) and $\partial_0 g_{bc} = 0$, then $\mathscr{L}_X g = 0$ in every coordinate system. [Vector fields with this property are called *Killing vectors*: they arise from symmetries of spacetime.]

Q7: Let ∇ be any torsion-free connection. Show that if X, Y are vector fields, then $\mathcal{L}_X Y$ has components $X^b\nabla_b Y^a - Y^b\nabla_b X^a$. What happens if ∇ has torsion?

Q8: Let ∇ be the Levi-Civita connection. Show that

$$\mathcal{L}_X g_{ab} = \nabla_a X_b + \nabla_b X_a.$$

34 Given a one-form, α, we define its *exterior derivative*, $d\alpha$, as

$$d\alpha_{ab} = \partial_a \alpha_b - \partial_b \alpha_a.$$

Show that $d\alpha$ is a tensor.
Show that, acting on arbitrary vector fields X, Y, we have

$$d\alpha(X, Y) = \mathcal{L}_X (\alpha(Y)) - \mathcal{L}_Y (\alpha(X)) - \alpha([X, Y]).$$

(Note that this allows us to define the exterior derivative of a one-form in terms of the Lie derivative.) Also, show that for arbitrary vector fields X, Y, we have

$$(\mathcal{L}_X \alpha)(Y) = d\alpha(X, Y) + Y(\alpha(X)).$$

Recall that if f is a function, then $df = \partial_i f\, dx^i$. Show that $d(df) = 0$, and

$$\mathcal{L}_X(df) = d(\mathcal{L}_X f).$$

35 The Heisenberg group is the group of matrices

$$G := \left\{ \begin{bmatrix} 1 & x & y \\ 0 & 1 & z \\ 0 & 0 & 1 \end{bmatrix} \middle| x, y, z \in \mathbb{R} \right\},$$

with the group operation given by matrix multiplication. Given a group element $g = \begin{bmatrix} 1 & a & b \\ 0 & 1 & c \\ 0 & 0 & 1 \end{bmatrix} \in G$, we define the operation of left-multiplication $L_g : G \to G$

by $h \mapsto gh$. Calculate the image of $h = \begin{bmatrix} 1 & x & y \\ 0 & 1 & z \\ 0 & 0 & 1 \end{bmatrix}$ under the map L_g, and the

push-forward map $(L_g)_*$, at h.
A vector field on G is *left-invariant* if $(L_g)_* v = v$ for all $g \in G$. Show that the vector fields

$$\mathbf{v}_1 = \partial_x, \qquad \mathbf{v}_2 = \partial_y, \qquad \mathbf{v}_3 = \partial_z + x\partial_y$$

are left-invariant vector fields on G. Calculate the flows of the vector fields $\mathbf{v}_1, \mathbf{v}_2, \mathbf{v}_3$ on G (i.e., the integral curves of the vector fields starting at a fixed point $g_0 \in G$).

36 Consider the metric

$$g = g_{\text{Schw}} - \frac{4J \sin^2 \theta}{r} dt\, d\varphi\,,$$

where g_{Schw} denotes the Schwarzschild metric.
1. Show that are geodesics for the metric g of the form $\gamma(\tau) = (t(\tau), r(\tau), \theta = 0)$.
2. Show that if $s^r = 0$ at some point, then it is zero along the whole such geodesic γ.
3. Write-down explicitly the gyroscope equation on γ assuming that $s^r = 0$. (Can you solve it?)

37 (a) Show that the function $f(t-r)/r$ solves the wave equation on four-dimensional Minkowski spacetime for $r > 0$.
(b) Show that $-\delta(t - r)/(4\pi r)$ is a Green function for the wave equation on four-dimensional Minkowski spacetime.

38 A distant galaxy has a redshift $z = (\lambda_{observed} - \lambda_{emitted})/\lambda_{emitted}$ of 0.2. According to Hubble's law, how far away was the galaxy when the light was emitted if the Hubble constant is 72 (km/s)/Mpc?

39 A Cepheid variable star is observed with an apparent magnitude of 22 (see http://outreach.atnf.csiro.au/education/senior/astrophysics/photometry_magnitude.html#magnapparent for the notion of the magnitude of a star) and a period of 28 days. Using data from http://hyperphysics.phy-astr.gsu.edu/hbase/astro/cepheid.html, determine the distance to this star.
[*Hint: The apparent magnitude m of the star is related to that of the Sun by the formula*

$$m - m_\odot = -2.5 \log(\ell/L_\odot)\,,$$

where ℓ is the apparent luminosity, $L_\odot$ is the luminosity of the Sun, and $\log$ is the logarithm at base ten. Use the fact that distance is inversely proportional to the square root of luminosity. Use $m_\odot = -26.5$, $d_\odot = 1$ astronomical unit = 149.60×10^6 km = 149.60×10^9 m = 4.8481×10^{-6} parsecs, 1 parsec = 3.26 light years = 3×10^{13} km.]

40 Let X be a Killing vector on a compact Riemannian manifold with non-positive Ricci tensor: $R_{ij} X^i X^j \leq 0$. Integrating by parts in the integral

$$\int X^i X^j{}_{;ij}$$

show that X is covariantly constant, and that the isometry group of compact Riemannian Einstein (this means that $R_{ij} = \frac{R}{n} g_{ij}$) manifolds with $R < 0$ is discrete. [*Hint: Use the equation satisfied by the second derivatives of a Killing vector. Also note that "integration by parts" is equivalent to the fact that on a compact manifold without boundary we have $\int Y^j{}_{;j} = 0$ for any differentiable vector field Y.*]

41 Show, by a direct calculation of the Christoffel symbols or otherwise, that the radial null rays in a FLRW metric are geodesics (perhaps, but not necessarily, affinely parameterized).

42 Find the geodesics of the maximally symmetric Riemannian and Lorentzian manifolds. [*Hint: it suffices to find a family with the property that all geodesics can be obtained from the members of the family by applying isometries.*]
Using this, or otherwise, show that in the embedded model where the maximally symmetric manifold is a submanifold S_a of $\mathbb{R}^3$ as described in the lectures, the geodesics are intersections of S_a with planes through the origin.

43 Verify that for FLRW models with $\rho + 3p \geq 0$ and with non-positive cosmological constant the scale factor R is a concave function of t (i.e., $\frac{d^2R}{dt^2} \leq 0$). Assuming that $R(t) \approx ct^\alpha$ as $t \to 0$, deduce from this that $1/H(t) \geq t$ for $t > 0$. [*Hint: Calculate the time derivative of $1/H - t$.*]

44 Suppose that the spatial volume of a closed, matter dominated, FLRW universe with spherical space sections and vanishing cosmological constant is 10^{12}Mpc^3 at the moment of maximum expansion. What is the duration of this universe from big bang to big crunch in years?

45 Show that solutions of the linearization of the Friedman equation in the radiation-and-dust universe at $\dot{R} = 0$ are unstable.

46 Recall that, given a one-form α, we defined its exterior derivative $d\alpha$ to be the two-form with components

$$(d\alpha)_{ab} = \partial_a\alpha_b - \partial_b\alpha_a.$$

Consider the manifold $M = \mathbb{R}^{2n}$, with coordinates (x^i, p_i), $i = 1, \ldots, n$. (We will also denote the coordinates by q^a, $a = 1, \ldots, 2n$, with $q^1 = x^1, \ldots, q^n = x^n, q^{n+1} = p_1, \ldots, q^{2n} = p_n$). Let

$$\alpha = \sum_{i=1}^{n} p_i dx^i.$$

(So the components of α are $\alpha_{x^i} = p_i, \alpha_{p_i} = 0$.)
Calculate $\omega := -d\alpha$. Check that the matrix ω_{ab} is invertible, find its inverse. (($\mathbb{R}^{2n}, \omega$) is an example of what is called a *symplectic manifold*.)
Let $H = H(x, p)$ be a function on M, hence its differential dH is

$$dH = \frac{\partial H}{\partial q^a}dq^a = \sum_{i=1}^{n}\left[\frac{\partial H}{\partial x^i}dx^i + \frac{\partial H}{\partial p_i}dp_i\right].$$

The *Hamiltonian vector field* corresponding to H is the vector field, X_H, on M defined by the equations

$$\omega_{ab}X_H^a = \frac{\partial H}{\partial q^b}, \qquad a, b = 1, \ldots, 2n.$$

Check that, in index-free notation, this is the same as $\omega(X_H, \cdot) = dH$.
Calculate the components of the Hamiltonian vector field X_H. Write down the equations for an integral curve of the vector field X_H. Do you recognize these equations?
Given two functions f and g, calculate $[X_f, X_g]$. What do you obtain if $f = p^i$ and $g = x^j$? Do you recognize this operation?

47 Calculate the total gravitational potential energy of a spherically symmetric New-tonian star with mass density ρ which is constant within a ball of radius R. [*Hint: Start by explaining why a shell at radius r of thickness dr contributes*

$$dU = -4\pi\rho r^2 \frac{m(r)}{r} dr$$

to the total potential energy U of the star.]

48 *The Lane-Emden equation* Recall from lectures that, for a Newtonian static fluid, p and ρ satisfy the equation

$$\frac{1}{r^2}\frac{d}{dr}\left(\frac{r^2}{\rho}\frac{dp}{dr}\right) = -4\pi\rho.$$

Assume that $\rho = \rho_c\theta(r)^n$ and $p = p_c\theta(r)^{n+1}$, where n, ρ_c, p_c are constants, and θ is a function to be determined. Check that

$$p = K\rho^{\frac{n+1}{n}},$$

with a constant K that you will determine. Show that the function θ satisfies the differential equation

$$\frac{K(n+1)}{4\pi}\rho_c^{\frac{1}{n}-1}\frac{1}{r^2}\frac{d}{dr}\left(r^2\frac{d\theta}{dr}\right) = -\theta^n.$$

Let $r_n > 0$ be defined by

$$r_n^2 = \frac{K(n+1)}{4\pi}\rho_c^{\frac{1}{n}-1},$$

and define a new variable ξ by $r = r_n\xi$. Deduce that

$$\frac{1}{\xi^2}\frac{d}{d\xi}\left(\xi^2\frac{d\theta}{d\xi}\right) = -\theta^n.$$

In the case $n = 0$ solve this equation with the boundary conditions

$$\theta(0) = 1, \quad \theta'(0) = 0. \tag{B.1.31}$$

In the case $n = 1$ solve again with the boundary conditions (B.1.31). *[Hint: When n = 1 use the substitution $\theta(\xi) = \psi(\xi)/\xi$.]*
Finally, verify that the function

$$\theta(\xi) = \frac{1}{\left(1 + \frac{1}{3}\xi^2\right)^{1/2}}$$

is a solution in the case $n = 5$.

Bibliography

1. B.P. Abbott et al., GW151226: observation of gravitational waves from a 22-solar-mass binary black hole coalescence. Phys. Rev. Lett. **116**, 241103 (2016). arXiv:1606.04855 [gr-qc]. https://doi.org/10.1103/PhysRevLett.116.241103
2. B.P. Abbott et al., Observation of gravitational waves from a binary black hole merger. Phys. Rev. Lett. **116**, 061102 (2016). arXiv:1602.03837 [gr-qc]. https://doi.org/10.1103/PhysRevLett.116.061102
3. B.P. Abbott et al., First low-frequency Einstein@Home all-sky search for continuous gravitational waves in Advanced LIGO data. Phys. Rev. **D96**(12), 122004 (2017). arXiv:1707.02669 [gr-qc]. https://doi.org/10.1103/PhysRevD.96.122004
4. B.P. Abbott et al., Gravitational waves and gamma-rays from a binary neutron star merger: GW170817 and GRB 170817A. Astrophys. J. **848**, L13 (2017). arXiv:1710.05834 [astro-ph.HE]. https://doi.org/10.3847/2041-8213/aa920c
5. B.P. Abbott et al., GW170104: observation of a 50-solar-mass binary black hole coalescence at redshift 0.2. Phys. Rev. Lett. **118**, 221101 (2017). arXiv:1706.01812 [gr-qc]. https://doi.org/10.1103/PhysRevLett.118.221101
6. B.P. Abbott et al., Multi-messenger observations of a binary neutron star merger. Astrophys. J. **848**, L12 (2017). arXiv:1710.05833 [astro-ph.HE]. https://doi.org/10.3847/2041-8213/aa91c9
7. B.P. Abbott et al., GWTC-1: a gravitational wave transient catalog of compact binary mergers observed by LIGO and Virgo during the first and second observing runs (2018). arXiv:1811.12907 [astro-ph.HE]
8. B.P. Abbott et al., Searches for continuous gravitational waves from fifteen Supernova remnants and Fomalhaut b with Advanced LIGO (2018). arXiv:1812.11656 [astro-ph.HE]
9. A. Albert et al., Search for high-energy neutrinos from binary neutron star merger GW170817 with ANTARES, IceCube, and the Pierre Auger Observatory. Astrophys. J. **850**, L35 (2017). https://doi.org/10.3847/2041-8213/aa9aed
10. L. Andersson, R. Beig, B.G. Schmidt, Rotating elastic bodies in Einstein gravity. Commun. Pure Appl. Math. **63**, 559–589 (2010). https://doi.org/10.1002/cpa.20302
11. N. Andersson et al., The transient gravitational-wave sky. Classical Quantum Gravity **30**, 193002 (2013). arXiv:1305.0816 [gr-qc]. https://doi.org/10.1088/0264-9381/30/19/193002
12. H. Andréasson, M. Kunze, G. Rein, Rotating, stationary, axially symmetric spacetimes with collisionless matter. Commun. Math. Phys. **329**, 787–808 (2012). arXiv:1212.5028 [gr-qc]. https://doi.org/10.1007/s00220-014-1904-5
13. J. Antoniadis et al., A massive pulsar in a compact relativistic binary. Science **340**, 6131 (2013). arXiv:1304.6875 [astro-ph.HE]. https://doi.org/10.1126/science.1233232
14. R. Beig, W. Simon, Proof of a multipole conjecture due to Geroch. Commun. Math. Phys. **78**, 75–82 (1980)
15. R. Beig, W. Simon, On the uniqueness of static perfect–fluid solutions in general relativity. Commun. Math. Phys. **144**, 373–390 (1992)
16. A.L. Besse, *Einstein Manifolds*. Ergebnisse der Mathematik und ihrer Grenzgebiete. 3. Folge, vol. 10 (Springer, Berlin/New York/Heidelberg, 1987)
17. M. Betoule et al., Improved cosmological constraints from a joint analysis of the SDSS-II and SNLS supernova samples. Astron. Astrophys. **568**, A22 (2014). arXiv:1401.4064 [astro-ph.CO]. https://doi.org/10.1051/0004-6361/201423413

© Springer Nature Switzerland AG 2019
P. T. Chruściel, *Elements of General Relativity*, Compact Textbooks in Mathematics,
https://doi.org/10.1007/978-3-030-28416-9

18. G. Beyerle, Visualization of Thomas-Wigner rotations. Symmetry **9**, 292 (2017). arXiv:1706.02755 [physics.class-ph]. https://doi.org/10.3390/sym9120292

19. I. Białynicki-Birula, Z. Białynicka-Birula, Gravitational waves carrying orbital angular momentum. New J. Phys. **18**, 023022 (2016). arXiv:1511.08909 [gr-qc]. https://doi.org/10.1088/1367-2630/18/2/023022

20. G.D. Birkhoff, *Relativity and Modern Physics* (Harvard University Press, Harvard, 1923)

21. V. Bonvin et al., H0LiCOW - V. New COSMOGRAIL time delays of HE 0435-1223: H_0 to 3.8 per cent precision from strong lensing in a flat ΛCDM model. Mon. Not. R. Astron. Soc. **465**, 4914–4930 (2017). arXiv:1607.01790 [astro-ph.CO]. https://doi.org/10.1093/mnras/stw3006. https://ui.adsabs.harvard.edu/#abs/2017MNRAS.465.4914B

22. H.A. Buchdahl, General relativistic fluid spheres. Phys. Rev. **116**(2), 1027–1034 (1959)

23. A. Buonanno, Gravitational waves, in *Les Houches Summer School - Session 86: Particle Physics and Cosmology: The Fabric of Spacetime Les Houches*, 31 July–25 August 2006 (2007). arXiv:0709.4682 [gr-qc]

24. T.E. Collett et al., A precise extragalactic test of general relativity. Science **360**, 1342–1346 (2018).

25. A. Conley et al., Supernova constraints and systematic uncertainties from the first 3 years of the supernova legacy survey. Astrophys. J. Suppl. **192**, 1 (2011). arXiv:1104.1443 [astro-ph.CO]. https://doi.org/10.1088/0067-0049/192/1/1

26. J. Creswell, S. von Hausegger, A.D. Jackson, H. Liu, P. Naselsky, On the time lags of the LIGO signals. J. Cosmol. Astropart. Phys. **1708**, 013 (2017). https://doi.org/10.1088/1475-7516/2017/08/013

27. P. Delva et al., Gravitational redshift test using eccentric Galileo satellites. Phys. Rev. Lett. **121**, 231101, 6 pp. (2018). arXiv:1812.03711 [gr-qc]. https://doi.org/10.1103/PhysRevLett.121.231101

28. P. Demorest, T. Pennucci, S. Ransom, M. Roberts, J. Hessels, Shapiro delay measurement of a two solar mass neutron star. Nature **467**, 1081–1083 (2010). arXiv:1010.5788 [astro-ph.HE]. https://doi.org/10.1038/nature09466

29. F. Dyson, *Interstellar Communication*, Chap. 12 (A.G. Cameron, New York, 1963)

30. J. Ehlers, *Über den Newtonschen Grenzwert der Einsteinschen Gravitationstheorie*. Fundamental Problems of Modern Physics (Bibliographisches Inst., Mannheim, 1981), pp. 65–84

31. J. Ehlers, R. Geroch, Equation of motion of small bodies in relativity. Ann. Phys. **309**, 232–236 (2004). https://doi.org/10.1016/j.aop.2003.08.020

32. L.P. Eisenhart, *Riemannian Geometry*. Princeton Landmarks in Mathematics (Princeton University Press, Princeton, NJ, 1997). Eighth printing, Princeton Paperbacks

33. Event Horizon Telescope Collaboration, First M87 Event Horizon Telescope results. VI. The shadow and mass of the central black hole. Astrophys. J. Lett. **875**, L6, 44 pp. (2019). https://doi.org/10.3847/2041-8213/ab1141. https://ui.adsabs.harvard.edu/abs/2019ApJ...875L...6E

34. C.W.F. Everitt et al., Gravity Probe B data analysis status and potential for improved accuracy of scientific results. Classical Quantum Gravity **25**, 114002 (2008). https://doi.org/10.1088/0264-9381/25/11/114002

35. C.W.F. Everitt et al., Gravity Probe B: final results of a space experiment to test general relativity. Phys. Rev. Lett. **106**, 221101 (2011). arXiv:1105.3456 [gr-qc]. https://doi.org/10.1103/PhysRevLett.106.221101

36. C. Fronsdal, Completion and embedding of the Schwarzschild solution. Phys. Rev. **116**, 778–781 (1959)

37. R. Geroch, J.O. Weatherall, The motion of small bodies in space-time. Commun. Math. Phys. **364**, 607–634 (2018). arXiv:1707.04222 [gr-qc]. https://doi.org/10.1007/s00220-018-3268-8

38. F. Giannoni, A. Masiello, P. Piccione, A variational theory for light rays in stably causal Lorentzian manifolds: regularity and multiplicity results. Commun. Math. Phys. **187**, 375–415 (1997)

39. J.C. Hafele, R.E. Keating, Around-the-world atomic clocks: observed relativistic time gains. Science **177**, 168–170 (1972). http://www.jstor.org/stable/1734834

40. J.C. Hafele, R.E. Keating, Around-the-world atomic clocks: predicted relativistic time gains. Science **177**, 166–168 (1972). http://www.jstor.org/stable/1734833

41. B.T. Hayden et al., The rise and fall of type Ia supernova light curves in the SDSS-II supernova survey. Astrophys. J. **712**, 350–366 (2010). https://doi.org/10.1088/0004-637X/712/1/350. http://adsabs.harvard.edu/abs/2010ApJ...712..350H

42. U. Heilig, On the existence of rotating stars in general relativity. Commun. Math. Phys. **166**, 457–493 (1995)

43. J.M. Heinzle, Bounds on $2m/r$ for static perfect fluids (2007). arXiv:0708.3352 [gr-qc]

44. S. Herrmann et al., Test of the gravitational redshift with Galileo satellites in an eccentric orbit. Phys. Rev. Lett. **121**, 231102 (2018). arXiv:1812.09161 [gr-qc]. https://doi.org/10.1103/PhysRevLett.121.231102

45. D. Hilbert, *Die Grundlagen der Physik*, Nachrichten von der Gesellschaft der Wissenschaften zu Göttingen. Mathematisch–physikalische Klasse (1915), pp. 395–407

46. G. Hinshaw et al., Nine-year Wilkinson Microwave Anisotropy Probe (WMAP) observations: cosmological parameter results. Astrophys. J. Suppl. Ser. **208**, 19, 25 pp. (2013). https://doi.org/10.1088/0067-0049/208/2/19

47. Hubble Legacy Archive, http://hla.stsci.edu/

48. E. Hubble, A relation between distance and radial velocity among extra-galactic nebulae. Proc. Natl. Acad. Sci. **15**, 168–173 (1929). https://www.pnas.org/content/15/3/168

49. J.T. Jebsen, On the general spherically symmetric solutions of Einstein's gravitational equations in vacuo. Gen. Relativ. Gravit. **37**, 2253–2259 (2005). Translation of Ark. Mat. Ast. Fys. (Stockholm) 15, nr.18 (1921). https://doi.org/10.1007/s10714-005-0168-y

50. J. Jezierski, Thermo-hydrodynamics as a field theory, in *Nonequilibrium Theory and Extremum Principles*, ed. by S. Sieniutycz, P. Salamon. Advances of Thermodynamics, vol. 3 (Taylor and Francis, New York, 1990), pp. 282–317

51. J. Jezierski, J. Kijowski, Une description hamiltonienne du frottement et de la viscosité. C. R. Acad. Sci. Paris Sér. II Méc. Phys. Chim. Sci. Univers Sci. Terre **301**, 221–224 (1985)

52. P. Karageorgis, J.G. Stalker, Sharp bounds on $2m/r$ for static spherical objects. Classical Quantum Gravity **25**, 195021, 14 pp. (2008). arXiv:0707.3632 [gr-qc]. https://doi.org/10.1088/0264-9381/25/19/195021

53. J. Kijowski, A. Smólski, A. Górnicka, Hamiltonian theory of self-gravitating perfect fluid and a method of effective deparametrization of Einstein's theory of gravitation. Phys. Rev. D **41**(3), 1875–1884 (1990). https://doi.org/10.1103/PhysRevD.41.1875

54. R. Kippenhahn, A. Weigert, *Stellar Structure and Evolution* (Springer, New York, 1994)

55. M. Kramer, Probing gravitation with pulsars. IAU Symp. **291**, 19–26 (2013). arXiv:1211.2457 [astro-ph.HE]. https://doi.org/10.1017/S1743921312023306X

56. M.D. Kruskal, Maximal extension of Schwarzschild metric. Phys. Rev. **119**, 1743–1745 (1960)

57. G. Lemaître, Un Univers homogène de masse constante et de rayon croissant rendant compte de la vitesse radiale des nébuleuses extra-galactiques. Ann. Soc. Sci. Brux. **A47**, 49–59 (1927). https://ui.adsabs.harvard.edu/abs/1927ASSB...47...49L

58. D. Lovelock, The uniqueness of the Einstein field equations in a four-dimensional space. Arch. Ration. Mech. Anal. **33**, 54–70 (1969). https://doi.org/10.1007/BF00248156

59. D. Lovelock, The four-dimensionality of space and the Einstein tensor. J. Math. Phys. **13**, 874–876 (1972)

60. A.K.M. Masood-ul Alam, Proof that static stellar models are spherical. Gen. Relativ. Gravit. **39**, 55–85 (2007). https://doi.org/10.1007/s10714-006-0364-4

61. J.-P. Nicolas, Dirac fields on asymptotically flat space-times. Dissertationes Math. (Rozprawy Mat.) **408**, 1–85 (2002)

62. T.A. Oliynyk, Post-Newtonian expansions for perfect fluids. Commun. Math. Phys. **288**, 847–886 (2009). https://doi.org/10.1007/s00220-009-0738-z

63. T.A. Oliynyk, A rigorous formulation of the cosmological Newtonian limit without averaging. J. Hyperbolic Differ. Equ. **7**, 405–431 (2010). https://doi.org/10.1142/S0219891610002189

64. W.J. Percival et al., Baryon acoustic oscillations in the sloan digital sky survey data release 7 galaxy sample. Mon. Not. R. Astron. Soc. **401**, 2148–2168 (2010). arXiv:0907.1660 [astro-ph.CO]. https://doi.org/10.1111/j.1365-2966.2009.15812.x

65. V. Perlick, On Fermat's principle in general relativity: I. The general case Classical Quantum Gravity **7**, 1319–1331 (1990)

66. S. Perlmutter et al., Cosmology from Type Ia supernovae. Bull. Am. Astron. Soc. **29**, 1351 (1997). arXiv:astro-ph/9812473

67. Planck Collaboration, Planck 2013 results. I. Overview of products and scientific results. Astron. Astrophys. **571**, A1 (2014). arXiv:1303.5062 [astro-ph.CO]. https://doi.org/10.1051/0004-6361/201321529

68. Planck Collaboration, Hubble's law and the expanding universe. Proc. Natl. Acad. Sci. **112**, 3173–3175 (2015). https://doi.org/10.1073/PNAS.1424299112

69. Planck Collaboration, Planck 2015 results. XIII. Cosmological parameters. Astron. Astrophys. **594**, A13 (2016). arXiv:1502.01589 [astro-ph.CO]. https://doi.org/10.1051/0004-6361/201525830

70. Planck Collaboration, Planck 2018 results. VI. Cosmological parameters. Astron. Astrophys. (2018), in press. arXiv:1807.06209 [astro-ph.CO]

71. R.D. Reasenberg et al., Viking relativity experiment - verification of signal retardation by solar gravity. Astrophys. J. Lett. **234**, L219–L221 (1979)

72. A.D. Rendall, The Newtonian limit for asymptotically flat solutions of the Vlasov-Einstein system. Commun. Math. Phys. **163**, 89–112 (1994). http://projecteuclid.org/euclid.cmp/1104270381

73. A.G. Riess et al., New Hubble Space Telescope discoveries of type Ia Supernovae at $z > 1$: narrowing constraints on the early behavior of dark energy. Astrophys. J. **659**, 98–121 (2007). arXiv:astro-ph/0611572

74. A.G. Riess et al., New parallaxes of galactic Cepheids from spatially scanning the Hubble Space Telescope: implications for the Hubble constant. Astrophys. J. **855**, 136 (2018). arXiv:1801.01120 [astro-ph.SR]. https://doi.org/10.3847/1538-4357/aaadb7. https://ui.adsabs.harvard.edu/#abs/2018ApJ...855..136R

75. C. Ripken, *Coordinate Systems in De Sitter Spacetime*, Bachelor thesis (2013), https://www.ru.nl/publish/pages/913454/thesis_chris_ripken.pdf

76. J. Sbierski, The C^0-inextendibility of the Schwarzschild spacetime and the spacelike diameter in Lorentzian Geometry. J. Differ. Geom. **108**, 319–378 (2018). arXiv:1507.00601 [gr-qc]. https://doi.org/10.4310/jdg/1518490820

77. P. Schneider, J. Ehlers, E.E. Falco, *Gravitational Lenses* (Springer, New York, 1993)

78. I.I. Shapiro et al., Fourth test of general relativity: new radar result. Phys. Rev. Lett. **26**, 1132–1135 (1971). https://doi.org/10.1103/PhysRevLett.26.1132

79. S.L. Shapiro, S.A. Teukolsky, *Black Holes, White Dwarfs, and Neutron Stars: The Physics of Compact Objects* (Wiley, New York, 1983)

80. M. Spradlin, A. Strominger, A. Volovich, Les Houches lectures on de Sitter space, in *Unity from Duality: Gravity, Gauge Theory and Strings. Proceedings*, NATO Advanced Study Institute, Les Houches, 30 July–31 August 2001, pp. 423–453. arXiv:hep-th/0110007 [hep-th]

81. M. Sullivan et al., SNLS3: constraints on dark energy combining the Supernova Legacy Survey three year data with other probes. Astrophys. J. **737**, 102 (2011). arXiv:1104.1444 [astro-ph.CO]. https://doi.org/10.1088/0004-637X/737/2/102

82. N. Suzuki et al., The hubble space telescope cluster supernova survey: V. Improving the dark energy constraints above $z > 1$ and building an early-type-hosted supernova sample. Astrophys. J. **746**, 85 (2012). aXiv:1105.3470 [astro-ph.CO]. https://doi.org/10.1088/0004-637X/746/1/85

83. J.L. Synge, The escape of photons from gravitationally intense stars. Mon. Not. R. Astron. Soc. **131**, 463–466 (1966)

84. G. Szekeres, On the singularities of a Riemannian manifold. Gen. Relativ. Gravit. **34**, 2001–2016 (2002). Reprinted from Publ. Math. Debrecen **7**, 285–301 (1960). https://doi.org/10.1023/A: 1020744914721

85. M.E. Taylor, *Partial Differential Equations III. Nonlinear Equations*. Applied Mathematical Sciences, vol. 117, 2nd edn. (Springer, New York, 2011)

86. K.S. Thorne, Multipole expansions of gravitational radiation. Rev. Mod. Phys. **52**, 299–339 (1980). https://doi.org/10.1103/RevModPhys.52.299

87. J.M. Weisberg, Y. Huang, Relativistic measurements from timing the binary pulsar PSR B1913+16. Astrophys. J. **829**, 55 (2016). arXiv:1606.02744 [astro-ph.HE]. https://doi.org/10.3847/0004-637X/ 829/1/55

88. W.M. Wood-Vasey et al., Observational constraints on the nature of the dark energy: first cosmological results from the ESSENCE Supernova Survey. Astrophys. J. **666**, 694–715 (2007). arXiv:astro-ph/0701041

89. N.M.J. Woodhouse, *General Relativity*. Springer Undergraduate Mathematics Series (Springer, London, 2007)

90. S. Yang, On the geodesic hypothesis in general relativity. Commun. Math. Phys. **325**, 997–1062 (2014). arXiv:1209.3985 [math.AP]. https://doi.org/10.1007/s00220-013-1834-7